Mechanics of Machines

Elementary Theory and Examples

Mechanics of Machines

Elementary Theory and Examples

J Hannah

B Sc (Eng), C Eng, F I Mech E, F I Prod E

R C Stephens

M Sc (Eng), C Eng, M I Mech E

Fourth Edition

Edward Arnold

First published 1958
by Edward Arnold (Publishers) Ltd,
41 Bedford Square, London WC1B 3DQ

Reprinted 1959, 1962
Second Edition 1963
Reprinted 1965, 1966, 1968
Third Edition 1970
Reprinted 1971, 1972, 1973, 1975, 1977, 1979
Fourth Edition 1984

British Library Cataloguing in Publication Data

Hannah, John
Mechanics of machines: elementary theory and
examples.—4th ed.
1. Machinery 2. Mechanics
I. Title II. Stephens, R. C.
621.8'11 TJ145

ISBN 0–7131–3471–2

Printed in Great Britain by
Thomson Litho Ltd, East Kilbride, Scotland

Preface

This book is intended for students taking the first year of an engineering degree or diploma course and the scope of the book is sufficiently wide to cover the variations in syllabuses and standards of the various universities, polytechnics, colleges and professional bodies.

Each chapter consists of a concise but thorough statement of the theory and a large number of examples and problems. The examples form a major part of the work and have been chosen to amplify the theory.

To meet requests for an enlarged edition, many additional examples and problems have been added, particularly to the general dynamics chapter. These have been taken from recent university and engineering institution examination papers and reflect the changing trends since this book was first published. All numerical work has been checked and some out-dated examples have been deleted.

The opportunity has been taken to include an introductory chapter on automatic control, since this topic now features in most degree and diploma courses. The treatment is elementary and is intended to do no more than introduce the student to the language and concepts of automatic control, thus paving the way for the study of specialized books on the subject.

The authors acknowledge with thanks the permission granted by the Senates of the Universities of London and Glasgow, the Institution of Mechanical Engineers and the Council of Engineering Institutions to use questions set at their examinations. These have been designated U. Lond., U. Glas., I. Mech. E. and C. E. I. respectively.

The authors are indebted to their colleagues and students for constructive criticism and help in checking the solutions to the examples.

J Hannah
R C Stephens

Contents

1

Dynamics

1.1 Velocity and acceleration

The *velocity* of a body is the rate of change of its displacement (linear or angular) with respect to time. Velocity is a vector quantity, and to specify it completely the magnitude, direction and sense must be known. The *speed* of a body is merely the magnitude of its velocity.

The *acceleration* of a body is the rate of change of its velocity (linear or angular) with respect to time. A body accelerates if there is a change in either the magnitude, direction or sense of its velocity and can thus accelerate without change in speed, as in the case of a body moving in a circular path with uniform speed.

1.2 Equations of uniformly accelerated motion

Let a body having linear motion accelerate uniformly from an initial velocity u to a final velocity v in time t; let the acceleration be f and the distance from the initial position be s.

Then
$$v = u + ft \tag{1.1}$$
$$s = ut + \tfrac{1}{2}ft^2 \tag{1.2}$$
and
$$v^2 = u^2 + 2fs \tag{1.3}$$

The corresponding equations for angular motion are
$$\omega_2 = \omega_1 + \alpha t \tag{1.4}$$
$$\theta = \omega_1 t + \tfrac{1}{2}\alpha t^2 \tag{1.5}$$
and
$$\omega_2^2 = \omega_1^2 + 2\alpha\theta \tag{1.6}$$
where ω_1 and ω_2 are the initial and final angular velocities respectively, θ is the angle turned through in time t and α is the angular acceleration.

1.3 Non-uniform acceleration

If the acceleration is a function of time, distance or velocity, it must be expressed in the form
$$f = \frac{dv}{dt} \quad \text{or} \quad \frac{d^2s}{dt^2} \quad \text{or} \quad v\frac{dv}{ds}$$
which then leads to a differential equation of motion.

The corresponding equations for angular motion are

$$\alpha = \frac{d\omega}{dt} \quad \text{or} \quad \frac{d^2\theta}{dt^2} \quad \text{or} \quad \omega\frac{d\omega}{d\theta}$$

If the acceleration is neither constant nor varying in a mathematical manner, the acceleration at any instant is represented by the slope of the velocity–time graph and the distance travelled is represented by the area under the graph, Fig. 1.1.

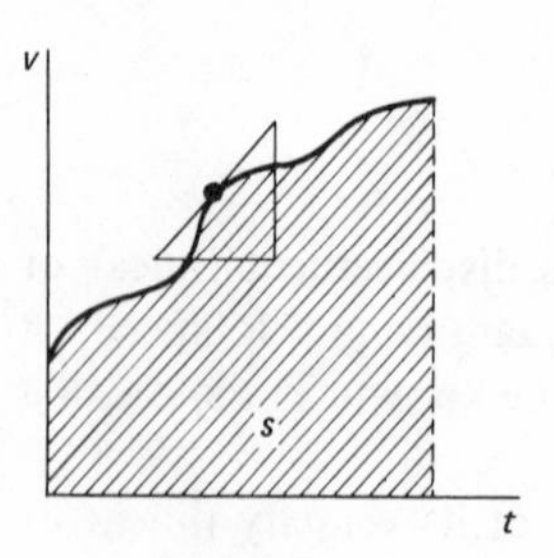

FIG. 1.1

1.4 Relations between linear and angular motion

If a point on the circumference of a rotating body of radius r, Fig. 1.2, moves a distance s, the angle subtended at the centre θ (measured in radians) is given by

$$\theta = s/r$$

or

$$s = \theta r$$

Then, by successive differentiation,

$$v = \frac{ds}{dt} = \frac{d\theta}{dt} r = \omega r \tag{1.7}$$

and

$$f = \frac{dv}{dt} = \frac{d\omega}{dt} r = \alpha r \tag{1.8}$$

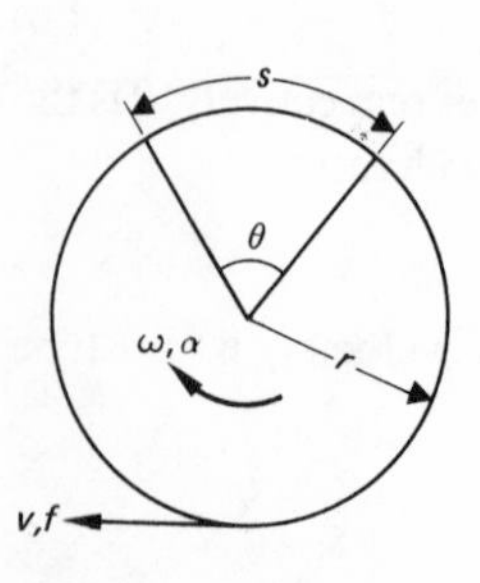

FIG. 1.2

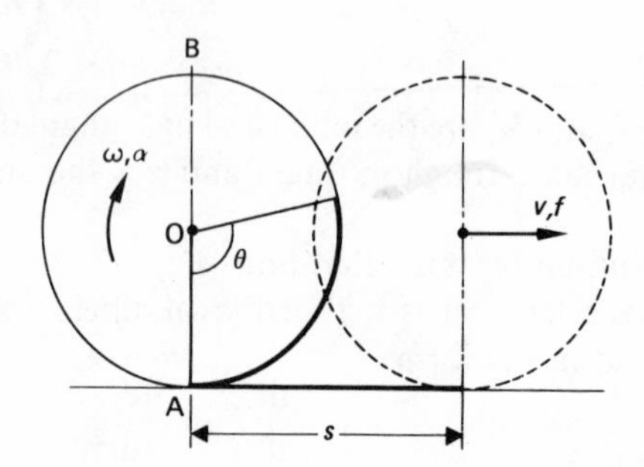

FIG. 1.3

In the case of a wheel or cylinder which rolls without slip on a flat surface, Fig. 1.3, the arc length s is the same as the distance moved by the centre O,

i.e. $$\text{movement of O} = \theta r$$

The velocity and acceleration of O are then given by

$$v = \omega r$$

and $$f = \alpha r$$

as before.

Since the point of contact A is instantaneously at rest, the values of the displacement, velocity and acceleration of the point B are twice those of the centre O, i.e. they are $2\theta r$, $2\omega r$ and $2\alpha r$ respectively.

1.5 Mass, force, weight and momentum

The *mass* of a body is determined by comparison with a standard mass, using a beam-type balance. Thus mass is independent of gravitational acceleration since any variation in g will have an equal effect on the standard mass.

Force is that which tends to change the state of rest or uniform motion of a body. Unit force is that required to give unit acceleration to unit mass.

The *weight* of a body is the force of attraction which the earth exerts upon it and is determined by a suitably-calibrated spring-type balance. Thus the weight varies from place to place as g varies but is standardized at a point where g has the value 9·806 65 m/s^2. For normal engineering purposes, however, g is taken as 9·81 m/s^2.

The *momentum* of a body is the product of its mass and velocity.

1.6 Newton's laws of motion

(1) Every body continues in its state of rest or uniform motion in a straight line, unless acted upon by an external force.
(2) The rate of change of momentum is proportional to the applied external force and takes place in the direction of the force.
(3) To every action there is an equal and opposite reaction.

From the second law,

$$\begin{aligned} \text{force} &\propto \text{rate of change of momentum} \\ &\propto \text{mass} \times \text{rate of change of velocity} \end{aligned}$$

i.e. $$P = kmf$$

where k is a constant.

The units of the quantities are chosen so as to make the value of k unity,

i.e. $$P = mf \tag{1.9}$$

In the *Système International d'Unités*, the fundamental quantities are mass, length and time, force being a derived quantity. The unit of mass is the kilogramme, the unit of length is the metre and the unit of force is the newton, which is the force required to give a mass of 1 kg an acceleration of 1 m/s^2. Then

$$P\ (\text{N}) = m\ (\text{kg}) \times f\ (\text{m/s}^2)$$

Since a body falling freely under the earth's gravitational force has an acceleration g, the weight

$$W = mg$$

1.7 Impulse

The *impulse* of a constant force P acting for a time t is the product Pt. If, during this time, the velocity changes from u to v, then

$$P = mf = m(v - u)/t$$

or

$$Pt = m(v - u) \tag{1.10}$$

i.e. impulse of force = change of momentum

A force which acts for a very short time is referred to as an *impulsive force*.

The units of impulse are N s or kg m/s.

If the force varies with time, as shown in Fig. 1.4, then

$$\text{impulse} = \int_{t_1}^{t_2} P\,\mathrm{d}t$$

$$= \text{area under graph}$$

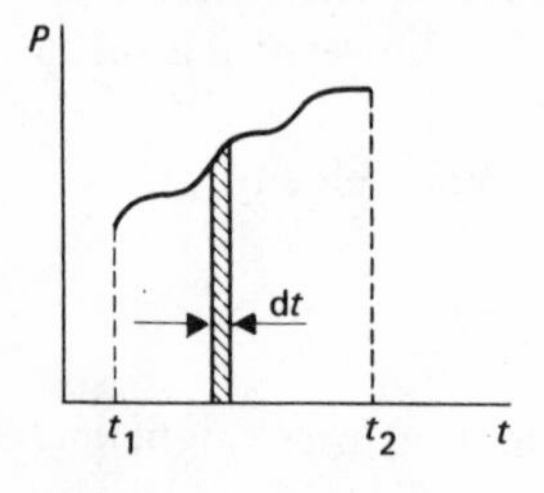

FIG. 1.4

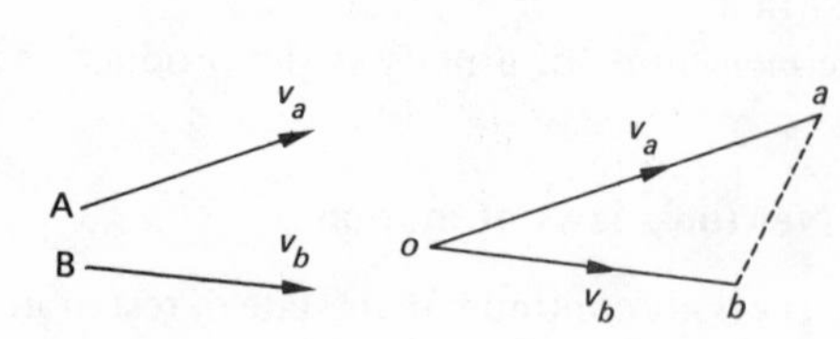

FIG. 1.5

1.8 Relative velocity

If two bodies A and B are moving with velocities v_a and v_b respectively, Fig. 1.5, then the relative velocity of one to the other is the vector difference of v_a and v_b, i.e. if vectors *oa* and *ob*, representing v_a and v_b in magnitude, direction and sense, are drawn from the same point *o*, then *ab* represents the velocity of B relative to A and *ba* the velocity of A relative to B.

If *oa* and *ob* represent the velocities of the same body at different times, then *ab* represents the change in its velocity.

1.9 Centripetal acceleration and centrifugal force

Consider a body of mass m moving in a circular path of radius r with constant speed v, Fig. 1.6. If it moves from A to B in time $\mathrm{d}t$ and the angle AOB is $\mathrm{d}\theta$,

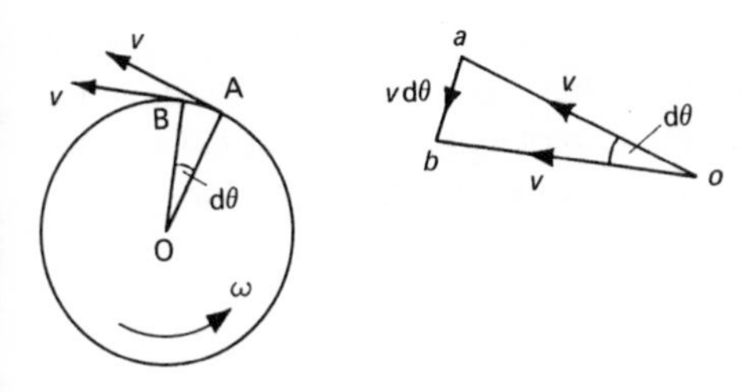

FIG. 1.6

then, from the relative velocity diagram, the change of velocity is represented by *ab*.

Thus $$\text{change of velocity} = v\,\mathrm{d}\theta$$

$$\therefore \text{acceleration} = v\frac{\mathrm{d}\theta}{\mathrm{d}t}$$

i.e. $$f = v\omega$$

where ω is the angular velocity of OA.

But $$v = \omega r \qquad \text{from equation 1.7}$$

$$\therefore f = \omega^2 r \quad \text{or} \quad v^2/r \tag{1.11}$$

This acceleration is directed towards the centre of rotation, O, and is called the *centripetal acceleration*. The radially inward, or centripetal, force required to produce this acceleration is given by

$$P = mf = m\omega^2 r \quad \text{or} \quad mv^2/r \tag{1.12}$$

If a body rotates at the end of an arm, this force is provided by the tension in the arm. The reaction to this force acts at the centre of rotation and is called the *centrifugal force*. It represents the inertia force of the body, resisting the change in the direction of its motion.

A common concept of centrifugal force in engineering problems is to regard it as the radially outward force which must be applied to a body to convert the dynamical condition to the equivalent static condition; this is known as *d'Alembert's Principle*.

Thus the system shown in Fig. 1.7(a) is equivalent to that in Fig. 1.7(b), which is frequently more easy to consider. This concept is particularly useful in problems on engine governors and balancing of rotating masses.

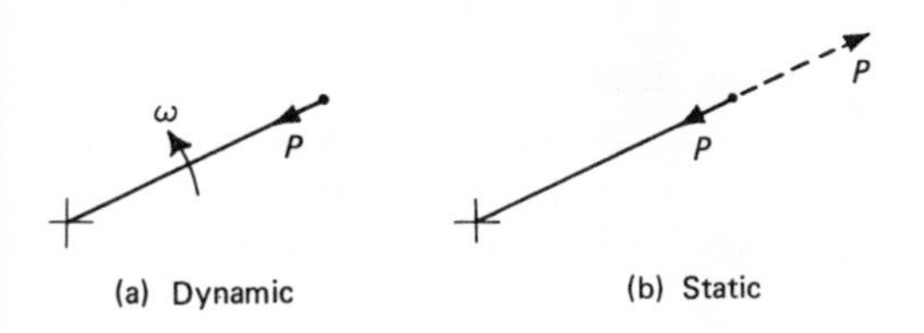

FIG. 1.7

1.10 Work and power

Work is the product of the average force and the distance moved in the direction of the force by its point of application. The unit of work is the joule (J), which is the work done by a force of 1 N moving through a distance of 1 m.

If a constant force P moves through a distance x,

$$\text{work done} = Px$$

If the force varies linearly from zero to a maximum value P,

$$\text{work done} = \tfrac{1}{2}Px$$

In the general case where $P = f(x)$,

$$\text{work done} = \int_0^x f(x)\,\mathrm{d}x$$

Power is the rate of doing work. The unit of power is the watt (W), which is 1 J/s or 1 N m/s. Thus the power developed by a force P N moving at v m/s is Pv W.

1.11 Energy

Energy is the capacity to do work, mechanical energy being equal to the work done on a body in altering either its position, velocity or shape.

The *potential energy* (P.E.) of a body is the energy it possesses due to its position and is equal to the work done in raising it from some datum level. Thus the P.E. of a body of mass m at a height h above datum level is mgh.

The *kinetic energy* (K.E.) of a body is the energy it possesses due to its velocity. If a body of mass m attains a velocity v from rest in time t under the influence of a force P and moves a distance s, then

$$\text{work done by } P = P \times s$$

i.e.

$$\text{K.E. of body} = mf \times v^2/2f = \tfrac{1}{2}mv^2 \qquad (1.13)$$

The *strain energy* of an elastic body is the energy stored when the body is deformed. In the simple case of a body which is strained from its natural position, the relation between the deformation x and the straining force P is a straight line passing through the origin, Fig. 1.8, and the work done is represented by the area of the triangle. Thus

$$\text{work done} = \text{strain energy} = \tfrac{1}{2}Px$$
$$= \tfrac{1}{2}Sx^2 \qquad (1.14)$$

where S is the stiffness (i.e. force per unit deflection).

In the more general case of a body with some initial distortion, e.g. a buffer spring with initial compression, the strain energy is represented by the area under the force–deformation graph, Fig. 1.9. If P_0 is the initial force when $x = 0$, then

$$\text{strain energy} = P_0x + \tfrac{1}{2}Sx^2 \qquad (1.15)$$

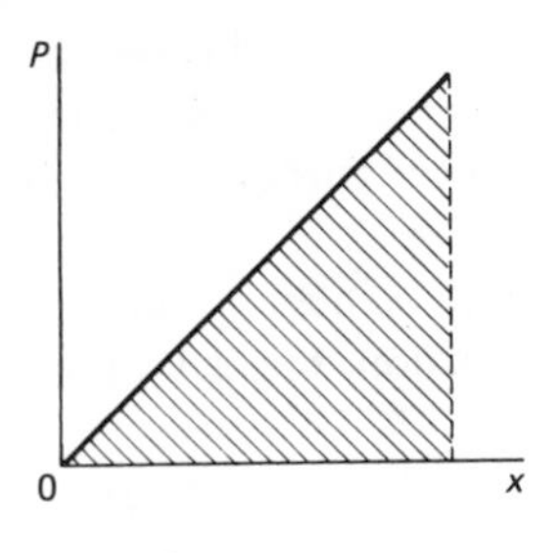

FIG. 1.8

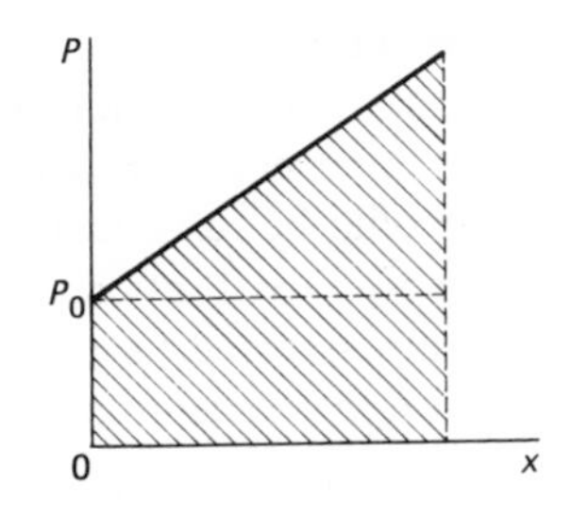

FIG. 1.9

The term $P_0 x$ represents the additional work done by the initial spring force moving through a distance x.

1.12 Principle of conservation of energy

Energy can neither be created nor destroyed. It may exist in a variety of forms, such as mechanical, electrical or heat energy, but a loss of energy in any one form is always accompanied by an equivalent increase in another form. When work is done on a rigid body, the work is converted into kinetic or potential energy or is used in overcoming friction. If the body is elastic, some of the work will also be stored as strain energy.

1.13 Principle of conservation of linear momentum

The total momentum of a system of masses in any one direction remains constant unless acted upon by an external force in that direction.

1.14 Collision of bodies moving in a straight line

When two bodies, moving in the same straight line, collide, each exerts an impulse on the other, these impulses being equal in magnitude and opposite in direction. Thus the changes in momenta of the two masses are equal and opposite, so that the total momentum before and after contact remains constant; this accords with the principle of conservation of momentum.

Let a body of mass m_1, moving with velocity u_1, collide with a body of mass m_2, moving with velocity u_2 in the same straight line, Fig. 1.10.

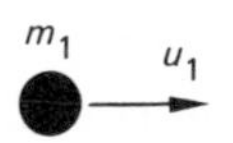

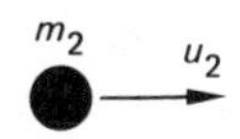

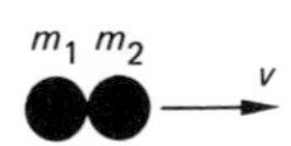

FIG. 1.10

If the bodies remain coupled together after impact and move on with a common velocity v, then

$$m_1 u_1 + m_2 u_2 = (m_1 + m_2)v \tag{1.16}$$

The loss of kinetic energy at impact is given by

$$\tfrac{1}{2}m_1 u_1^2 + \tfrac{1}{2}m_2 u_2^2 - \tfrac{1}{2}(m_1 + m_2)v^2 \tag{1.17}$$

If the bodies are completely inelastic, the energy is lost to the system, appearing mainly as heat energy, but if the bodies are partially or completely elastic, some or all of this energy is stored in the form of strain energy, such as in buffer springs, and could be subsequently recovered.

If the bodies move apart after collision with velocities v_1 and v_2 respectively, which are assumed to be in the same direction, Fig. 1.11, then from the conservation of momentum,

$$m_1 u_1 + m_2 u_2 = m_1 v_1 + m_2 v_2 \tag{1.18}$$

FIG. 1.11

If the bodies are perfectly elastic, the strain energy at impact will be recovered and so the kinetic energy after impact will be the same as that before impact,

i.e. $$\tfrac{1}{2} m_1 u_1^2 + \tfrac{1}{2} m_2 u_2^2 = \tfrac{1}{2} m_1 v_1^2 + \tfrac{1}{2} m_2 v_2^2 \tag{1.19}$$

From equations 1.18 and 1.19,

$$v_1 - v_2 = -(u_1 - u_2) \tag{1.20}$$

Thus, when two perfectly elastic bodies collide, the relative velocity of separation is equal to the relative velocity of approach but in the opposite direction.

If the two bodies are not perfectly elastic, there will always be some loss of energy at impact and the relative velocity of separation will be less than the relative velocity of approach. The ratio of these relative velocities is called the *coefficient of restitution* and is denoted by e, so that

$$v_1 - v_2 = -e(u_1 - u_2) \tag{1.21}$$

This is known as *Newton's Law of Impact*.

The value of e is constant for any given pair of bodies, ranging from zero for completely inelastic impact to unity for perfectly elastic impact.

1.15 Moment of inertia

If the mass of every particle of a body is multiplied by the square of its distance from an axis, the summation of these quantities for the whole body is termed the moment of inertia of the body about the axis and is denoted by I.

The moment of inertia of a particle of mass dm, Fig. 1.12, at a distance l from an axis through O perpendicular to the plane of the paper is d$m\, l^2$. Hence the moment of inertia of the whole body about O,

$$I = \int \mathrm{d}m\, l^2 \tag{1.22}$$

If the total mass of the body is m, this may be written $I = mk^2$; k is termed the *radius of gyration* and is the radius at which the mass would have to be concentrated to give the same value of I.

The units of I are $kg\,m^2$.

1.16 Theorem of parallel axes

The moment of inertia of a body about any axis is equal to the moment of inertia of the body about a parallel axis through the centre of mass together with the product of the mass and the square of the distance between the axes.

Let I_G be the moment of inertia of a body about an axis through the centre of mass G, Fig. 1.13. It is required to find the moment of inertia about a parallel axis through O, which is a distance h from G.

$$\begin{aligned}\text{Moment of inertia of particle about O} &= dm\,l^2\\ &= dm\{x^2+(h+y)^2\}\\ &= dm\{x^2+h^2+2hy+y^2\}\\ &= dm\{r^2+h^2+2hy\}\end{aligned}$$

Therefore the moment of inertia of the body about O,

$$\begin{aligned}I_O &= \int dm\,r^2 + \int dm\,h^2 + 2h\int dm\,y\\ &= I_G + mh^2 + 2h \times (\text{total moment of the mass about XX})\end{aligned}$$

Since XX passes through the centre of mass G, the total moment of the mass about XX is zero. Hence

$$I_O = I_G + mh^2 \tag{1.23}$$

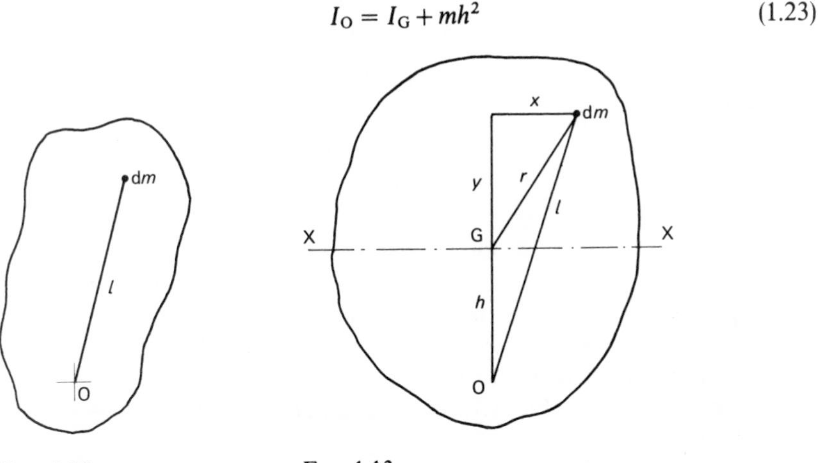

FIG. 1.12 FIG. 1.13

1.17 Moments of inertia for common cases

(a) Uniform disc or cylinder, radius r. I about the central axis is $mr^2/2$ and $k = r/\sqrt{2}$.

(b) Hollow disc or cylinder, external and internal radii R and r respectively. I about the central axis is $m(R^2+r^2)/2$.

(c) Thin disc, radius r. I about a diameter is $mr^2/4$ and $k = r/2$.
(d) Thin uniform rod, length l. I about an axis through the centre perpendicular to the length is $ml^2/12$; I about a parallel axis through one end is $ml^2/3$.
(e) Solid cylinder, radius r, and length l. I about an axis through the centre perpendicular to the length is $m(r^2/4 + l^2/12)$.
(f) Solid sphere, radius r. I about a diametral axis is $\frac{2}{5}mr^2$.

1.18 Torque and angular acceleration

Let the body shown in Fig. 1.14 rotate about an axis through O and let the angular acceleration produced by a torque T be α. Then

$$\text{acceleration of particle of mass } \mathrm{d}m \text{ at radius } l = \alpha l$$

$$\therefore \text{force required to accelerate particle} = \mathrm{d}m\,\alpha l$$

$$\therefore \text{torque required to accelerate particle} = \mathrm{d}m\,\alpha l^2$$

$$\therefore \text{total torque required to accelerate the body} = \int \mathrm{d}m\,\alpha l^2$$

i.e.

$$T = I_O\alpha = mk_O^2\alpha \tag{1.24}$$

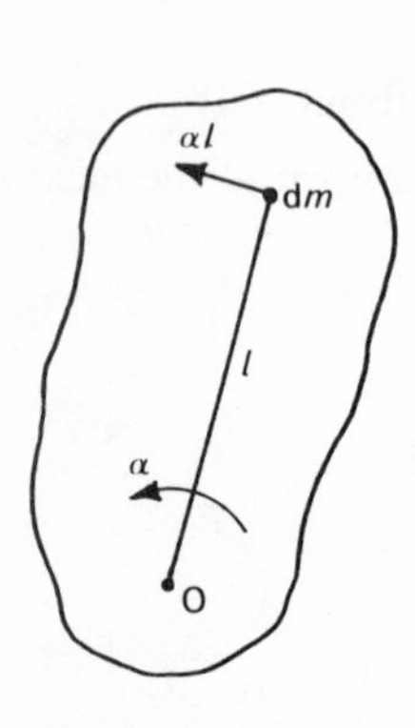

FIG. 1.14

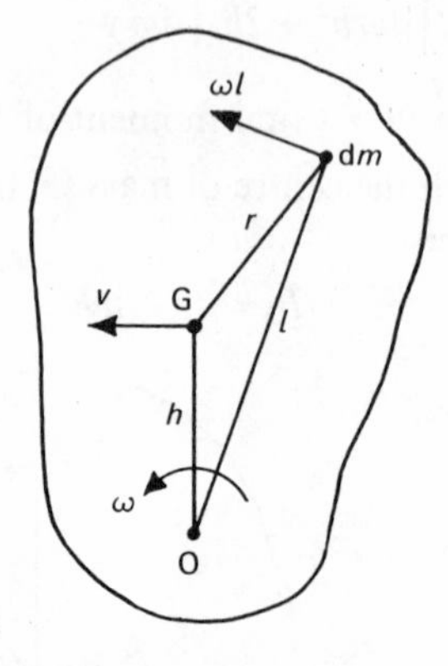

FIG. 1.15

1.19 Angular momentum or moment of momentum

The *angular momentum* of a body about its axis of rotation is the moment of its linear momentum about that axis.

If the body shown in Fig. 1.15 is rotating about an axis through O with angular velocity ω, then

$$\text{velocity of particle of mass } \mathrm{d}m \text{ at radius } l = \omega l$$

$$\therefore \text{momentum of particle} = \mathrm{d}m\,\omega l$$

$$\therefore \text{moment of momentum of particle about O} = \mathrm{d}m\,\omega l^2$$

$$\therefore \text{total moment of momentum of body about O} = \int \mathrm{d}m\,\omega l^2$$

$$= I_O\omega \tag{1.25}$$

If G is the centre of mass, then $I_O = I_G + mh^2$

so that moment of momentum about O $= I_G\omega + mh^2\omega$

$$= I_G\omega + mvh \qquad (1.26)$$

i.e. the angular momentum about any axis O is equal to the angular momentum about a parallel axis through G, together with the moment of the linear momentum about O.

The units of linear and angular momentum are different, being kg m/s and kg m^2/s respectively, so that, for a body possessing both linear and angular momentum, these quantities are not additive.

1.20 Principle of conservation of angular momentum

The total angular momentum of a system of masses about any one axis remains constant unless acted upon by an external torque about that axis.

1.21 Angular impulse

The *angular impulse* of a constant torque T acting for a time t is the product Tt. If, during this time, the angular velocity changes from ω_1 to ω_2, then

$$T = I\alpha = I\frac{(\omega_2 - \omega_1)}{t}$$

or

$$Tt = I(\omega_2 - \omega_1) \qquad (1.27)$$

i.e. impulse of torque = change of angular momentum

A torque which acts for a very short time is referred to as an *impulsive torque.*

The units of angular impulse are N m s or kg m^2/s.

If an impulsive force acts on a body, causing rotation, the angular momentum of the body about the line of action of the force, immediately before and after impact, remains unchanged, since the force has no moment about that line.

1.22 Sudden meshing of rotating wheels

If two rotating gear wheels or friction discs are suddenly meshed together, the same impulsive force acts at the circumference of both wheels, but if the radii are different, the wheels are subjected to different impulsive torques and thus the changes in angular momentum of the wheels are different. The external impulsive torque causing the change in total momentum is supplied by the reactions at the bearings. When the wheels have equal radii, however, the impulsive torques are equal and there is no change in the total angular momentum.

1.23 Work done by a torque

If a constant torque T moves through an angle θ,

$$\text{work done} = T\theta$$

If the torque varies linearly from zero to a maximum value T,

$$\text{work done} = \tfrac{1}{2}T\theta$$

In the general case where $T = f(\theta)$,

$$\text{work done} = \int_0^\theta f(\theta)\,d\theta$$

The power developed by a torque T N m moving at ω rad/s is given by

$$\text{power} = T\omega \quad \text{or} \quad \frac{2\pi NT}{60}\ \text{W}$$

where N is the speed in rev/min.

1.24 Angular kinetic energy

Let a body rotate about O, Fig. 1.15, with an angular velocity ω. Then

$$\text{K.E. of particle of mass } dm = \frac{dm}{2}(\omega l)^2$$

$$\therefore \text{ total angular K.E. of body} = \frac{\omega^2}{2}\int dm\, l^2$$

$$= \tfrac{1}{2} I_O \omega^2 \qquad (1.28)$$

$$= \tfrac{1}{2}(I_G + mh^2)\omega^2$$

$$= \tfrac{1}{2} I_G \omega^2 + \tfrac{1}{2} mv^2 \qquad (1.29)$$

1.25 Total K.E. and rate of change of K.E.

If the centre of mass of a body is moving with linear velocity v and the body is also rotating about the centre of mass with angular velocity ω, then

$$\text{total K.E.} = \text{K.E. of translation} + \text{K.E. of rotation}$$

$$= \tfrac{1}{2} mv^2 + \tfrac{1}{2} I\omega^2$$

where I is the moment of inertia of the body about the centre of mass.

$$\text{Rate of change of K.E.} = \frac{d}{dt}\{\tfrac{1}{2} mv^2 + \tfrac{1}{2} I\omega^2\}$$

$$= \tfrac{1}{2} m\, 2v \frac{dv}{dt} + \tfrac{1}{2} I\, 2\omega \frac{d\omega}{dt}$$

$$= mfv + I\alpha\omega$$

$$= Pv + T\omega \qquad (1.30)$$

$$= \text{work done in unit time}$$

1.26 Equivalent mass of a rotating body

Consider a body of mass m rotating about an axis through O, Fig. 1.16; let the radius of gyration about this axis be k. If a tangential force P, acting at radius r, produces an angular acceleration α, then the equation of angular motion of the body is

$$P \times r = I_O \alpha = mk^2 f/r$$

where f is the linear acceleration at radius r.

$$\therefore P = m(k/r)^2 f \qquad (1.31)$$

which is the equation of linear motion of the body, assumed concentrated at a radius r.

The quantity $m(k/r)^2$ is the equivalent mass of the body, referred to the line of action of P. In problems concerning both linear and angular accelerations, such as vehicle dynamics, this equivalent mass may be added to the actual mass in order to obtain the total equivalent mass having linear acceleration only.

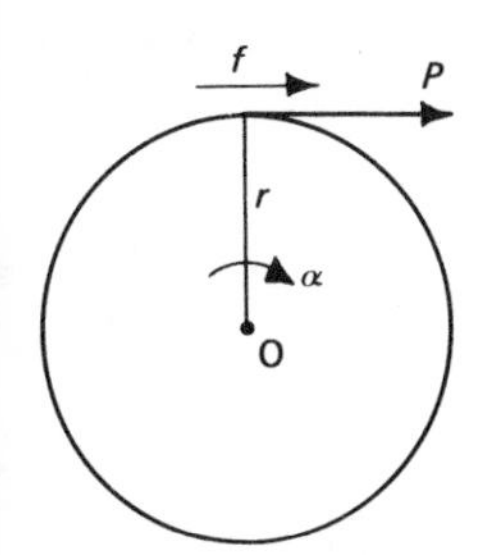

FIG. 1.16

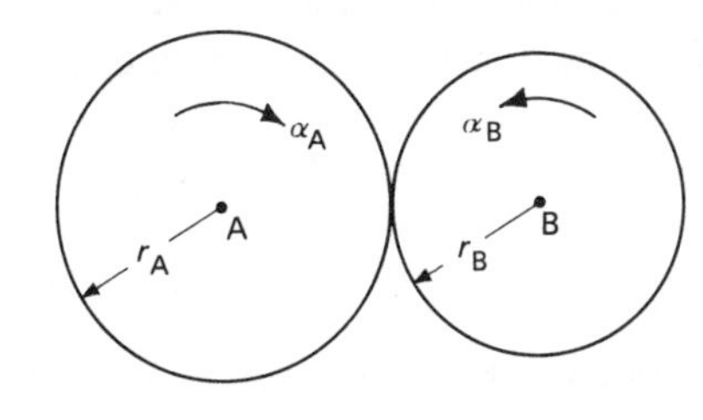

FIG. 1.17

1.27 Acceleration of a geared system

Let two gear wheels A and B, having moments of inertia I_A and I_B respectively, mesh with a speed ratio $\omega_B/\omega_A = n$, Fig. 1.17. If a torque T is applied to wheel A to accelerate the system and α_A and α_B are the accelerations of A and B respectively, then

$$\text{torque required on B to accelerate B} = I_B\alpha_B = I_B n\alpha_A$$

$$\therefore \text{torque required on A to accelerate B} = n^2 I_B\alpha_A$$

since the torques are inversely proportional to the speeds.

$$\text{Torque required on A to accelerate A} = I_A\alpha_A$$

$$\therefore \text{total torque required on A to accelerate A and B} = I_A\alpha_A + n^2 I_B\alpha_A$$

i.e.

$$T = (I_A + n^2 I_B)\alpha_A \tag{1.32}$$

The quantity $I_A + n^2 I_B$ may be regarded as the equivalent moment of inertia of the gears referred to wheel A. This principle may be extended to any number of wheels geared together, the moment of inertia of each wheel in the train being multiplied by the square of its gear ratio relative to the reference wheel. Thus, in problems on hoists, the moments of inertia of the various gears may be reduced to an equivalent moment of inertia of the motor shaft.

The above result can also be obtained from the general principle that the net energy supplied to a system in unit time is equal to the rate of change of its kinetic energy (see Section 1.25). Thus

$$T\omega_A = \frac{d}{dt}\{\tfrac{1}{2}I_A\omega_A^2 + \tfrac{1}{2}I_B\omega_B^2\}$$

$$= \tfrac{1}{2}\{I_A + n^2 I_B\} \times \frac{d}{dt}(\omega_A)^2$$

$$= \tfrac{1}{2}\{I_A + n^2 I_B\} \times 2\omega_A\alpha_A$$

i.e.

$$T = \{I_A + n^2 I_B\}\alpha_A$$

This torque is in addition to any torque required on A to overcome external resisting torques applied to A and/or B.

If P is the tangential force between the teeth, then

$$P = \frac{I_B\alpha_B}{r_B} \tag{1.33}$$

or

$$P = \frac{T - I_A\alpha_A}{r_A}$$

where r_A and r_B are the pitch circle radii* of A and B respectively.

1.28 Equivalent dynamical system

It is required to replace a rigid body by a dynamically equivalent system of concentrated masses. The necessary conditions are as follows:

(1) The total mass must be the same in each case.
(2) The position of the centre of mass must be the same in each case.
(3) The moment of inertia about an axis through the centre of mass must be the same in each case.

Thus, if a body of mass m and radius of gyration k about its centre of mass is to be replaced by an equivalent two-mass system, Fig. 1.18, then

$$m_1 + m_2 = m \tag{1.34}$$

$$m_1 a = m_2 b \tag{1.35}$$

$$m_1 a^2 + m_2 b^2 = mk^2 \tag{1.36}$$

From equations 1.34 and 1.35,

$$m_1 = \frac{b}{a+b}m \quad \text{and} \quad m_2 = \frac{a}{a+b}m$$

Substituting in equation 1.36 gives the essential condition for the placing of the masses, i.e. $ab = k^2$. Either a or b can be chosen arbitrarily and the other term obtained from this relation.

1.29 Maximum acceleration of vehicles

Assuming that sufficient power is available, the maximum possible acceleration of a vehicle is limited by the adhesion between the tyres and road. The friction force available depends on whether the power is supplied to the rear, front or all wheels.

* See Section 11.1.

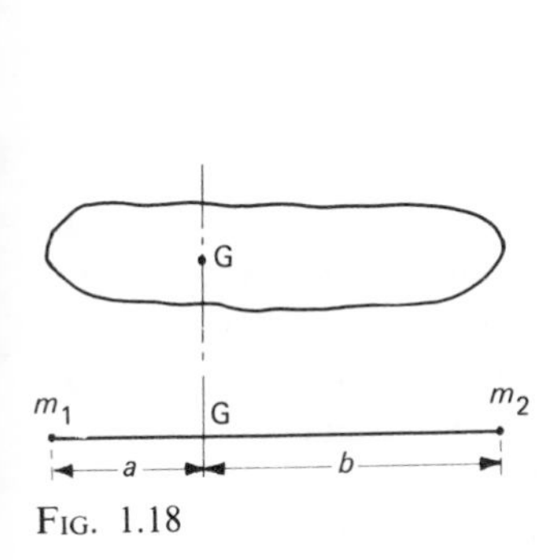

Fig. 1.18

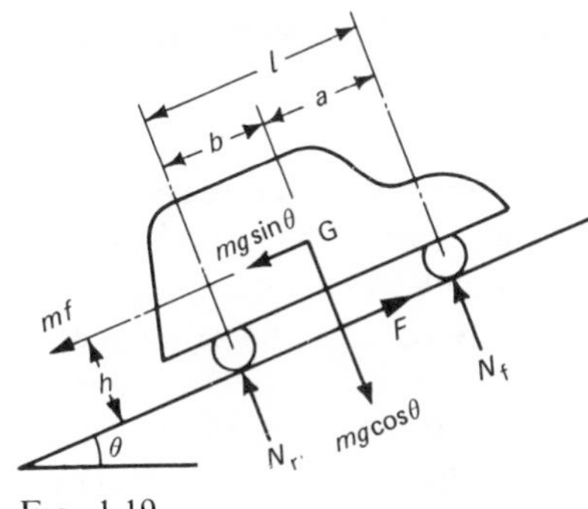

Fig. 1.19

When a car is accelerating or hill-climbing, the rear wheel reaction is increased due to the inertia force and weight component down the slope and the front wheel reaction is correspondingly decreased. Thus, in these circumstances, the wheel grip is increased for rear wheel drives but reduced for front wheel drives.

In the case of four wheel drives, slipping would only occur simultaneously at the rear and front wheels if the wheel reactions were equal but, since these will vary with the acceleration and gradient of slope, this situation is very unlikely to occur and the maximum friction force is governed by the limiting friction at the slipping wheels.

Consider a vehicle of mass m accelerating up a gradient which is inclined at an angle of θ to the horizontal, Fig. 1.19. Let the tractive force be F, irrespective of the type of drive, and the reactions at the rear and front wheels be N_r and N_f respectively.

Resolving forces perpendicular and parallel to the incline,

$$N_f + N_r = mg\cos\theta \tag{1.37}$$

and

$$F = mg\sin\theta + mf \tag{1.38}$$

Taking moments about the centre of mass, G,

$$Fh = N_r b - N_f a \tag{1.39}$$

For rear wheel drives,

$$F = \mu N_r$$

For front wheel drives,

$$F = \mu N_f$$

In the case of four wheel drives, the arrangement of the drive ensures that equal torques are applied to the rear and front wheels, so that the tractive force at the rear and front axles are equal. From equations 1.37, 1.38 and 1.39, it is possible to determine which wheels will reach their limiting adhesion first. If $N_f < N_r$, the front wheels will slip first, giving a limiting friction force of μN_f and, since the same torque is applied to the rear wheels, the total accelerating force is

$$F = 2\mu N_f$$

Similarly, if $N_r < N_f$, the rear wheels will slip first, in which case

$$F = 2\mu N_r$$

1.30 Maximum retardation of vehicles

The analysis of the maximum retardation is identical with that for the maximum acceleration, except that F is now the braking force instead of the accelerating force and, consequently, acts in the opposite direction.

The brakes are normally applied to all four wheels and, if they were arranged to apply equal braking torques at the rear and front, the treatment would be the same as that for the four wheel drive considered in Section 1.29. If, however, the brakes are designed so that the rear and front wheels will slip simultaneously, the braking force can be increased to

$$F = \mu(N_r + N_f) = \mu mg \cos\theta$$

Examples: general dynamics

Example 1.1 The velocity of a train travelling at 100 km/h decreases by 10 per cent in the first 40 s after application of the brakes.

(a) Calculate the velocity at the end of a further 80 s, assuming that, during the whole period of 120 s, the retardation is proportional to the velocity.

(b) Derive an expression for the retarding force in N/tonne mass of the train.

(c) Find the power being dissipated at the end of the whole period, if the train has a mass of 500 tonnes. (U. Lond.)

Solution

(a) Since the retardation is proportional to the velocity,

$$f = -kv = \frac{dv}{dt} \qquad \text{where } k \text{ is a constant}$$

i.e.

$$\frac{dv}{v} = -k\,dt$$

i.e.

$$\ln v = -kt + C$$

When $t = 0$, $v = 100$ km/h,

$$\therefore \ln 100 = C$$

so that

$$\ln(v/100) = -kt$$

or

$$v = 100e^{-kt}$$

When $t = 40$ s, $v = 0{\cdot}9 \times 100 = 90$ km/h

$$\therefore k = 0{\cdot}002\,635$$

Therefore, when $t = 120$ s,

$$v = 100\,e^{-0{\cdot}002\,635 \times 120} = \underline{72{\cdot}9 \text{ km/h}}$$

(b)

$$P = mf = 10^3 kv = \underline{2{\cdot}635v \text{ N/tonne}}$$

where v is in m/s.

(c) At the end of 120 s,

$$\text{total retarding force} = 500 \times 2{\cdot}635 \times 72{\cdot}9 \times 10^3/3600 = 25{\cdot}7 \times 10^3 \text{ N}$$

$$\therefore \text{power} = 25{\cdot}7 \times 10^3 \times 72{\cdot}9 \times 10^3/3600 \text{ W} = \underline{520 \text{ kW}}$$

Example 1.2 Cylinder A in Fig. 1.20 moves towards cylinder B at a speed of 7 m/s along a frictionless rod. Cylinder B is attached to a spring of negligible mass which is undeformed in the position shown. If the coefficient of restitution is 0·9, what is the maximum deflection of the spring, given that the masses of A and B, m_A and m_B, are 5 kg and 20 kg respectively, and the stiffness of the spring, S, is 2000 N/m? (U. Lond.)

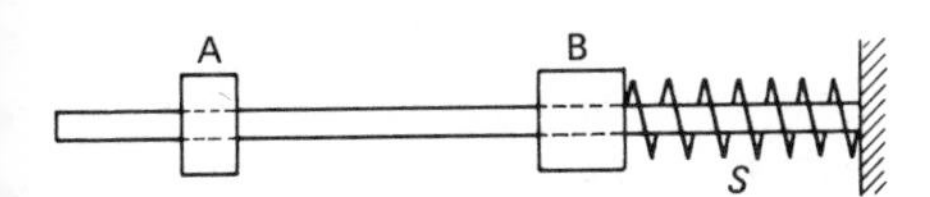

Fig. 1.20

Solution

If the initial velocities of A and B are u_A and u_B respectively, with velocities immediately after impact of v_A and v_B, then, since the total momentum remains constant,

$$m_A v_A + m_B v_B = m_A u_A + m_B u_B \qquad \text{from equation 1.18}$$

i.e.
$$5v_A + 20v_B = 5 \times 7 + 20 \times 0$$

or
$$v_A + 4v_B = 7 \qquad (1)$$

Also
$$v_A - v_B = -e(u_A - u_B) \qquad \text{from equation 1.20}$$
$$= -0{\cdot}9(7-0)$$
$$= -6{\cdot}3 \qquad (2)$$

Hence, from equations (1) and (2),

$$v_B = 2{\cdot}66 \text{ m/s}$$

When B comes to rest, the energy stored in the spring is equal to the loss of kinetic energy of B,

i.e.
$$\tfrac{1}{2}Sx^2 = \tfrac{1}{2}m_B v_B^2$$

i.e.
$$2000x^2 = 20 \times 2{\cdot}66^2$$

from which
$$x = 0{\cdot}266 \text{ m} = \underline{266 \text{ mm}}$$

Example 1.3 Show that the efficiency of a pile driver (i.e. useful work done in driving pile/initial energy of monkey), when employed for driving a given size of pile, increases with the mass of the monkey, assuming no rebound from the head of the pile.

A pile of mass 500 kg is driven by a monkey of mass 800 kg falling a distance of 3 m on to the top of the pile. The pile is driven a distance of 6 mm and there is no rebound. Find the penetration resistance of the pile, assumed constant, and also the amount of energy expended in deforming the head. (U. Lond.)

Solution

Let m_1 = mass of monkey
m_2 = mass of pile
v_1 = velocity of the monkey just before impact
v = common velocity of the monkey and pile after impact
h = height above the pile from which the monkey falls
x = penetration of the pile
R = resistance of the ground

The monkey must be raised back to its initial position after each blow and the energy input is the potential energy of the monkey in its highest position, i.e. $m_1 gh$ or $\frac{1}{2}m_1 v_1^2$, neglecting the effect of the distance penetrated at each blow.

At impact, the monkey and pile move on together with a common velocity v which is obtained by equating the momenta before and after impact,

i.e
$$m_1 v_1 = (m_1 + m_2)v$$

or
$$v = \frac{m_1 v_1}{m_1 + m_2} \tag{1}$$

The kinetic energy of the monkey and pile moving with this common velocity, together with the potential energy lost when descending a distance x, is equal to the work done against the resistance R,

i.e
$$\tfrac{1}{2}(m_1 + m_2)v^2 + (m_1 + m_2)gx = Rx$$

If the potential energy term is neglected, then

$$Rx = \tfrac{1}{2}(m_1 + m_2)v^2 \tag{2}$$

The efficiency, η, is given by

$$\eta = \frac{\text{useful work done}}{\text{initial energy of monkey}}$$

$$= \frac{Rx}{\frac{1}{2}m_1 v_1^2}$$

$$= \frac{\frac{1}{2}(m_1 + m_2)v^2}{\frac{1}{2}m_1 v_1^2} \quad \text{from equation (2)}$$

Substituting for v from equation (1)

$$\eta = \frac{m_1}{m_1 + m_2} \quad \text{which increases with } m_1$$

Note that this expression is approximate since (a) the efficiency varies at each blow as the height to which the monkey has to be raised increases, and (b) the potential energy term, $(m_1 + m_2)gx$, has been ignored.

In this case,

$$m_1 = 800 \text{ kg}$$
$$m_2 = 500 \text{ kg}$$

and $$v_1 = \sqrt{(2gh)} = \sqrt{(2 \times 9{\cdot}81 \times 3)} = 7{\cdot}67 \text{ m/s}$$

Therefore, from equation (1),

$$v = \frac{800}{1300} \times 7{\cdot}67 = 4{\cdot}72 \text{ m/s}$$

and $$\text{work done by } R = Rx = \tfrac{1}{2}(m_1 + m_2)v^2 + (m_1 + m_2)gx$$

or, substituting,

$$R \times 0{\cdot}006 = \tfrac{1}{2} \times 1300 \times 4{\cdot}72^2 + 1300 \times 9{\cdot}81 \times 0{\cdot}006$$
$$= 14\,480 + 77$$
$$= 14\,557 \text{ N}$$
$$\therefore R = \underline{2{\cdot}43 \text{ MN}}$$

Energy expended in deforming the head

= initial K.E. of monkey − K.E. of monkey and pile after impact

$= \tfrac{1}{2} \times 800 \times 7{\cdot}67^2 - 14\,480$ J

$= \underline{9{\cdot}05 \text{ kJ}}$

Note that the loss of potential energy of the monkey and pile in this example is negligible, which justifies the derivation of the formula for efficiency. However, in cases where the penetration is larger in comparison with the drop of the hammer, the loss of potential energy is appreciable and must be taken into account.

Example 1.4 Figure 1.21 shows a tilt hammer, hinged at O, with its head A resting on top of the pile B. The hammer, including the arm OA, has a mass of

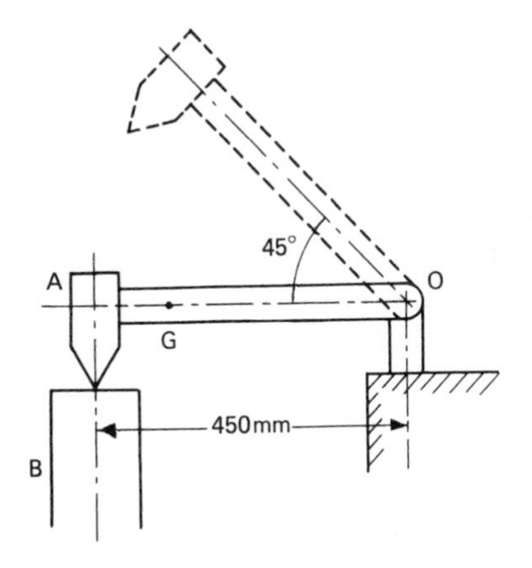

FIG. 1.21

25 kg. Its centre of mass G is 400 mm horizontally from O, and its radius of gyration about an axis through G parallel to the axis of the pin O is 75 mm. The pile has a mass of 135 kg. The hammer is raised through 45° to the position shown in dotted lines, and released. On striking the pile there is no rebound.

Find the angular velocity of the hammer immediately before impact and the linear velocity of the pile immediately after impact. Neglect any impulsive resistance offered by the earth into which the pile is being driven.

(U. Lond.)

Solution

Let ω_1 be the angular velocity of OA before impact, ω_2 be the angular velocity of OA after impact, and v be the initial velocity of pile after impact.

$$\text{Moment of inertia of hammer about O} = mk_0^2 = \frac{25(75^2+400^2)}{10^6} = 4{\cdot}14\ \text{kg m}^2$$

Potential energy in raised position = kinetic energy in lowest position

i.e.
$$25 \times 0{\cdot}4 \sin 45° = \frac{4{\cdot}14}{2} \times \omega_1^2$$

from which
$$\omega_1 = \underline{5{\cdot}79\ \text{rad/s}}$$

Neglecting the impulsive resistance of the earth, the total moment of momentum of the system about O is constant, the impulsive reaction of the hinge having no moment about that point,

i.e. initial moment of momentum about O

= final moment of momentum about O

i.e.
$$4{\cdot}14 \times 5{\cdot}79 = 4{\cdot}14 \times \omega_2 + 135 \times v \times 0{\cdot}45$$

But since the hammer and pile remain in contact after impact, $v = 0{\cdot}45\omega_2$.

$$\therefore\ 24 = 4{\cdot}14 \times \omega_2 + 135 \times (0{\cdot}45\omega_2) \times 0{\cdot}45$$

$$\therefore\ \omega_2 = 0{\cdot}762\ \text{rad/s}$$

$$\therefore\ v = 0{\cdot}762 \times 0{\cdot}45 = \underline{0{\cdot}343\ \text{m/s}}$$

Example 1.5 A vertical pile, 7·6 m long, weighs 650 N/m and is driven by a drop hammer weighing 15 kN. With the last blow, the hammer dropped freely from 5 m and the pile penetrated 80 mm deeper into the ground. The coefficient of restitution of the pile is 0·5. Calculate the carrying capacity of the pile, assuming that frictional forces are constant during penetration. (U. Lond.)

Solution

$$\text{Mass of hammer} = 15 \times 10^3/9{\cdot}81 = 1529\ \text{kg}$$

$$\text{Mass of pile} = 7{\cdot}6 \times 650/9{\cdot}81 = 503{\cdot}6\ \text{kg}$$

The velocity of the hammer at impact is given by

$$u_1 = \sqrt{(2 \times 9{\cdot}81 \times 5)} = 9{\cdot}9\ \text{m/s}$$

If the velocities of the hammer and pile immediately after impact are v_1 and v_2 respectively, the total momentum remains constant,

i.e. $$1529 \times 9{\cdot}9 = 1529v_1 + 503{\cdot}6v_2 \quad (1)$$

Also $$v_1 - v_2 = -e(u_1 - u_2) \quad \text{from equation 1.20}$$
$$= -0{\cdot}5(9{\cdot}9 - 0) = -4{\cdot}95 \quad (2)$$

Hence, from equations (1) and (2),

$$v_2 = 11{\cdot}17 \text{ m/s}$$

Let the resistance of the pile be R, which is its carrying capacity. Then the work done against R is equal to the loss of kinetic and potential energy of the pile which it possesses immediately after impact,

i.e. $$R \times 0{\cdot}08 = \tfrac{1}{2} \times 503{\cdot}6 \times 11{\cdot}17^2 + 503{\cdot}6 \times 9{\cdot}81 \times 0{\cdot}08$$
$$\therefore \; R = \underline{398 \text{ MN}}$$

Example 1.6 The two buffers at one end of a truck each require a force of 0·7 MN/m of compression and engage with similar buffers on a truck which it overtakes on a straight horizontal track. The truck has a mass of 10 tonnes and its initial speed is 1·8 m/s, while the second truck has a mass of 15 tonnes with initial speed 0·6 m/s, in the same direction.

Find: (a) the common speed when moving together during impact, (b) the kinetic energy then lost to the system and the compression of each buffer spring to store this, and (c) the velocity of each truck on separation if only half of the energy stored in the springs is returned. (I. Mech. E.)

Solution

(a) Momentum before impact = momentum at instant of common velocity

i.e. $$10 \times 1{\cdot}8 + 15 \times 0{\cdot}6 = 25v$$
$$\therefore \text{ common velocity, } v = \underline{1{\cdot}08 \text{ m/s}}$$

(b) $$\text{K.E. lost} = [\tfrac{1}{2}(10 \times 1{\cdot}8^2) + \tfrac{1}{2}(15 \times 0{\cdot}6^2) - \tfrac{1}{2}(25 \times 1{\cdot}08^2)] \times 10^3$$
$$= 4350 \text{ J} \quad \text{or} \quad \underline{4{\cdot}35 \text{ kJ}}$$

If x is the compression of each spring, in metres,

strain energy stored in springs = K.E. lost in impact

i.e $$4 \times \frac{0{\cdot}7 \times 10^6 x^2}{2} = 4350 \quad \text{from equation 1.14}$$
$$\therefore \; x = 0{\cdot}0557 \text{ m} \quad \text{or} \quad \underline{55{\cdot}7 \text{ mm}}$$

(c) Final K.E. after separation = K.E. at instant of common velocity
+ $\tfrac{1}{2}$ × strain energy stored in springs

i.e. $$\tfrac{1}{2}(10 \times 10^3 \times v_1^2) + \tfrac{1}{2}(15 \times 10^3 \times v_2^2) = \tfrac{1}{2}(25 \times 10^3 \times 1{\cdot}08^2) + \tfrac{1}{2} \times 4350$$

i.e. $$10v_1^2 + 15v_2^2 = 33{\cdot}55 \quad (1)$$

where v_1 and v_2 are the final velocities of the 10- and 15-tonne trucks respectively.

Also, the initial and final momenta must be the same,

i.e. $$10v_1 + 15v_2 = 25 \times 1{\cdot}08 = 27 \quad (2)$$

Therefore, from equations (1) and (2),

$$v_1 = \underline{0{\cdot}6 \text{ m/s}} \quad \text{and} \quad v_2 = \underline{1{\cdot}4 \text{ m/s}}$$

Example 1.7 Two parallel shafts can be connected by a gear wheel of 65 mm radius sliding on one shaft A to engage with a wheel of 130 mm radius fixed to the other shaft B. Before engagement the shaft speeds of A and B are respectively 1400 rev/min and 900 rev/min in opposite directions. If the inertia of A is equivalent to 9 kg at 75 mm radius and that of B is equivalent to 36 kg at 130 mm radius, calculate the speed of A after engagement and the tangential impulse measured in N s, experienced at the wheel teeth. (I. Mech. E.)

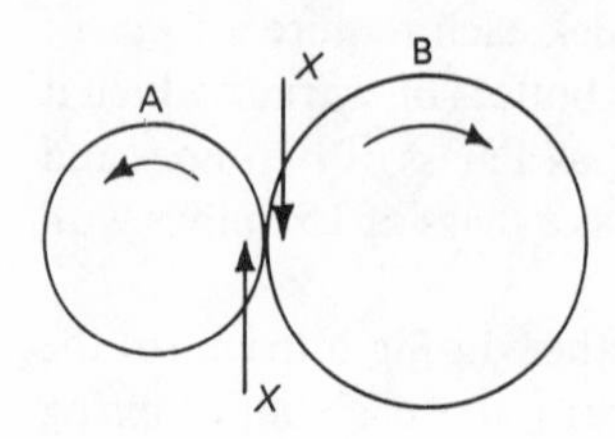

FIG. 1.22

Solution

Referring to Fig. 1.22,

$$I_A = 9 \times 0{\cdot}075^2 = 0{\cdot}0506 \text{ kg m}^2$$
$$I_B = 36 \times 0{\cdot}13^2 = 0{\cdot}608 \text{ kg m}^2$$

Let n_1 and n_2 be the initial and final speeds of A in rev/min, N_1 and N_2 be the initial and final speeds of B in rev/min, and X be the tangential impulse between the teeth.

Then X acts so as to increase the speed of A and decrease the speed of B.

Angular impulse = change of angular momentum

i.e. $$Xr_A = I_A(n_2 - n_1) \times \frac{2\pi}{60} \quad (1)$$

and $$Xr_B = -I_B(N_2 - N_1) \times \frac{2\pi}{60} \quad (2)$$

$$\therefore \frac{r_A}{r_B} = -\frac{I_A\,(n_2 - n_1)}{I_B(N_2 - N_1)}$$

i.e. $$\frac{65}{130} = -\frac{0{\cdot}0506\,(n_2 - 1400)}{0{\cdot}608\,(N_2 - 900)} \quad (3)$$

Also
$$N_2 = \frac{65}{130} \times n_2 = \frac{n_2}{2} \tag{4}$$

Therefore, from equations (3) and (4), the speed of A after engagement,
$$n_2 = \underline{1700 \text{ rev/min}}$$

Substituting in equation (1),
$$0{\cdot}065\,X = 0{\cdot}0506\,(1700 - 1400) \times \frac{2\pi}{60}$$
$$\therefore\ X = \underline{24{\cdot}45 \text{ N s}}$$

Example 1.8 A rigid uniform beam AB, 6 m long, is supported vertically with the end B resting on the ground. End A is released and the beam is allowed to fall. It turns about the end B which remains in its original position. A point C on the beam, 3·6 m from B, strikes the edge of a horizontal step. After the impact, the beam rotates about the edge of the step without slipping. Determine the height of the step if the beam comes momentarily to rest in the horizontal position. (U. Lond.)

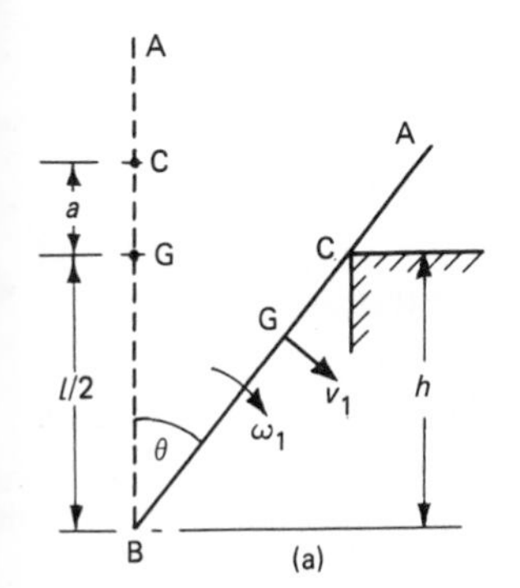

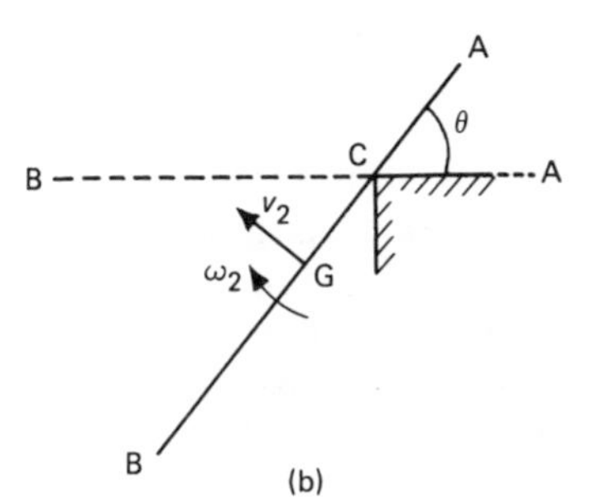

Fig. 1.23

Solution

Referring to Fig. 1.23, let ω_1 and ω_2 be the angular velocities of the beam immediately before and after impact, and v_1 and v_2 be the linear velocities of G immediately before and after impact.

In falling from the vertical position,
$$\text{loss of P.E.} = \text{gain of K.E.}$$
i.e.
$$mg\frac{l}{2}(1-\cos\theta) = \tfrac{1}{2}I_B\omega_1^2 = \tfrac{1}{2}\frac{ml^2}{3}\omega_1^2$$
$$\therefore\ \omega_1 = \sqrt{\left(\frac{3g}{l}(1-\cos\theta)\right)} \tag{1}$$

In moving to the horizontal position after impact,
$$\text{gain of P.E.} = \text{loss of K.E.}$$

i.e.
$$mga\cos\theta = \tfrac{1}{2}I_C\omega_2^2 = \tfrac{1}{2}m\left(\frac{l^2}{12}+a^2\right)\omega_2^2$$

$$\therefore\ \omega_2 = \sqrt{\left(\frac{24ga\cos\theta}{l^2+12a^2}\right)} \tag{2}$$

During impact, the moment of momentum of the beam about C remains constant since the impulsive force at C has no moment about that point,

i.e.
$$I_G\omega_1 - mv_1a = I_G\omega_2 + mv_2a$$

taking clockwise momentum as positive.

But
$$v_1 = \omega_1\frac{l}{2} \quad\text{and}\quad v_2 = \omega_2 a$$

$$\therefore\ m\frac{l^2}{12}\omega_1 - m\omega_1\frac{l}{2}a = m\frac{l^2}{12}\omega_2 + m\omega_2a^2$$

Substituting for ω_1 and ω_2 from equations (1) and (2),

$$\left(\frac{l^2}{12}-\frac{la}{2}\right)\sqrt{\left(\frac{3g}{l}(1-\cos\theta)\right)} = \left(\frac{l^2}{12}+a^2\right)\sqrt{\left(\frac{24ga\cos\theta}{l^2+12a^2}\right)}$$

Therefore, when $l = 6$ m and $a = 0{\cdot}6$ m, $\cos\theta = 0{\cdot}1515$

$$\therefore\ \text{height of step, } h = 3{\cdot}6\cos\theta = \underline{0{\cdot}546\text{ m}}$$

Example 1.9 Two uniform slender bars AB and BC are pin-jointed at B and BC is pinned to a frame at C, as shown in Fig. 1.24. Bar AB is 4 m long and has a mass of 60 kg; bar BC is 2 m long and has a mass of 30 kg.

If a horizontal impulse of $F = 180$ N s is applied to A as shown over such a short period of time that the bars can be considered not to move during the impulse, calculate the angular velocities of the two bars immediately after the impulse. (U. Lond.)

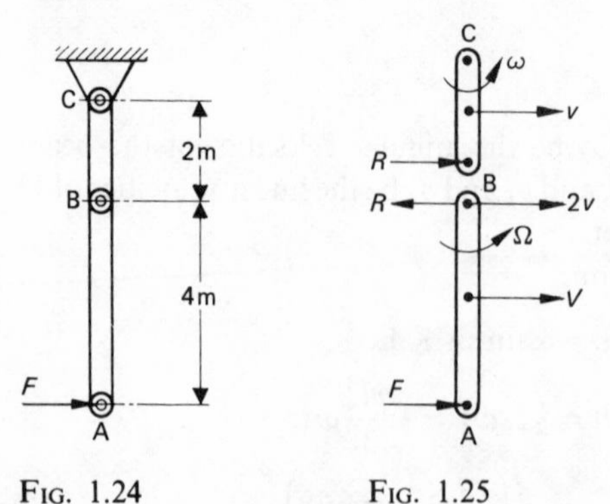

FIG. 1.24

FIG. 1.25

Solution

Let the linear velocities of the centres of AB and BC be V and v respectively, Fig. 1.25, with corresponding angular velocities Ω and ω. It will be assumed that both angular velocities are in an anticlockwise direction. If the impulsive reaction at B is R, this will be to the right on BC to produce the angular velocity ω and in the opposite direction (i.e. to the left) on BA.

For AB,

$$\text{angular impulse} = \text{change of angular momentum}$$

i.e.
$$F \times 4 = I_B\,\Omega$$

i.e.
$$180 \times 4 = \frac{60 \times 4^2}{3}\Omega$$

$$\therefore\ \Omega = \underline{2{\cdot}25 \text{ rad/s}}$$

The velocity of B will be twice that of the centre of BC, so that

$$V - 2v = \Omega \times 2 = 4{\cdot}5 \text{ rad/s}$$

Hence
$$V = 4{\cdot}5 + 2v$$

$$\text{Impulse} = \text{change of linear momentum}$$

i.e.
$$F - R = mV$$

i.e.
$$180 - R = 60(4{\cdot}5 + 2v) = 270 + 120v$$

i.e.
$$R = -120v - 90$$
$$= -120\omega - 90$$

since $v = \omega \times 1$.

For BC,

$$\text{angular impulse} = \text{change of angular momentum}$$

i.e.
$$R \times 2 = I_C\,\omega$$

i.e
$$(-120\omega - 90) \times 2 = \frac{30 \times 2^2}{3}\omega$$

from which
$$\omega = \underline{-0{\cdot}643 \text{ rad/s}}$$

The rotation of BC is therefore in the opposite direction to that initially assumed.

Example 1.10 A railway truck of mass 20 tonnes, moving at 6·5 km/h, is brought to rest by a buffer stop. The buffer exerts a force of 22·5 kN initially, and this force increases uniformly by 60 kN for each 1 m compression of the buffer. Neglecting any loss of energy at impact, find the maximum compression of the buffer and the time required for the truck to be brought to rest.

(U. Lond.)

Solution

When the buffer is compressed a distance x m,

$$\text{restoring force} = 22{\cdot}5 + 60x \text{ kN}$$

$$\therefore\ (22{\cdot}5 + 60x) \times 10^3 = -20 \times 10^3 \frac{d^2x}{dt^2}$$

i.e
$$\frac{d^2x}{dt^2} + 3x = -1{\cdot}125$$

The solution is*

$$x = A \cos 1{\cdot}732t + B \sin 1{\cdot}732t - 0{\cdot}375$$

When $t = 0$, $x = 0$

$$\therefore\ 0 = A - 0{\cdot}375 \qquad \text{i.e. } A = 0{\cdot}375$$

When $t = 0$, $dx/dt = (6{\cdot}5 \times 10^3)/3600 = 1{\cdot}805$ m/s

$$\therefore\ 1{\cdot}805 = 1{\cdot}732B \qquad \text{i.e. } B = 1{\cdot}042$$

$$\therefore\ x = 0{\cdot}375 \cos 1{\cdot}732t + 1{\cdot}042 \sin 1{\cdot}732t - 0{\cdot}375$$

and
$$\frac{dx}{dt} = -0{\cdot}65 \sin 1{\cdot}732t + 1{\cdot}805 \cos 1{\cdot}732t$$

When the truck is brought to rest, $dx/dt = 0$

$$\therefore\ 0{\cdot}65 \sin 1{\cdot}732t = 1{\cdot}805 \cos 1{\cdot}732t$$

i.e.
$$\tan 1{\cdot}732t = 2{\cdot}78$$

from which
$$\underline{t = 0{\cdot}707 \text{ s}}$$

The maximum compression of the buffer is given by

$$x = 0{\cdot}375 \cos (1{\cdot}732 \times 0{\cdot}707) + 1{\cdot}042 \sin (1{\cdot}732 \times 0{\cdot}707) - 0{\cdot}375 = \underline{0{\cdot}733 \text{ m}}$$

Example 1.11 A frictionless flexible chain, of total length 3 m, hangs over the edge of a table by 1 m and is held in that position.

Determine the time taken for the chain to just slide off the table if released.

(U. Lond.)

Solution

Let m be the mass of the chain per unit length, Fig. 1.26.

When a length x m has slid off the table,

$$\text{accelerating force} = mg(1 + x)$$

$$\therefore\ m \times 9{\cdot}81(1 + x) = 3m\frac{d^2x}{dt^2}$$

i.e.
$$\frac{d^2x}{dt^2} - 3{\cdot}27x = 3{\cdot}27$$

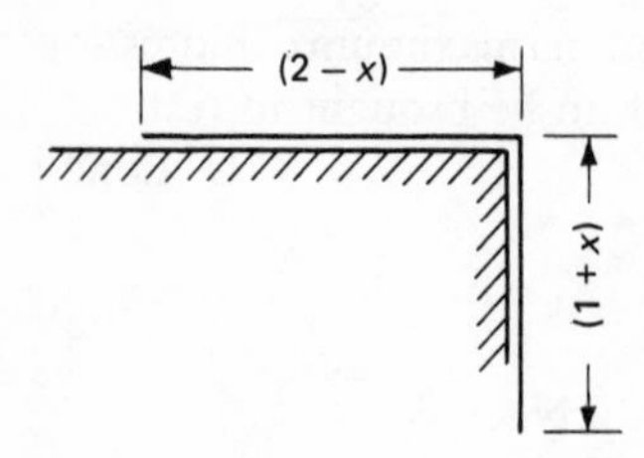

Fig. 1.26

* See the Appendix.

The solution is*

$$x = A\cosh 1{\cdot}81t + B\sinh 1{\cdot}81t - 1$$

When $t = 0$, $x = 0$

$$\therefore\ 0 = A - 1 \qquad \text{i.e. } A = 1$$

When $t = 0$, $\mathrm{d}x/\mathrm{d}t = 0$

$$\therefore\ 0 = 1{\cdot}81B \qquad \text{i.e. } B = 0$$

$$\therefore\ x = 1(\cosh 1{\cdot}81t - 1)$$

When $x = 2\,\text{m}$, $\quad 2 = 1(\cosh 1{\cdot}81t - 1)$

from which $\quad \underline{t = 0{\cdot}972\,\text{s}}$

Alternative Solution

If the chain moves a distance $\mathrm{d}x$ from the position shown,

$$\text{work done} = \text{loss of P.E.} = mg(1+x)\,\mathrm{d}x$$

$$\therefore\ \text{total work done for a movement } x = \int_0^x mg(1+x)\,\mathrm{d}x$$

$$= mg\left(x + \frac{x^2}{2}\right) + C$$

When $x = 0$, work done $= 0$, $\therefore\ C = 0$

Also $\quad \text{gain in K.E.} = \dfrac{3m}{2}v^2$

$$\therefore\ \frac{3m}{2}v^2 = mg\left(x + \frac{x^2}{2}\right)$$

$$\therefore\ v^2 = \tfrac{2}{3} \times 9{\cdot}81\left(x + \frac{x^2}{2}\right)$$

$$\therefore\ v = \frac{\mathrm{d}x}{\mathrm{d}t} = 2{\cdot}56\sqrt{\left(x + \frac{x^2}{2}\right)} = 1{\cdot}805\sqrt{(2x + x^2)}$$

$$\therefore\ \mathrm{d}t = \frac{\mathrm{d}x}{1{\cdot}81\sqrt{(2x+x^2)}}$$

$$\therefore\ t = \int_0^2 \frac{\mathrm{d}x}{1{\cdot}81\sqrt{\{(x+1)^2 - 1^2\}}}$$

$$= \frac{1}{1{\cdot}81}\left[\cosh^{-1}(x+1)\right]_0^2$$

$$= \frac{1}{1{\cdot}81}\left[\cosh^{-1} 3 - \cosh^{-1} 1\right]$$

$$= \frac{1}{1{\cdot}81} \times 1{\cdot}7615 = \underline{0{\cdot}972\,\text{s}}$$

* See the Appendix.

Example 1.12 An electric motor drives a machine through a speed-reducing gear of ratio 9:1. The motor armature, with its shaft and gear wheel, has moment of inertia 0·5 kg m². The rotating part of the driven machine has moment of inertia 40 kg m². The driven machine has a constant resisting torque of 100 N m and the efficiency of the reduction gear is 95 per cent.
(a) What power must the motor develop to drive the machine at a uniform speed of 160 rev/min?
(b) The torque developed on the motor armature, in starting from rest, is 30 N m. What time will be required for the speed of the machine to increase from zero to 60 rev/min?
(c) If the gear ratio were altered so as to give the machine the greatest possible angular acceleration in starting from rest, what would then be the gear ratio? The starting torque of the motor is 30 N m as before. (U. Lond.)

Solution

(a)

$$\text{Input power} = \frac{\text{output power}}{0{\cdot}95} = \frac{2\pi \times 100 \times 160}{60 \times 0{\cdot}95}$$
$$= 1765\text{ W} \quad \text{or} \quad \underline{1{\cdot}765\text{ kW}}$$

(b) Let A and B refer to the input and output shafts respectively, and let the gear ratio $\omega_A/\omega_B = n$.

$$\text{Torque available at gear box} = 30 - I_A\alpha_A$$
$$= 30 - 0{\cdot}5\, n\alpha_B \text{ N m}$$
$$\therefore \text{ torque available at machine} = (30 - 0{\cdot}5\, n\alpha_B) \times n \times 0{\cdot}95$$
$$= 28{\cdot}5n - 0{\cdot}475n^2\alpha_B \text{ N m}$$

The equation of motion of the machine is therefore

$$28{\cdot}5n - 0{\cdot}475n^2\alpha_B = 100 + I_B\alpha_B$$
$$= 100 + 40\alpha_B$$

from which

$$\alpha_B = \frac{28{\cdot}5n - 100}{40 + 0{\cdot}475n^2}$$

When $n = 9$,

$$\alpha_B = 2 \text{ rad/s}^2$$
$$\therefore 2 = \frac{2\pi}{60} \times \frac{60}{t}$$
$$\therefore t = \underline{3{\cdot}14 \text{ s}}$$

(c) For maximum acceleration, $d\alpha_B/dn = 0$

i.e

$$(40 + 0{\cdot}475n^2) \times 28{\cdot}5 = (28{\cdot}5 - 100) \times 0{\cdot}95n$$

i.e.

$$n^2 - 7n - 84 = 0$$

from which

$$n = \underline{13{\cdot}32}$$

Example 1.13 A ventilating fan is driven by an electric motor and normally operates at a constant speed of 600 rev/min, with the motor delivering a torque of 400 N m. Determine the time required for the fan to reach 540 rev/min from rest, assuming that the resisting torque of the fan is proportional to the square of the speed and that the motor torque is constant. The moment of inertia of the fan and other rotating parts is 12 kg m^2. (C.E.I.)

Solution

At a constant speed of 600 rev/min, the motor torque is completely absorbed in overcoming the resisting torque,

i.e. $$400 = k\left(600 \times \frac{2\pi}{60}\right)^2 \quad \text{where } k \text{ is a constant}$$

from which $$k = \frac{1}{\pi^2}$$

At a speed of ω,

$$\text{accelerating torque} = \text{driving torque} - \text{resisting torque}$$
$$= 400 - \frac{1}{\pi^2}\omega^2$$

Hence $$400 - \frac{\omega^2}{\pi^2} = I\alpha = 12\frac{d\omega}{dt}$$

or $$dt = \frac{12\pi^2\, d\omega}{400\pi^2 - \omega^2} = \frac{12\pi^2\, d\omega}{(20\pi)^2 - \omega^2}$$

$$\therefore\ t = 12\pi^2 \int_0^{540 \times 2\pi/60} \frac{d\omega}{(20\pi)^2 - \omega^2}$$

$$= \frac{12\pi^2}{40\pi}\left[\ln\frac{20\pi + \omega}{20\pi - \omega}\right]_0^{18\pi}$$

$$= 0{\cdot}3\pi\left[\ln\frac{38\pi}{2\pi} - \ln\frac{20\pi}{20\pi}\right]$$

$$= 0{\cdot}3\pi \ln 19$$

i.e. $$t = \underline{2{\cdot}775 \text{ s}}$$

Example 1.14 A cord is wound round the inner drum, radius 0·1 m, of a cylindrical spool of outer radius 0·15 m and is pulled horizontally by a force of 200 N, Fig. 1.27. The mass of the spool is 50 kg and the radius of gyration about its axis is 0·12 m. If the spool is supported on a rough horizontal surface, determine the minimum coefficient of friction, μ, between the spool and surface for the spool to roll without slip.

If $\mu = 0{\cdot}2$, calculate the acceleration of the centre of the spool and its angular acceleration. (U. Lond.)

Solution

Assuming that the spool rolls without slip to the right, the linear acceleration f and the angular acceleration α are as shown in Fig. 1.28 and the friction force is to the left.

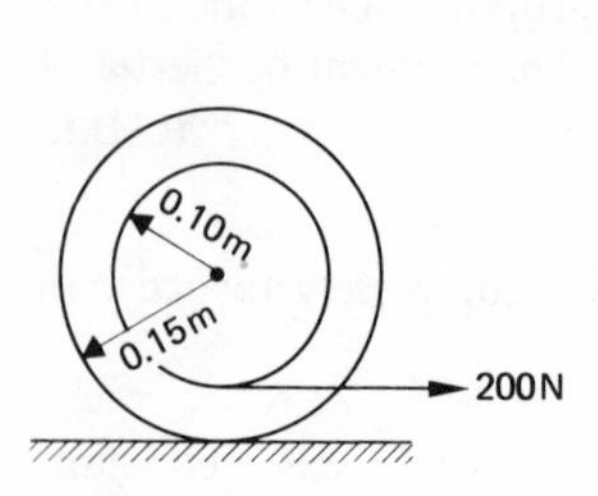

FIG. 1.27

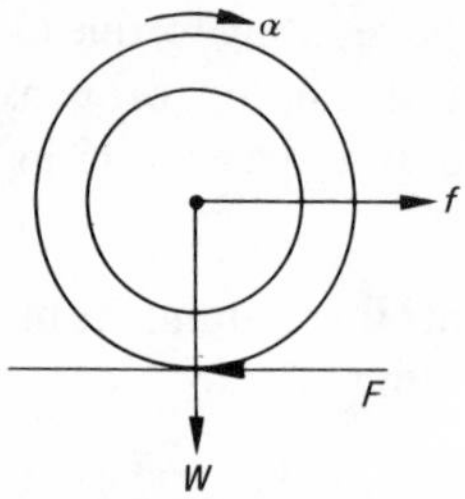

FIG. 1.28

Then, for the linear motion,

$$200 - F = 50f \tag{1}$$

and, for the angular motion,

$$F \times 0{\cdot}15 - 200 \times 0{\cdot}1 = 50 \times 0{\cdot}12^2 \alpha \tag{2}$$

If the spool rolls without slip,

$$\alpha = f/0{\cdot}15$$

and so equation (2) reduces to

$$3F - 400 = 96f \tag{3}$$

From equations (1) and (3),

$$F = 159{\cdot}3 \text{ N}$$

This is the minimum value of the friction force for no slipping and, since $F = \mu W$, the minimum value of μ is given by

$$\mu = 159{\cdot}3/(50 \times 9{\cdot}81) = \underline{0{\cdot}325}$$

When $\mu = 0{\cdot}2$, the spool slips and $\alpha \neq f/0{\cdot}15$.

$$F = 0{\cdot}2 \times 50 \times 9{\cdot}81 = 98{\cdot}1 \text{ N}$$

Hence $f = \underline{2{\cdot}04 \text{ m/s}^2}$ from equation (1)

and $\alpha = \underline{-7{\cdot}3 \text{ rad/s}^2}$ from equation (2)

These values show that f is still to the right but the angular acceleration is now anticlockwise.

Example 1.15 A solid ball of radius r, having an angular velocity ω_0 and no translational velocity, is placed on a rough, inclined plane of angle θ and coefficient of friction μ and released, so that it moves up the plane. The axis of rotation is transverse to the slope.

Describe the initial motion of the ball and determine the time at which the ball starts to roll without slipping up the plane. In addition, derive an expression for the distance through which the ball subsequently rolls without slipping up the plane. (U. Lond.)

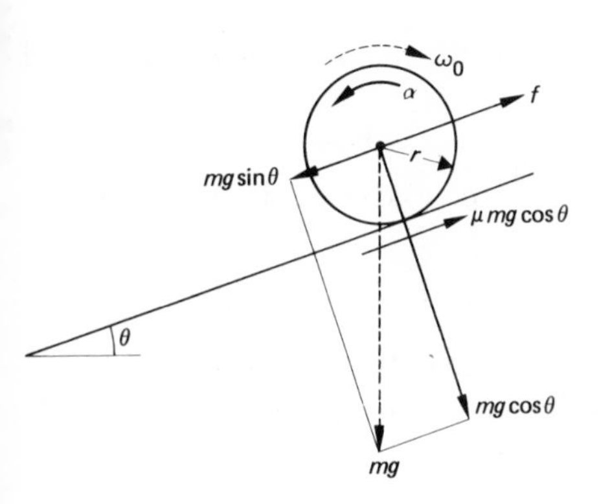

FIG. 1.29

Solution

When the sphere touches the plane, slipping takes place until the linear motion becomes compatible with the angular velocity, i.e. when $v = \omega r$.

The forces acting on the sphere during the *slipping* stage are as shown in Fig. 1.29.

Thus, for the linear motion,

$$\mu mg \cos\theta - mg \sin\theta = mf$$

or
$$\mu g \cos\theta - g \sin\theta = f \tag{1}$$

and, for the angular motion,

$$\mu mg \cos\theta \times r = I\alpha = \tfrac{2}{5} mr^2 \alpha$$

or
$$\mu g \cos\theta = \tfrac{2}{5} r\alpha \tag{2}$$

Let the time of slipping be t and the corresponding linear and angular velocities be v and ω respectively at this time. Then

$$f = v/t$$

$$\alpha = (\omega_0 - \omega)/t$$

and
$$v = \omega r$$

Thus equations (1) and (2) become

$$\mu g \cos\theta - g \sin\theta = \frac{v}{t} = \frac{\omega r}{t}$$

and
$$\mu g \cos\theta = \frac{2}{5} r \left(\frac{\omega_0 - \omega}{t} \right) = \frac{2}{5} \left(\frac{\omega_0 r}{t} - \frac{\omega r}{t} \right)$$

$$= \frac{2}{5} \left(\frac{\omega_0 r}{t} - \mu g \cos\theta + g \sin\theta \right)$$

from which
$$t = \frac{2\omega_0 r}{g(7\mu \cos\theta - 2 \sin\theta)} \tag{3}$$

From equations (1) and (3), the velocity of the ball when slipping has ceased is given by

$$v = ft = (\mu g \cos\theta - g \sin\theta) \times \frac{2\omega_0 r}{g(7\mu \cos\theta - 2 \sin\theta)}$$

$$= 2\omega_0 r\left(\frac{\mu \cos\theta - \sin\theta}{7\mu \cos\theta - 2 \sin\theta}\right) \quad (4)$$

The ball then continues to roll up the incline until the gain in potential energy is equal to the loss of kinetic energy. Let the distance travelled be s, Fig. 1.30.

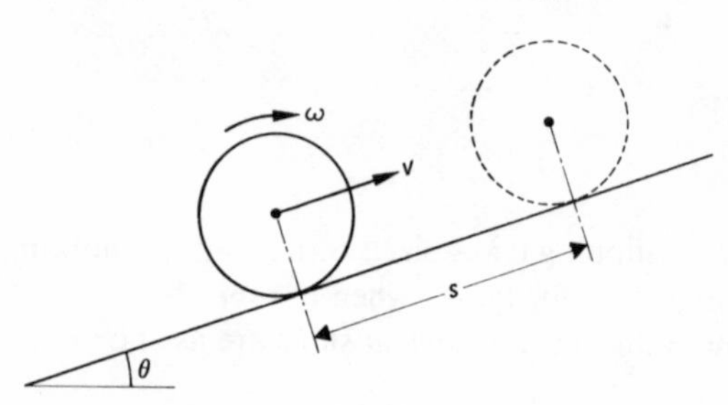

FIG. 1.30

Then

$$mgs \sin\theta = \tfrac{1}{2}mv^2 + \tfrac{1}{2}I\omega^2$$

i.e.

$$2mgs \sin\theta = mv^2 + \tfrac{2}{5}mr^2(v/r)^2$$

Thus

$$s = \frac{7v^2}{10g \sin\theta}$$

$$= \frac{7}{10g \sin\theta}\left\{2\omega_0 r\left(\frac{\mu \cos\theta - \sin\theta}{7\mu \cos\theta - 2 \sin\theta}\right)\right\}^2 \quad \text{from equation (3)}$$

i.e.

$$s = \frac{14\omega_0^2 r^2}{5g \sin\theta}\left(\frac{\mu \cos\theta - \sin\theta}{7\mu \cos\theta - 2 \sin\theta}\right)^2$$

Problems: general dynamics

1. A wheel of mass m consists of a thin disc of constant radius R about its axis of rotation. The centre of mass of the disc is situated a distance r from the axis. The wheel is caused to roll along the ground so that the axis moves at uniform speed v.

Develop an expression, in terms of the given quantities, for the normal force N exerted by the wheel on the ground. Hence find the minimum speed at which the wheel will tend to lose contact with the ground, given that $R = 390$ mm and $r = 130$ mm. Sketch a curve of N against time for this condition. (C.E.I.)
(*Answer:* $mg - m(v/R)^2 r \sin\theta$, where θ = angle of eccentricity to horizontal; 3·39 m/s)

2. A toy car track shown in Fig. 1.31 consists of a sloping section AB, a circular loop-the-loop section having a radius r, and a straight horizontal section BD. Neglecting

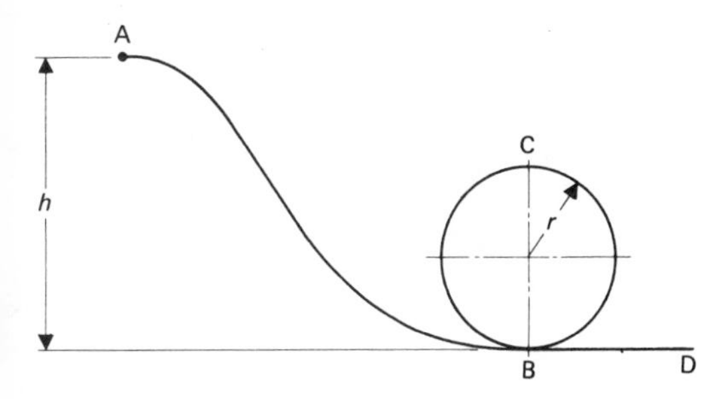

FIG. 1.31

friction, what is the least height h, of A above B, at which the car can start from rest without falling off the track at C?

If there is a constant force resisting the motion of the car along the track equal to 0·2 N, what is the velocity of the car at D when BD = 350 mm, h = 450 mm, r = 150 mm, the length of the AB section of the track is 630 mm, and the car has a mass of 500 g? (U. Lond.) (*Answer:* $2{\cdot}5r$; 2·7 m/s)

3. A uniform pole AB of length l and mass m is initially held vertically with end A resting on level ground. The pole is then disturbed from this position and topples. It is observed that the pole first pivots about end A until θ, the angle of inclination of the pole from the vertical, reaches a value of 18°, after which slipping occurs. Find the value of the coefficient of friction between the pole and the ground. (U. Lond.)

(*Answer:* 0·237)

4. Show that the acceleration f of a point moving along a straight line may be written in the form

$$f = v\frac{\mathrm{d}v}{\mathrm{d}s}$$

where v is the velocity of the point at any instant and s is the corresponding distance measured from some datum.

An aircraft of mass 50 Mg lands on a level runway at a speed of 195 km/h with a negligible vertical component of velocity. Immediately following contact with the ground, a horizontal retarding force of $(55\,000 + 4V^2)$ N, where V is the forward speed in m/s, is applied to the aircraft, bringing it to rest. Calculate the total length of the landing run. (C.E.I.) (*Answer:* 1·206 km)

5. The inlet valve of a 4-stroke petrol engine operates while the crank turns through 180°, the lift in mm being given by $3(1-\cos 2\omega t)$ where ω, the crank velocity, corresponds to an engine speed of 3600 rev/min.

Find the minimum force to be applied by the valve-spring to maintain contact, if the mass of the valve with its attachments is 0·22 kg and its axis is horizontal. (U. Lond.)

(*Answer:* 376 N)

6. A truck of mass 6 t and moving at a speed of 16 km/h runs into a stationary truck of mass 12 t which is in contact with a pair of spring buffers. The combined stiffness of the buffer springs is 800 kN/m and the buffers are already compressed 50 mm before impact occurs. Neglecting the effects of friction and the mass of the buffers, find how much further they will be compressed. (U. Lond.) (*Answer:* 177·8 mm)

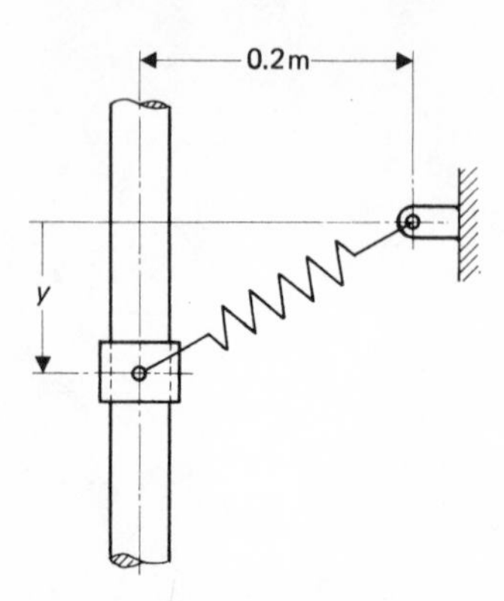

FIG. 1.32

7. The collar in Fig. 1.32 has a mass of 2 kg and slides freely on the fixed vertical guide. It is released from rest from the position $y = 0$. In this position, the spring, which has a stiffness of 500 N/m and a negligible mass, is compressed 0·1 m. Determine the velocity of the collar when $y = 0·2$ m. (U. Lond.) (*Answer:* 2·485 m/s)

8. Define the quantities impulse and momentum and state the principle of the conservation of momentum.

A stationary truck, of total mass 9 t, is set in motion by the action of a shunting locomotive which provides an impulse of 30 kN s. The truck travels freely along a level track for a period of 15 s when it collides with a truck of mass 12 t which is moving at 0·6 m/s in the same direction. The track resistance in 65 N/t. Both trucks move on together after the collision. Determine their common speed and the loss of energy at impact. (I. Mech. E.) (*Answer:* 1·353 m/s; 8 kJ)

9. A towing van of mass 2300 kg is attached to an automobile of mass 1350 kg by a wire rope of 4·5 m effective length. The initial distance between the fixing points for this rope is 3 m so that the van may move 1·5 m forward before the rope tightens. The van is accelerated from the initial at-rest position under a uniform tractive effort of 2·25 kN.

Determine the van speed: (a) just before the rope starts to tauten and (b) at the instant when the rope ceases to stretch.

Determine also the average impulsive force on the van during this stretching period, assuming that the rope takes up the strain in 0·1 s.

Ignore frictional losses and neglect the effect of the tractive effort during the tensioning period of 0·1 s. (U. Lond.) (*Answer:* 1·713 m/s; 1·08 m/s; 14·55 kN)

10. A mass of 700 kg falling 0·2 m is used to drive a pile of mass 500 kg into the ground. Assuming there is no rebound, find the common velocity of the driver and pile at the end of the blow and the loss of kinetic energy. If the resistance of the ground is constant, find its value if the pile is driven 75 mm. (U. Lond.)
(*Answer:* 1·155 m/s; 572 J; 22·45 kN)

11. A mass of 275 kg is allowed to fall vertically through 0·9 m on to the top of a pile of mass 450 kg. Assuming that the falling mass and pile remain in contact after impact and that the pile is moved 150 mm at each blow, find, allowing for the action of gravity after impact: (a) the average resistance against the pile, and (b) the energy lost in the blow. (U. Lond.) (*Answer:* 13·25 kN; 1·51 kJ)

12. A pile of mass 0·75 tonne can just carry a steady load of 40 tonnes without subsidence. The load is removed and the pile driven to a greater depth by blows of a

2-tonne hammer dropping on to the top of the pile from a height of 1·22 m. The hammer does not rebound from the top of the pile.
(a) Calculate the penetration per blow, assuming that the ground resistance is constant.
(b) What is the efficiency of the pile-driving operation? Would this efficiency be increased or decreased if a heavier hammer were used? (U. Lond.)

(*Answer:* 46·7 mm; 72·8 per cent; increased)

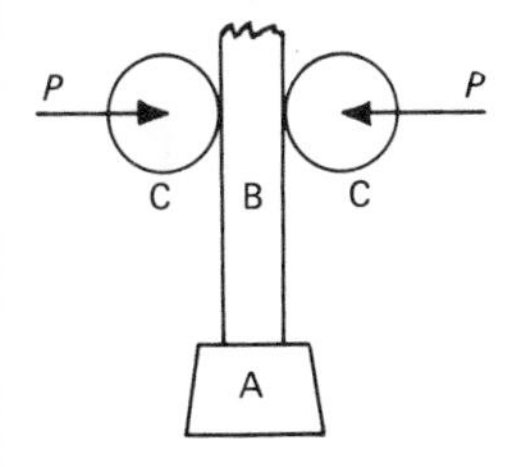

FIG. 1.33

13. The head, A, of a drop hammer is attached to a wooden board, B, which is raised when two rotating rollers, CC, are pressed against it with a force P, Fig. 1.33. The hammer is raised 1·5 m and then falls freely to strike the die. If the net mass is 180 kg and the penetration of the hot metal is 6 mm, what is the average force of the blow?

The diameter of the rollers is 200 mm and they are rotating at a constant speed of 86 rev/min. The force P is 4 kN and the coefficient of friction between the rollers and the board is 0·25. Find the time taken to raise the hammer 1·5 m. If blows were delivered at the rate of 15/min, what would be the average power of the driving motor? Determine also the maximum power that the motor would have to give if no flywheel effect were present. (U. Glas.) (*Answer:* 441 kN; 2·01 s; 0·68 kW; 1·8 kW)

14. A pile driver of mass 300 kg is used to drive a pile of mass 500 kg vertically into the ground. The pile driver falls freely through a distance of 4·0 m, rebounding with a velocity relative to the pile equal to the relative velocity immediately before impact.

Determine: (a) the velocity of the driver immediately before impact, (b) the velocity of the pile immediately after impact, (c) the depth of penetration of the pile after impact, given that the ground resisting force is constant and equal to 115 kN, and (d) the time taken for the penetration. (C.E.I.) (*Answer:* 8·86 m/s; 6·64 m/s; 0·1 m; 0·03 s)

15. A drop-forge consists of a hammer of mass m_1 and an anvil of mass m_2. The anvil is supported on four springs, each of stiffness S and arranged in parallel. The hammer is to fall vertically a distance h before striking the anvil. The coefficient of restitution between the hammer and anvil is e.

Assuming the hammer is stationary after impacting the anvil, derive expressions for: (a) the mass of the anvil, (b) the energy dissipated on impact, and (c) the maximum deflection of the springs, in terms of m_1, e, h, S and g, the gravitational acceleration. (U. Lond.)

$$\left(\textit{Answer: } \frac{m_1}{e};\ m_1gh(1-e);\ \frac{m_1 g}{4eS}\left[1+\sqrt{\left(1+\frac{8e^3hS}{m_1 g}\right)}\right]\right)$$

16. Body A is released at the position shown in Fig. 1.34 and swings down to strike body B. Given that the mass of A is 3 kg and that of B is 5 kg, what are the maximum angles the bodies will reach after the first impact if the coefficient of restitution is 0·8? Neglect the mass of the cables. (U. Lond.) (*Answer:* 5·5°; 30°)

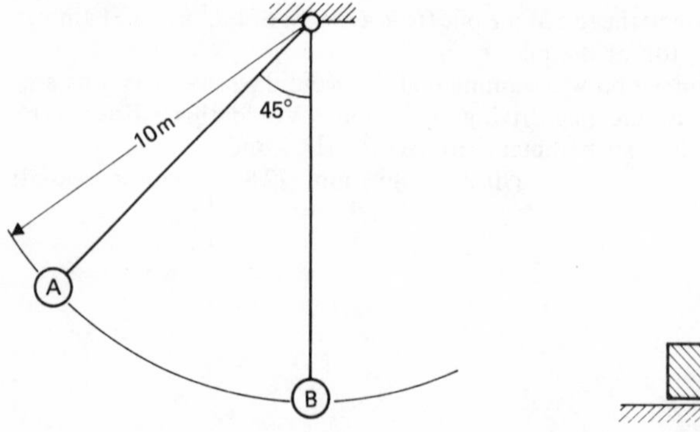

FIG. 1.34

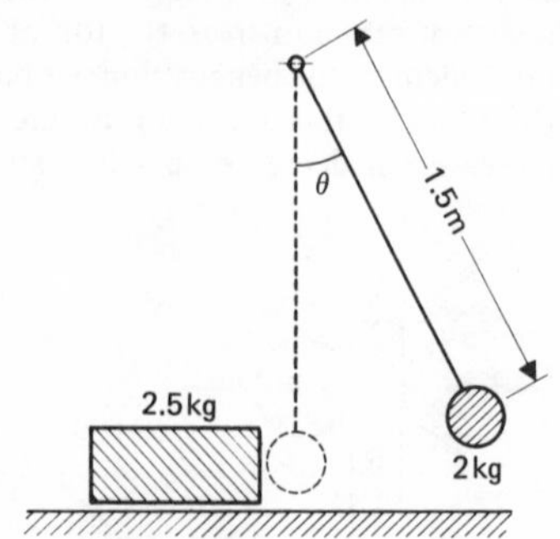

FIG. 1.35

17. A 2 kg sphere attached to a string strikes a 2·5 kg block as shown in Fig. 1.35. The sphere is initially at rest with $\theta = 60°$. It is observed that after impact the velocity of the sphere is zero, and that the block moves 1 m before coming to rest.

Determine: (a) the coefficient of restitution between the sphere and the block, and (b) the coefficient of sliding friction between the block and the horizontal surface, assuming it to be constant. (U. Lond.) (*Answer:* 0·8; 0·48)

18. A uniform rod of mass $6m$ and length l is pivoted to a fixed point at one of its ends and is free to swing in a vertical plane. The rod is held at an angle of 60° to the vertical and released. On reaching its vertical position, the lower end of the rod strikes a small body of mass m, which is at rest on a horizontal smooth table. The small body is projected across the table and the rod continues to swing through an angle of 25° before coming momentarily to rest. Determine the coefficient of restitution between the rod and the body.

Given that m is 1 kg and l is 300 mm, determine the energy dissipated at the impact. (U. Lond.) (*Answer:* 0·7; 0·75 J)

19. A thin straight uniform rod, of length l and mass m_2, is suspended freely at one end from a block, of mass m_1, so that it may move in a vertical plane. The block rests on a horizontal plane and is free to slide thereon without friction.

Starting with the rod and block in equilibrium, a horizontal impulse I is applied at the centre point of the rod in the plane in which the rod is free to move. Determine the velocity of the block and the angular velocity of the rod immediately after application of the impulse. (C.E.I.) (*Answer:* $I/4(m_1+m_2)$; $3I/2m_2l$)

20. A rocking lever of mass 70 kg has a radius of gyration of 225 mm about its centre of mass, which is 75 mm from the axis of oscillation of the lever. Find completely the equivalent two-mass system, if one of the masses is placed at the axis of oscillation.

What torque is required to give the lever an angular acceleration of 150 rad/s^2? (I. Mech. E) (*Answer:* 63 kg at axis; 7 kg at 675 mm from c.m.; 591 N m)

21. Two cylinders, each of mass 20 kg, external radius 200 mm, and radius of gyration 150 mm, rotate freely in parallel fixed bearings which are at the same level.

One cylinder is revolving at 60 rev/min and the other is revolving at 90 rev/min when a plank of mass 40 kg is laid gently upon them. Calculate the velocity attained by the plank and the time for this velocity to be reached. Assume that each cylinder supports one-half of the weight of the plank and that the coefficient of friction between plank and cylinders is 0·3. (U. Lond.) (*Answer:* 0·57 m/s; 0·252 s)

22. The inertia starter for a small aero-engine consists of a hand-driven crank geared to a small flywheel running at 30 times the crank speed. To start the engine, the handle is used to wind the flywheel up to a high speed and the handle is then pushed in to clutch with the engine. The energy stored in the flywheel then turns the engine to start it.

The flywheel rim is 50 mm wide at inner and outer radii of 80 mm and 105 mm respectively, and is made of steel of density 7·8 Mg/m^3. Calculate the kinetic energy stored when the crank is being turned at 130 rev/min. If the mean hand-turning effort is 9 N m, how long will it take to get up to this speed and what is the torque given to the engine if the handcrank speed drops at the rate of 75 rev/min per second on being clutched to the engine? (U. Glas.) (*Answer:* 4·1 kJ; 67·2 s; 11·64 N m)

23. A flywheel of mass 50 kg is mounted on a 75 mm diameter shaft in two bearings, one on either side of the wheel. Under the action of the friction on the bearings, the speed of the flywheel falls from 200 rev/min to 150 rev/min in 14 s with uniform deceleration.

A plain cast iron ring of 450 mm outside diameter and 350 mm inside diameter, and 75 mm thick, is now bolted on to the side of the flywheel, concentrically with it. The density of cast iron is 7·2 Mg/m^3. The effect of bearing friction now is to reduce the speed uniformly from 200 rev/min to 150 rev/min in 20 s.

Assuming that the coefficient of friction is constant, find its value. Find also the radius of gyration of the flywheel. (U. Lond.) (*Answer:* 0·02; 140 mm)

24. Figure 1.36 shows a cylinder in contact with a plane whose inclination, α, to the horizontal may be varied. The coefficient of friction between the cylinder and the plane is 0·2 and the cylinder is solid and of homogeneous material.

Determine the value of α for the cylinder to just slip before it rolls when first released. What is the acceleration of the cylinder under these conditions? (U. Lond.)

(*Answer:* 31°; 3·37 m/s^2)

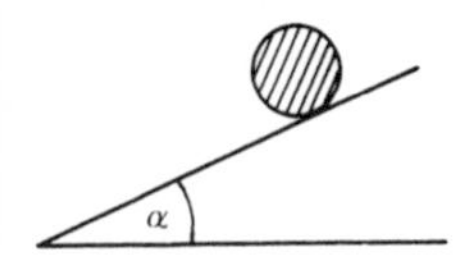

FIG. 1.36

25. A steel bar, 2·4 m long and 75 mm diameter, starts from rest and rolls without slipping, with its axis perpendicular to the track of an inclined plane.

Find: (a) the velocity of translation and (b) the kinetic energy of rotation, at the instant when the bar has rolled 6 m down the track. The mass of the bar is 88 kg and the angle of inclination of the plane is 10°. (I. Mech. E.) (*Answer:* 3·69 m/s; 300 J)

26. Two shafts A and B are parallel to each other and are connected by a pair of gear wheels so that the shaft A rotates at four times the speed of B. B carries a flywheel whose moment of inertia is 500 kg m^2. On A there are various rotating parts whose total moment of inertia is 160 kg m^2. Find the total K.E. of the system when B rotates at 200 rev/min. Find the driving torque required on shaft B to accelerate the system uniformly from rest so that the speed of B is 200 rev/min after 30 s.

If the shafts are 0·75 m apart, what is the tangential force between the teeth on the two gear wheels during this acceleration? (U. Lond.)

(*Answer:* 671 kJ; 2·14 kN m; 2·98 kN)

27. A short, heavy shaft is being turned in a lathe which is driven by a motor giving 2·25 kW at 1400 rev/min, the speed reduction between the motor and the lathe spindle being 10 to 1. The friction torque at the lathe spindle is 17·5 N m. The moment of inertia of the rotating parts of the motor is 0·08 kg m^2, that of the lathe face-plate and the piece being turned is 1·2 kg m^2.

If the turning tool is suddenly given an excessively heavy cut which stops the shaft in one revolution, calculate the total force on the tip of the tool if it is cutting at a radius of 140 mm. Assume the force to be uniform during the decelerating period. (U. Glas.)

(*Answer:* 2·1 kN)

28. A uniform disc of radius R, rolling without slipping along a horizontal plane with velocity V, encounters a step of height $3R/4$ perpendicular to its plane of motion. Assuming that no slipping occurs, show that the disc will surmount the step if $V^2 > 4Rg$. (C.E.I.)

29. A rigid beam AB of uniform cross-section and of mass 40 kg, is hinged at A to a fixed support and is maintained in a horizontal position by a vertical helical spring attached to B, AB being 1·8 m. A mass of 2·5 kg is allowed to fall on to the beam with a striking velocity of 3 m/s and the point of impact is 1·2 m from A.

Assuming the mass and beam move together, determine the angular velocity of the beam immediately after impact. (U. Lond.) (*Answer:* 0·1924 rad/s)

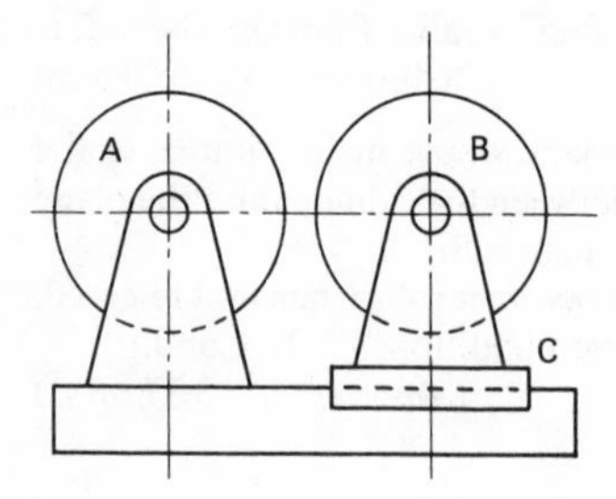

FIG. 1.37

30. Figure 1.37 shows a flywheel A of mass 12·5 kg, of outside diameter 450 mm, and radius of gyration 175 mm, initially rotating at 300 rev/min. A second flywheel B of mass 7·5 kg, outside diameter 380 mm, and radius of gyration 150 mm, is carried in bearings mounted on a slider C. Initially B is not rotating. The slider is now pushed to the left so that wheel B presses against A, until the wheels roll on each other without slipping.

Neglecting bearing friction, calculate the angular velocities of the two flywheels after slipping has ceased.

If the coefficient of friction between the wheels is 0·12, and the horizontal force between them is constant at 60 N, calculate the time from the instant when the wheels initially make contact until slipping ceases. (U. Lond.)

(*Answer:* A, 185·4 rev/min; B, 219·5 rev/min; 2·84 s)

31. A drum A of mass 200 kg, external diameter 380 mm, and a radius of gyration 150 mm, rotates on frictionless bearings with a speed of 250 rev/min; a stationary drum B of mass 50 kg, external diameter 200 mm, and radius of gyration 80 mm, mounted on a frictionless axis parallel to that of A, is brought into contact with A, the two being pressed together with a force of 90 N.

Assuming a coefficient of friction of 0·25, determine: (a) the time of slipping and the final speeds of A and B, and (b) the time of slipping if a torque is applied to A to maintain a constant speed of 250 rev/min. (U. Lond.)

(*Answer:* 5·63 s; 199 rev/min; 378 rev/min; 7·07 s)

32. Two gear wheels A and B are mounted on parallel shafts so that they may revolve separately or may be meshed together externally.

Wheel	*Mass (kg)*	*Number of teeth*	*Radius of gyration (mm)*	*Original rev/min*
A	5	100	300	300
B	6	50	250	100

The wheels were originally turning freely in the *same* direction.

Find: (a) the speed and direction of rotation of wheel A, if the wheels are suddenly meshed, assuming that there is no back-lash of the teeth and (b) the loss of energy due to impact. (U. Lond.) (*Answer:* 30·76 rev/min in original direction; 232·8 J)

33. A valve of mass 0·23 kg closes horizontally under the action of a spring. In the closed position the spring is compressed 12 mm, and the maximum opening of the valve is 6 mm. If the spring stiffness is 3·5 kN/m, find the time required for the valve to close and the velocity with which it strikes the seat. (I. Mech. E)

(*Answer:* 0·0068 s; 1·66 m/s)

34. A valve is opened vertically downwards by a cam and then released by a trip gear, to be shut by a close-coiled helical spring concentric with the valve stem. The maximum valve opening is 16 mm, the stiffness of the spring is 25 kN/m, and the mass of the valve and attached details is 3 kg. The spring compression when the valve is closed is 20 mm.

Neglecting all frictional resistances and the gravitational force on the valve and associated details, determine: (a) the time taken for the valve to close and (b) the velocity at the moment of impact. (U. Lond.) (*Answer:* 0·0108 s; 2·735 m/s)

35. The moving table of a machine tool slides upon horizontal guides. The table has a mass of 100 kg and the frictional force opposing its motion is 180 N. At a time when the speed of the table is 0·9 m/s, the driving mechanism is disconnected and the table is then brought to rest in a distance of 40 mm by collision with a spring buffer attached to the fixed frame of the machine. Calculate the time required for the spring to attain its greatest compression. Assume that the spring is initially free from stress and neglect any loss of energy in the initial impact. (U. Lond.) (*Answer:* 0·0723 s)

36. A massive uniform circular disc, of radius r, is rigidly attached normally to a light axle passing through its centre of mass. With the axle horizontal and rotating at angular velocity ω_0, the assembly is lowered on to a rough horizontal surface so that the initial velocity of translation, v, of the axle is zero. The limiting coefficient of friction between the disc and the surface is μ.

Develop expressions for v before and after slipping ceases. Find the ratio of the kinetic energy of the disc after slipping has ceased to the original kinetic energy of the disc. (C.E.I.) (*Answer:* $\mu g t$; $\omega_0 r/3$; 1/3)

37. A solid uniform sphere of mass m and radius r is released from rest on a rough slope with a coefficient of friction μ. Determine the maximum angle of inclination of the slope to the horizontal for the sphere to roll down the incline without slipping.

If the sphere does roll without slipping and acquires a speed v after moving a distance x down the slope, express v in terms of x and θ, the angle of the slope to the horizontal.

Finally, if $\mu = 0$, describe the motion of the sphere down the slope and once more express v in terms of x and θ. (U. Lond.)

(*Answer:* $\tan^{-1} 7\mu/2$; $v = \sqrt{(10gx \sin\theta/7)}$; $v = \sqrt{(2gx \sin\theta)}$)

38. An electric motor exerts a torque equal to $(1{\cdot}5 - 2{\cdot}4 \times 10^{-3}\,\omega)$ N m, where ω is the angular velocity in rad/s. The motor accelerates a gear train whose equivalent moment of inertia is $3{\cdot}33 \times 10^{-3}$ kg m^2 at the motor coupling. Assuming that friction and motor inertia are negligible, determine: (a) the speed at which the power is a maximum and the magnitude of the maximum power, and (b) the time taken to accelerate the gear train from rest to the speed for maximum power. (C.E.I.)

(*Answer:* 312·5 rad/s; 0·96 s)

Examples: hoists

Example 1.16 A load of mass 230 kg is lifted by means of a rope which is wound several times round a drum and which then supports a balance mass of 140 kg. As the load rises, the balance mass falls. The drum has a diameter of 1·2 m, a radius of gyration of 530 mm, and its mass is 70 kg. The frictional resistance to the movement of the load is 110 N and that to the movement of the balance mass 90 N. The frictional torque on the drum shaft is 80 N m.

Find the torque required on the drum, and also the power required, at an instant when the load has an upward velocity of 2·5 m/s and an upward acceleration of 1·2 m/s². (U. Lond.)

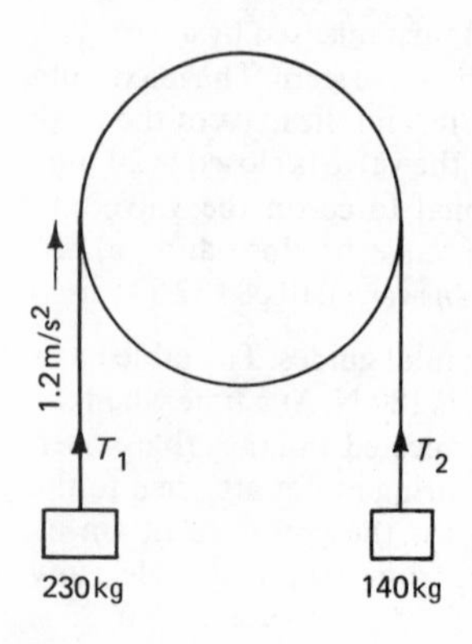

Fig. 1.38

Solution

Let T_1 and T_2 be the tensions in the rope, Fig. 1.38. Then

$$T_1 = 230 \times 9{\cdot}81 + 230 \times 1{\cdot}2 + 110$$

and

$$T_2 = 140 \times 9{\cdot}81 - 140 \times 1{\cdot}2 - 90$$

$$\therefore\ T_1 - T_2 = 1527 \text{ N}$$

$$\therefore \text{ torque to accelerate loads} = 1527 \times 0{\cdot}6 = 916{\cdot}2 \text{ N m}$$

$$\text{Torque to accelerate drum} = 70 \times 0{\cdot}53^2 \times (1{\cdot}2/0{\cdot}6) + 80 = 119{\cdot}3 \text{ N m}$$

$$\therefore \text{ total torque} = 916{\cdot}2 + 119{\cdot}3 = \underline{1035{\cdot}5 \text{ N m}}$$

$$\text{Power} = 1035{\cdot}5 \times 2{\cdot}5/0{\cdot}6 \text{ W} = \underline{4{\cdot}32 \text{ kW}}$$

Alternative Solution

$$\text{Equivalent mass of drum at rope} = \frac{mk^2}{r^2} = \frac{70 \times 0{\cdot}53^2}{0{\cdot}6^2} = 54{\cdot}7 \text{ kg}$$

$$\text{Total mass to be accelerated} = 230 + 140 + 54{\cdot}7 = 424{\cdot}7 \text{ kg}$$

Total friction force to be overcome at rope, $P = 90 + 110 + \dfrac{80}{0{\cdot}6} = 333{\cdot}3$ N

Then

$$\text{equivalent tension in rope} = 333{\cdot}3 - 140 \times 9{\cdot}81 + 230 \times 9{\cdot}81 + 424{\cdot}7 \times 1{\cdot}2$$
$$= 1725{\cdot}8\ \text{N}$$
$$\therefore\ \text{torque} = 1725{\cdot}8 \times 0{\cdot}6 = \underline{1035{\cdot}5\ \text{N m}}$$
$$\text{Power} = Pv = 1725{\cdot}8 \times 2{\cdot}5\ \text{W} = \underline{4{\cdot}32\ \text{kW}}$$

Example 1.17 A hoist with a 1·2 m diameter drum lifts a cage by means of a wire rope that winds on to the drum. The drum is driven by an electric motor through a double reduction gear.

Determine the motor torque to give a cage an acceleration upwards of 0·9 m/s², given the following data:

	Speed (rev/min)	*Mass (kg)*	*k (mm)*
Rotor of motor and pinion	1440	230	115
Intermediate gear	340	360	240
Low speed gear	75	900	650
Drum and shaft	75	1100	550
Cage	—	1800	—
Rope (rising)	—	450	—

(U. Lond.)

Solution

$$\text{Total mass to be accelerated} = 1800 + 450 = 2250\ \text{kg}$$
$$\therefore\ \text{tension in rope} = 2250 \times 9{\cdot}81 + 2250 \times 0{\cdot}9 = 24\,125\ \text{N}$$
$$\therefore\ \text{torque on drum to produce this tension} = 24\,125 \times 0{\cdot}6 = 14\,480\ \text{N m}$$

Equivalent moment of inertia of rotating system referred to drum,

$$I_e = 1100 \times 0{\cdot}55^2 + 900 \times 0{\cdot}65^2$$
$$+ 360 \times 0{\cdot}24^2 \times (340/75)^2 + 230 \times 0{\cdot}115^2 \times (1440/75)^2$$
$$= 2262\ \text{kg m}^2$$

$$\text{Angular acceleration of drum} = 0{\cdot}9/0{\cdot}6 = 1{\cdot}5\ \text{rad/s}^2$$
$$\therefore\ \text{torque on drum to accelerate drum, gears and motor} = 2262 \times 1{\cdot}5$$
$$= 3393\ \text{N m}$$

$$\therefore\ \text{total drum torque} = 14\,480 + 3393 = 17\,873\ \text{N m}$$
$$\therefore\ \text{motor torque} = 17\,873 \times 75/1440$$
$$= \underline{930\ \text{N m}}$$

Example 1.18 A cage used for raising coal in a mine shaft has a mass of 0·75 t and carries a load of 1·25 t; the shaft is 245 m deep and the haulage rope has a mass of 4·5 kg/m. The maximum power which can be applied to the rope at the winding drum is 330 kW at a rope speed of 9 m/s.

Find the minimum time in which the speed of 9 m/s is attained.

(U. Lond.)

Solution

$$\text{Mass of cage and load} = (0{\cdot}75 + 1{\cdot}25) \times 1000 = 2000\,\text{kg}$$

$$\text{Mass of cable being accelerated} = 245 \times 4{\cdot}5 = 1102{\cdot}5\,\text{kg}$$

When the cage has risen a distance h m,

$$\text{mass of cable being raised} = (245 - h) \times 4{\cdot}5 = (1102{\cdot}5 - 4{\cdot}5h)\,\text{kg}$$

$$\therefore \text{ tension in cable} = (2000 + 1102{\cdot}5 - 4{\cdot}5h) \times 9{\cdot}81 + (2000 + 1102{\cdot}5)f$$
$$= 30\,420 - 44{\cdot}2h + 3102{\cdot}5f\,\text{N}$$

$$\text{Tension in cable at } 9\,\text{m/s} = 330 \times 1000/9 = 36\,670\,\text{N}$$

If the motor torque is assumed constant, then, since the inertia of the drum is not given, the tension in the cable may be assumed constant. The equation of motion is therefore

$$36\,670 = 30\,420 - 44{\cdot}2h + 3102{\cdot}5f$$

i.e.
$$6250 = 3102{\cdot}5\,\frac{d^2h}{dt^2} - 44{\cdot}2h$$

or
$$\frac{d^2h}{dt^2} - 0{\cdot}1193^2h = 2{\cdot}015$$

The solution is*

$$h = A\cosh 0{\cdot}1193t + B\sinh 0{\cdot}1193t - 141{\cdot}6$$

When $t = 0$, $h = 0$

$$\therefore 0 = A - 141{\cdot}6 \qquad \text{i.e. } A = 141{\cdot}6$$

When $t = 0$, $dh/dt = 0$

$$\therefore 0 = 0{\cdot}1193B \qquad \text{i.e. } B = 0$$

$$\therefore h = 141{\cdot}6(\cosh 0{\cdot}1193t - 1)$$

$$\therefore \frac{dh}{dt} = 16{\cdot}9\sinh 0{\cdot}1193t$$

Therefore when $dh/dt = 9$ m/s, $t = \underline{4{\cdot}28\text{ s}}$

Example 1.19 A wagon of mass 14 t is hauled up an incline of 1 in 20 by a rope which is parallel to the incline and is being wound round a 1 m diameter drum. The drum, in turn, is driven through a 40 to 1 reduction gear by an electric motor. The frictional resistance to the movement of the wagon is 1·2 kN, and the efficiency of the gear drive is 85 per cent. The bearing friction at the drum and motor shafts may be neglected. The rotating parts of the drum have a mass

* See the Appendix.

of 1·25 t, with radius of gyration 450 mm, and the rotating parts on the armature shaft have a mass of 110 kg, with radius of gyration 125 mm.

At a certain instant, the wagon is moving up the slope with a velocity of 1·8 m/s and an acceleration of 0·1 m/s². Calculate the motor torque and the power being developed. (U. Lond.)

Solution

Tension in rope = component of weight down slope + inertia force + friction force

$$= 14 \times 1000 \times 9{\cdot}81/20 + 14 \times 1000 \times 0{\cdot}1 + 1200 = 9467\ \text{N}$$

$$\therefore \text{ torque on drum to accelerate load} = 9467 \times 0{\cdot}5 = 4733{\cdot}5\ \text{N m}$$

$$\text{Moment of inertia of drum} = 1{\cdot}25 \times 1000 \times 0{\cdot}45^2 = 253\ \text{kg m}^2$$

$$\text{Angular acceleration of drum} = 0{\cdot}1/0{\cdot}5 = 0{\cdot}2\ \text{rad/s}^2$$

$$\therefore \text{ torque on drum to accelerate drum} = 253 \times 0{\cdot}2 = 50{\cdot}6\ \text{N m}$$

∴ torque on armature to accelerate drum and load

$$= (4733{\cdot}5 + 50{\cdot}6) \times \frac{1}{40} \times \frac{100}{85} = 140{\cdot}8\ \text{N m}$$

$$\text{Moment of inertia of armature} = 110 \times 0{\cdot}125^2 = 1{\cdot}72\ \text{kg m}^2$$

$$\text{Angular acceleration of armature} = \frac{0{\cdot}1}{0{\cdot}5} \times 40 = 8\ \text{rad/s}^2$$

$$\therefore \text{ torque on armature to accelerate armature} = 1{\cdot}72 \times 8 = 13{\cdot}76\ \text{N m}$$

$$\therefore \text{ total armature torque} = 140{\cdot}8 + 13{\cdot}76 = \underline{154{\cdot}6\ \text{N m}}$$

$$\text{Power developed by motor} = T\omega = 154{\cdot}6 \times \left(\frac{1{\cdot}8}{0{\cdot}5} \times 40\right)\text{W} = \underline{22{\cdot}25\ \text{kW}}$$

Note that the efficiency of the gearing does not affect the torque required to accelerate the motor armature.

Example 1.20 A single-axle bucket A is moved up and down a 30° inclined slope by the counterweight B and pulley system C shown in Fig. 1.39. The bucket (including wheels) has a mass of 500 kg and its wheels have a total axial

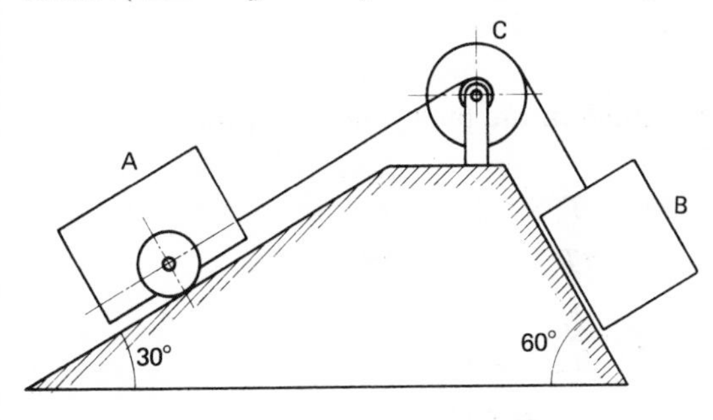

Fig. 1.39

moment of inertia of 50 kg m^2. The twin pulley block C has a total axial moment of inertia of 75 kg m^2. The counterweight weighs 20 kN and the coefficient of friction between it and the 60° slope on which it slides is 0·2. The wheels of A have a radius of 0·5 m, while the two pulleys which constitute C have radii of 0·2 m and 0·9 m.

Determine the acceleration with which the bucket will ascend the slope and find the tension in each of the two ropes, assuming these to remain parallel to the respective surfaces. (U. Lond.)

Solution

Motion of bucket From Section 1.26,

$$\text{equivalent mass of bucket for acceleration purposes} = 500 + (50/0{\cdot}5^2) = 700 \text{ kg}$$

Let the rope tension be T_1 and the acceleration of the bucket be f_1, Fig. 1.40. Then

$$T_1 - 500\,g \sin 30° = 700 f_1$$

or

$$T_1 = 700 f_1 + 2453 \qquad (1)$$

Motion of counterweight Let the rope tension be T_2 and the acceleration of the counterweight be f_2, Fig. 1.41. Then

$$20 \times 10^3 \sin 60° - T_2 - 0{\cdot}2 \times 20 \times 10^3 \cos 60° = \frac{20 \times 10^3}{9{\cdot}81} f_2$$

or

$$T_2 = 15\,321 - 2039 f_2 \qquad (2)$$

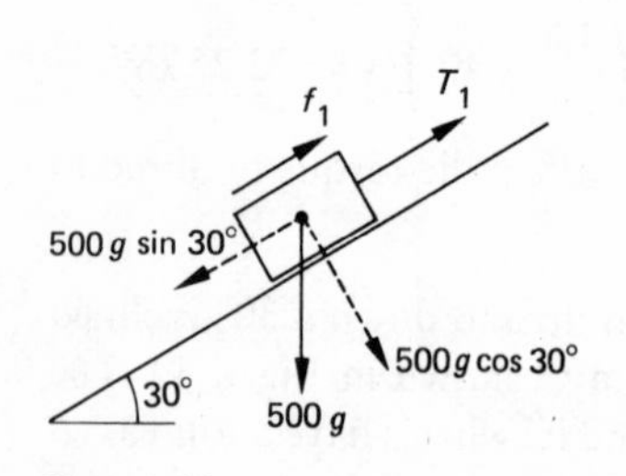

FIG. 1.40

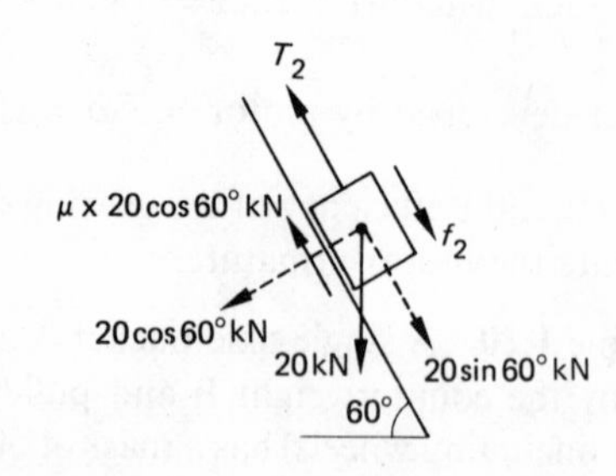

FIG. 1.41

Motion of pulley The forces acting on the pulley are as shown in Fig. 1.42. Hence

$$T_2 \times 0{\cdot}9 - T_1 \times 0{\cdot}2 = 75\alpha$$

or

$$4{\cdot}5\,T_2 - T_1 = 375\alpha \qquad (3)$$

Also

$$\alpha = \frac{f_1}{0{\cdot}2} = \frac{f_2}{0{\cdot}9}$$

so that

$$f_2 = 4{\cdot}5 f_1 \qquad (4)$$

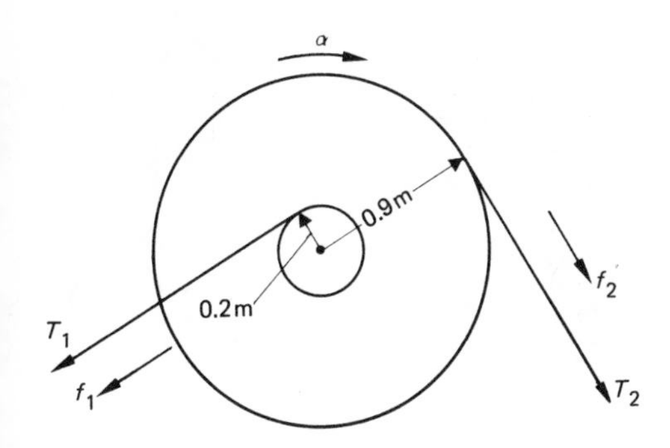

FIG. 1.42

Thus equations (1), (2) and (3) become

$$T_1 = 700 f_1 + 2453$$

$$T_2 = 15\,321 - 9176 f_1$$

and

$$4{\cdot}5\,T_2 - T_1 = 1875 f_1$$

from which

$$f_1 = \underline{1{\cdot}516\,\text{m/s}^2}$$

$$T_1 = \underline{3514\,\text{N}}$$

and

$$T_2 = \underline{1410\,\text{N}}$$

Example 1.21 The lifting system shown in Fig. 1.43 consists of a drum D of mass $4m$ and radius $2r$, a pulley P of mass m and radius r, and a load L of mass $50m$. A rope passes around the pulley and is wound onto the drum. The load is freely attached to the centre of the pulley. A motor having a constant torque output M and a maximum power output P is used to drive the drum and raise the load.

Taking the drum and the pulley to be uniform discs, and assuming that the rope does not slip on the pulley, show that the time taken for the load to be raised from rest to a maximum upwards speed is given by

$$t = \frac{119\,Pmr^2}{2M(M - 51mrg)} \qquad \text{(U. Lond.)}$$

Solution

Let the instantaneous velocity of the load be v, Fig. 1.44(a). Then

$$\text{angular velocity of pulley} = \frac{v}{r}$$

and

$$\text{angular velocity of drum} = \frac{2v}{2r} = \frac{v}{r}$$

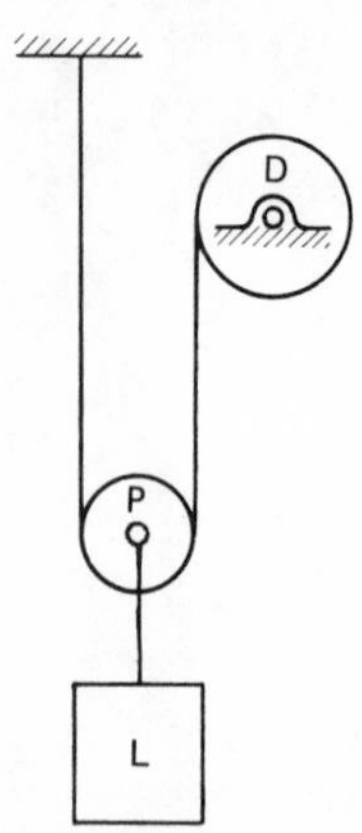

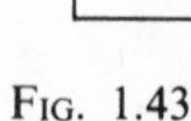

FIG. 1.43

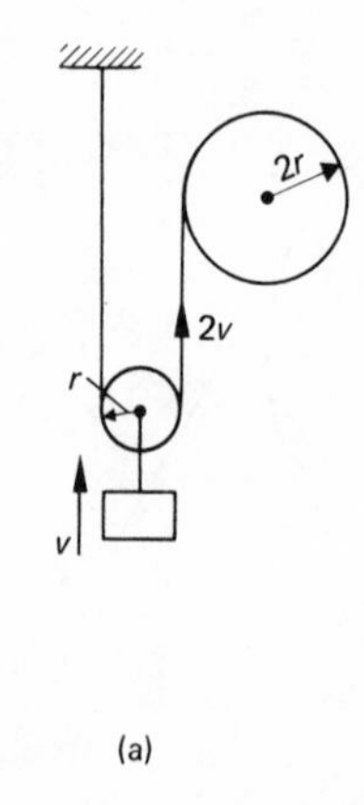

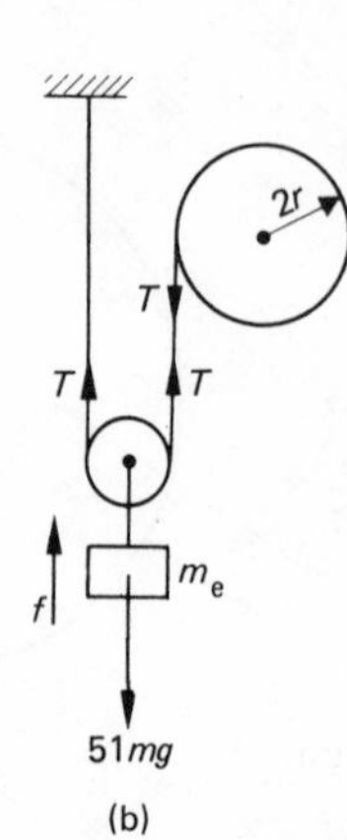

FIG. 1.44

$$\text{Total K.E. of system} = \text{K.E. of load} + \text{K.E. of pulley} + \text{K.E. of drum}$$

$$= \tfrac{1}{2}(50m)v^2 + \tfrac{1}{2}\left(\frac{mr^2}{2}\right)\left(\frac{v}{r}\right)^2 + \tfrac{1}{2}\left[\frac{4m(2r)^2}{2}\right]\left(\frac{v}{r}\right)^2$$

$$= \tfrac{1}{2}(58{\cdot}5m)v^2$$

It is therefore evident that the equivalent mass of the system, m_e, for acceleration purposes is $58{\cdot}5m$. Let the tension in the rope be T, Fig. 1.44(b). Then

$$\text{upward force on pulley and load} = 2T$$

$$\text{Weight of pulley and load} = (m + 50m)g = 51\,mg$$

Therefore the equation of motion of the equivalent mass is

$$2T - 51mg = m_e f = 58{\cdot}5mf$$

Since the motor torque is M, $T = M/2r$, and so

$$2\left(\frac{M}{2r}\right) - 51mg = 58{\cdot}5mf$$

or

$$f = \frac{2}{119mr}(M - 51mrg)$$

The maximum velocity is governed by the power available, P,

i.e.

$$P = M\omega = M\frac{v}{r}$$

or

$$v = \frac{Pr}{M}$$

Therefore $$t = \frac{v}{f} = \frac{Pr}{M} \bigg/ \frac{2}{119mr}(M - 51mrg)$$

i.e. $$t = \frac{119Pmr^2}{2M(M - 51mrg)}$$

Problems: hoists

39. The cage of a goods hoist has a total mass when loaded of 1200 kg. The rope passes over a drum at the top of the shaft and thence to a balancing mass of 500 kg. Cage and balance mass move on parallel guides and the force of friction in each line may be taken as the same and constant at 500 N. The drum has an effective diameter, between rope centres, of 1·5 m, a total mass, allowing for the rope, of 600 kg, and a radius of gyration of 0·6 m. The maximum acceleration attained is 1·5 m/s² which occurs at a speed of 2·5 m/s. The maximum speed is 4·5 m/s and retardation is at a uniform rate from that speed to zero in the last 4·5 m of travel.

Determine: (a) the power required to drive the system at the condition of maximum acceleration, and (b) the rope tensions in both lines during retardation. (I. Mech. E.)
(*Answer:* 27·5 kW; 9575 N and 5535 N)

40. A hoist has a winding drum of 0·9 m effective diameter and a radius of gyration of 0·35 m, the mass of the drum being 100 kg. A load of 320 kg is to be raised 36 m, the mass of the lifting rope being 1 kg/m. If the acceleration is 1·8 m/s² until a constant velocity of 6 m/s is reached, find the power necessary just at the end of the acceleration. (U. Lond.) (*Answer:* 24·9 kW)

41. An electrically driven capstan, at the top of an incline, hauls a truck of mass 4 t up a track having an inclination of 1 in 12. The resistances are constant and amount to 320 N. If the truck starts from rest and is accelerated uniformly to attain a speed of 24 km/h after an interval of 30 s, what is the tension in the cable? Find the tension in the cable if the speed drops to 8 km/h in the next 36 m length of track. Assuming the cable had broken when the speed had reached 24 km/h, how far would the truck proceed up the track? (I. Mech. E) (*Answer:* 4476 N; 1392 N; 24·75 m)

42. A winding drum raises a cage of mass 500 kg through a height of 120 m. The winding drum has a mass of 250 kg, an effective radius of 0·5 m, and a radius of gyration of 0·36 m. The mass of the rope is 3 kg/m. The cage has at first an acceleration of 1·5 m/s² until a velocity of 9 m/s is reached, after which the velocity is constant until the cage nears the top, when the final retardation is 6 m/s².

Find: (a) the time taken for the cage to reach the top, (b) the torque which must be applied to the drum at starting, and (c) the power at the end of the acceleration period. (U. Lond.) (*Answer:* 17·08 s; 4961 N m; 82·2 kW)

43. A train of loaded trucks has a mass of 20 t and is pulled up an incline 450 m long with a gradient of 1 in 40, by a wire rope wound round the drum of a winding engine at the top of the incline. The frictional resistance is 100 N/t. The drum has a mass of 3 t, its diameter is 2 m, and its radius of gyration is 0·9 m. If the tension in the rope must not be more than 12·5 kN, and if the speed of the train is not to exceed 50 km/h, find the minimum time in which the train can ascend the incline and the maximum power required. (U. Lond.) (*Answer:* 57·26 s; 183 kW)

44. The mass of the main girders and end cradles of an electric travelling crane is 7·5 t. The cross carriage with lifting crab has a mass of 2·5t. The lifting gear has an efficiency of

75 per cent when raising a load of 10 t, and the efficiency of the travelling gear is 65 per cent. A 10-t load is being raised with an acceleration of 0·15 m/s^2, while the crane is being accelerated forward at the same rate. At the instant when the hoisting speed is 30 m/min and the travelling speed is 55 m/min, the hoisting motor is developing eight times the power developed by the travelling motor.

Find: (a) the power developed by each motor, (b) the rail resistance, in N/t, and (c) the kinetic energy of the load. (U. Lond.) (*Answer:* 66·4 kW; 8·3 kW; 144·3 N/t; 5450 J)

45. The main shaft of a winding gear carries two drums and a brake ring. The rope from one drum raises the loaded cage, while that from the other lowers the empty one. The total moment of inertia of the shaft masses, inclusive of the constant length of rope wound on the drums, is 500 kg m^2 about the axis of rotation. Each cage has a mass of 2300 kg and the rising cage carries a load of 1150 kg. The total depth is 270 m, the winding drum diameter is 3 m and the rope has a mass of 3 kg/m.

The cages are to be uniformly accelerated to the full speed of 7·5 m/s in the first 21 m and uniformly retarded during the final 4 s. Frictional effects are neglected and the full mass of 270 m of rope may be assumed as constant on the ascending side during acceleration and on the descending side during retardation.

Determine: (a) the total time per wind, (b) the driving torque on the drum shaft during acceleration and the braking torque during retardation, and (c) the power developed just after acceleration ceases. (I. Mech. E.)

(*Answer:* 40·8 s; 42·5 kN m; 14·07 kN m; 144 kW)

46. A lift of mass 900 kg is connected to a rope which passes over a drum of 1 m diameter, and a balance mass of 450 kg is attached to the other end of the rope. The moment of inertia of the drum is 100 kg m^2 and it is driven by a motor through a reduction gear (25 to 1) of 90 per cent efficiency. Neglecting the inertia of the gears, calculate the motor torque for a lift acceleration of 3 m/s^2. If the maximum output of the motor is 15 kW, what will be the maximum uniform speed of the lift? (U. Lond.)

(*Answer:* 215 N m; 3·06 m/s)

47. A large winding drum with a direct-coupled motor is used to operate a cage in a mine 1·35 km deep. The diameter of the main part of the drum is 10 m. One end is conical, the smaller diameter being 5 m, and the rope begins to wind on this part when the cage is beginning to ascend. The drum and rotor of the motor together have a mass of 500 t, and the radius of gyration may be taken as 3 m. The mass of the loaded cage is 12 t and the mass of the whole length of rope is 9 t. If the motor can give a starting torque of 1 MN m, find the acceleration of the cage at the beginning of the ascent. If the maximum power of the motor is 3·75 MW, find the maximum possible speed of the cage when it is 600 m from the bottom of the shaft and the rope is winding on the large diameter of the drum. (U. Glas.) (*Answer:* 0·262 m/s^2; 22·5 m/s)

48. A haulage gear is driven by a motor which develops 15 kW at the full speed of 750 rev/min. The haulage drum is 1·5 m diameter and the total reduction ratio of the gearing between motor and drum is 18·5. The efficiency of the gear is 75 per cent, the gradient is 1 in 25 and the rail friction is 135 N per tonne of load. The masses of the gear are equivalent to a total mass of 2·3 Mg, radius of gyration of 0·5 m, at the drum shaft. Calculate the total load that can be hauled at the full speed.

If the motor can give an excess torque of 10 per cent of that at full speed, calculate the acceleration possible, assuming that efficiency and rail friction values remain unaltered. (U. Lond.) (*Answer:* 6·7 t; 0·0457 m/s^2)

49. A truck of mass 5 t is hauled up an incline of 1 in 15 by a rope parallel to the track. The rope is wound on a drum, 1 m in diameter, driven by an electric motor. The drum has a mass of 1 t and its radius of gyration is 0·4 m. The efficiency of the drive from motor to drum is 88 per cent and the frictional resistance to the motion of the truck is 1·3 kN.

At an instant when the truck is moving up the incline at a speed of 3 m/s and an acceleration of 0·3 m/s^2, find the power output from the electric motor. (U. Lond.)

(*Answer:* 21·4 kW)

50. A loaded mine skip has a mass of 13·5 Mg. When at rest at the bottom of the pit, it is supported by a length of 1·5 km of rope of mass 10 kg/m. The moment of inertia of the rotating parts of the winding gear is 500 Mg m^2 and the radius of the winding drum is 2·4 m. During winding, the skip has a constant acceleration of 0·9 m/s^2 at the beginning and the same retardation at the end, with a constant velocity of 15 m/s between these periods.

Find the maximum power required from the driving motor. (U. Lond.)

(*Answer:* 5·55 MW)

Examples: vehicle dynamics

Example 1.22 A self-propelled truck, of total mass 25 t and wheel diameter 750 mm, runs on a track for which the resistance is 180 N/t. The engine develops 60 kW at its maximum speed of 2400 rev/min and drives the axle through a gear box.

Determine: (a) the time to reach full speed from rest on the level if the gear reduction ratio is 10 to 1, assuming the engine torque to be constant and a gearing efficiency of 94 per cent, and (b) the gear ratio required to give an acceleration of 0·15 m/s^2 on an up gradient of 1 in 70, assuming a gearing efficiency of 90 per cent. (I. Mech. E.)

Solution

(a)
$$\text{Engine torque} = \frac{\text{power} \times 60}{2\pi N} = \frac{60 \times 1000 \times 60}{2\pi \times 2400} = 239 \text{ N m}$$

$$\therefore \text{ axle torque} = 239 \times 10 \times 94/100 = 2245 \text{ N m}$$

$$\therefore \text{ tractive force} = 2245/0{\cdot}375 = 5980 \text{ N}$$

$$\text{Resistance to motion} = 25 \times 180 = 4500 \text{ N}$$

$$\therefore \text{ net tractive force} = 5980 - 4500 = 1480 \text{ N}$$

The equation of motion of the truck is therefore

$$1480 = 25 \times 1000f$$

$$\therefore f = 0{\cdot}0592 \text{ m/s}^2$$

$$\text{Final speed} = \frac{2\pi}{60} \times \frac{2400}{10} \times 0{\cdot}375 = 9{\cdot}43 \text{ m/s}$$

$$\therefore t = \frac{9{\cdot}43}{0{\cdot}0592} = \underline{159 \text{ s}}$$

(b) If the required gear ratio is n, then

$$\text{tractive force} = 239 \times n \times \frac{90}{100} \times \frac{1}{0{\cdot}375} = 574n \text{ N}$$

$$\text{Component of weight down incline} = (25 \times 1000 \times 9{\cdot}81)/70 = 3503\ \text{N}$$

$$\therefore \text{net tractive force} = 574n - 4500 - 3503 = 574n - 8003$$

$$\therefore 574n - 8003 = 25 \times 1000 \times 0{\cdot}15$$

$$\therefore n = \underline{20{\cdot}5}$$

Example 1.23 A motor car has a mass of 1000 kg and the road wheels have a radius of 360 mm. The engine rotating parts have a moment of inertia of 0·3 kg m^2 and the four road wheels together have a moment of inertia of 3·6 kg m^2. The gear ratio from engine to back axle is 5·5 and, when the car speed is v m/s, the engine torque available for propulsion is $(88 - 0{\cdot}044v^2)$ N m and the resistance to motion is $(245 + 0{\cdot}96v^2)$ N.

Determine the acceleration of the car at a speed of 12 m/s on a level road and calculate the time required to increase the speed of the car from 12 m/s to 20 m/s. (U. Lond.)

Solution

$$\text{Equivalent moment of inertia of wheels} = 3{\cdot}6 + 5{\cdot}5^2 \times 0{\cdot}3 = 12{\cdot}675\ \text{kg m}^2$$

$$\therefore \text{total equivalent mass to be accelerated} = 1000 + \frac{12{\cdot}675}{0{\cdot}36^2} = 1097{\cdot}7\ \text{kg}$$

$$\text{Tractive force at road} = (88 - 0{\cdot}044v^2) \times 5{\cdot}5/0{\cdot}36$$

$$\therefore \text{net accelerating force} = (88 - 0{\cdot}044v^2) \times 5{\cdot}5/0{\cdot}36 - (245 + 0{\cdot}96v^2)$$

$$= 1100 - 1{\cdot}632v^2\ \text{N}$$

The equation of motion of the car is therefore

$$1100 - 1{\cdot}632v^2 = 1097{\cdot}7f$$

$$\therefore f = 1{\cdot}002 - 0{\cdot}001\,487v^2 \qquad (1)$$

When $v = 12$ m/s,

$$f = 1{\cdot}002 - 0{\cdot}001\,487 \times 144 = \underline{0{\cdot}788\ \text{m/s}^2}$$

Equation (1) may be written in the form

$$\frac{\mathrm{d}v}{\mathrm{d}t} = 1{\cdot}002 - 0{\cdot}001\,487v^2$$

or

$$\mathrm{d}t = \frac{\mathrm{d}v}{1{\cdot}002 - 0{\cdot}001\,487v^2} = \frac{672{\cdot}5\,\mathrm{d}v}{25{\cdot}9^2 - v^2}$$

Hence

$$t = \int_{12}^{20} \frac{672{\cdot}5\,\mathrm{d}v}{25{\cdot}9^2 - v^2}$$

$$= \frac{672{\cdot}5}{2 \times 25{\cdot}9}\left[\ln \frac{25{\cdot}9 + v}{25{\cdot}9 - v}\right]_{12}^{20}$$

$$= 12{\cdot}98 \ln 2{\cdot}853$$

i.e.

$$t = \underline{13{\cdot}6\ \text{s}}$$

Example 1.24 A motor vehicle of total mass 1360 kg has road wheels of 640 mm effective diameter. The engine develops a maximum torque of 200 N m when running at uniform speed, and the transmission efficiency is 85 per cent.

Given that the total moment of inertia of the road wheels and axles is 7·6 kg m^2, and that of the engine and flywheel is 0·84 kg m^2, calculate the value of the gear ratio that will give maximum acceleration up a slope of 1 in 8 with a road resistance of 270 N under maximum torque conditions. (I. Mech. E.)

Solution

Let n be the gear ratio and f be the acceleration of the vehicle in m/s^2. Then the tractive force required to overcome resistance, component of weight down slope and inertia force is given by

$$\text{tractive force} = 270 + (1360 \times 9{\cdot}81/8) + 1360f = (1937 + 1360f)\ \text{N}$$

$$\text{Torque to accelerate wheels} = 7{\cdot}6 \times f/0{\cdot}32 = 23{\cdot}7f\ \text{N m}$$

$$\therefore \text{ total torque on wheels} = 23{\cdot}7f + (1937 + 1360f) \times 0{\cdot}32$$

$$= 459f + 620\ \text{N m}$$

$$\therefore \text{ engine torque required} = (459f + 620) \times \frac{1}{n} \times \frac{100}{85} = \frac{540f + 729}{n}\ \text{N m}$$

$$\text{Torque to accelerate engine parts} = 0{\cdot}84 \times \frac{f}{0{\cdot}32} \times n = 2{\cdot}625nf$$

$$\therefore \text{ total engine torque} = \frac{540f + 729}{n} + 2{\cdot}625nf = 200\ \text{N m}$$

$$\therefore f = \frac{200n - 729}{540 + 2{\cdot}625n^2}$$

For maximum acceleration, $\mathrm{d}f/\mathrm{d}n = 0$

i.e. $$(540 + 2{\cdot}625n^2) \times 200 = (200n - 729) \times 5{\cdot}25n$$

from which $$n = \underline{18{\cdot}45}$$

Note that the transmission efficiency does not affect the torque required to accelerate the engine parts.

Example 1.25 A haulage vehicle has a mass of 40 t and is capable of carrying a load of mass 200 t. The vehicle has ten axles, the moment of inertia of the masses rotating with each axle being 150 kg m^2. The effective diameter of the wheels on the axles is 1 m and the moment of inertia of the masses rotating with the engine shaft is 10 kg m^2. The efficiency of transmission between the engine shaft and the driving axles is 0·80. The vehicle, when in bottom gear, attains a speed of 6 km/h at a maximum engine speed of 3000 rev/min and is capable of climbing, at a steady speed of 6 km/h, roads inclined at the angle $\sin^{-1} 0{\cdot}125$ to the horizontal when it is fully loaded. Frictional resistances to motion amount to 8000 N.

Determine: (a) the torque which the engine must be capable of developing,

(b) the greatest acceleration the fully loaded vehicle can attain in bottom gear along a level road, and (c) the greatest power the engine can develop.

(U. Lond.)

Solution

From Section 1.26, the effective mass of lorry and wheels for acceleration purposes is given by

$$m_e = 240 \times 10^3 + 10 \times 150/0{\cdot}5^2 = 246 \times 10^3 \text{ kg}$$

At constant speed up incline,

maximum tractive force = component of weight down incline + resistance

$$= 240 \times 10^3 \times 9{\cdot}81 \times 0{\cdot}125 + 8000$$

$$= 302\,300 \text{ N}$$

$$\text{Speed at which this force is developed} = \frac{6 \times 1000}{3600} = \frac{1}{0{\cdot}6} \text{ m/s}$$

Therefore $\quad$ output power required $= 302\,300/0{\cdot}6 = 503\,833$ W

and $\quad$ engine power required $= \dfrac{503\,833}{0{\cdot}8} = 629\,800 \text{ W} = \underline{629{\cdot}8 \text{ kW}}$

The engine speed,

$$\omega = 3000 \times 2\pi/60 = 100\pi \text{ rad/s}$$

Therefore $\quad$ engine torque $= 629{\cdot}8/100\pi = \underline{2 \text{ kN m}}$

If the acceleration of the lorry is f,

$$\text{acceleration of wheels} = f/0{\cdot}5 = 2f \text{ rad/s}^2$$

$$\text{Wheel speed} = \frac{1/0{\cdot}6}{0{\cdot}5} = \frac{1}{0{\cdot}3} \text{ rad/s}$$

$$\therefore \text{ gear ratio} = \frac{100\pi}{1/0{\cdot}3} = 30\pi$$

$$\therefore \text{ acceleration of engine} = 2f \times 30\pi = 60\pi f \text{ rad/s}^2$$

$$\therefore \text{ torque to accelerate engine parts} = 10 \times 60\pi f = 1885f \text{ N m}$$

$$\therefore \text{ output torque from engine} = 2000 - 1885f \text{ N m}$$

$$\therefore \text{ torque on wheels} = (2000 - 1885f) \times 30\pi \times 0{\cdot}8$$

$$= 150\,796 - 142\,126f \text{ N m}$$

$$\therefore \text{ tractive effort available} = \frac{150\,796 - 142\,126f}{0{\cdot}5}$$

$$= 301\,592 - 284\,252f \text{ N}$$

$$\text{The accelerating force} = m_e f$$

i.e. $\quad 301\,592 - 284\,252f = 246 \times 10^3 f$

from which $\quad f = \underline{0{\cdot}57 \text{ m/s}^2}$

Example 1.26 The resistance to motion of a train on a horizontal track is $35+0{\cdot}1v^2$ N/t where v is the velocity in m/s. The draw-bar pull is assumed constant and equal to 225 N per tonne mass of train. If the train approaches a gradient of 1 in 100 at a speed of 18 m/s, find the speed when it has travelled 1500 m up the gradient. (U. Lond.)

Solution
If the mass of the train is m tonnes,

$$\text{component of weight down the gradient} = \frac{m \times 1000 \times 9.81}{100} = 98{\cdot}1m$$

$$\therefore \text{ accelerating force} = 225m - m(35+0{\cdot}1v^2) - 98{\cdot}1m$$
$$= m(91{\cdot}9 - 0{\cdot}1v^2) \text{ N}$$

The equation of motion of the train is therefore

$$m(91{\cdot}9-0{\cdot}1v^2) = m \times 1000v\frac{dv}{ds}$$

$$\therefore \text{ ds} = \frac{1000v\,dv}{91{\cdot}9-0{\cdot}1v^2}$$

$$\therefore \ s = -\frac{1000}{0{\cdot}2}\int\frac{-0{\cdot}2v\,dv}{91{\cdot}9-0{\cdot}1v^2}$$
$$= -5000\ln(91{\cdot}9-0{\cdot}1v^2)+C$$

When $s=0$, $v=18$ m/s, and so

$$0 = -5000\ln(91{\cdot}9-32{\cdot}4)+C$$

i.e. $C = 20\,430$

Therefore, when $s = 1500$ m,

$$1500 = -5000\ln(91{\cdot}9-0{\cdot}1v^2)+20\,430$$

from which $\underline{v = 21{\cdot}83 \text{ m/s}}$

Example 1.27 A motor car has a wheelbase of 2·4 m with a centre of mass 0·84 m above the ground and 1 m behind the front axle. The coefficient of friction between the tyre and the ground is 0·5.

Calculate the maximum possible accelerations when the vehicle is: (a) driven on all four wheels, (b) driven on the front wheels only, and (c) driven on the rear wheels only. (I. Mech. E.)

Solution
From Section 1.29,

$$N_f + N_r = mg \quad (1)$$
$$F = mf \quad (2)$$

and
$$F \times 0{\cdot}84 = N_r \times 1{\cdot}4 - N_f \times 1 \quad (3)$$

(a) If the rear wheels are assumed to slip first

$$F = 2\mu N_r = 2 \times 0{\cdot}5 N_r = N_r$$

Substituting in equation (3) gives $N_f = 0{\cdot}56 N_r$ and, since $N_f < N_r$, the rear wheels will *not* slip first.

If the front wheels are assumed to slip first,

$$F = 2\mu N_f = N_f$$

Substituting in equation (3) gives $N_f = 0{\cdot}76 N_r$, which confirms that the assumption is correct.

Hence, from equation (1),

$$N_f + N_f/0{\cdot}76 = mg$$

or

$$N_f = 0{\cdot}432mg$$

so that, from equation (2),

$$0{\cdot}432mg = mf$$

$$f = \underline{4{\cdot}24 \text{ m/s}^2}$$

(b) For front wheel drive,

$$F = 0{\cdot}5N_f$$

$$\therefore \; N_r = 1{\cdot}014N_f \qquad \text{from equations (3) and (5)}$$

$$\therefore \; N_f = mg/2{\cdot}014 \qquad \text{from equation (1)}$$

$$\therefore \; 0{\cdot}5 \times mg/2{\cdot}014 = mf$$

$$\therefore \; f = \underline{2{\cdot}43 \text{ m/s}^2}$$

(c) For rear wheel drive,

$$F = 0{\cdot}5N_r$$

$$\therefore \; N_f = 0{\cdot}98N_r \qquad \text{from equations (3) and (5)}$$

$$\therefore \; N_r = mg/1{\cdot}98 \qquad \text{from equation (1)}$$

$$\therefore \; 0{\cdot}5 \times mg/1{\cdot}98 = mf$$

$$\therefore \; f = \underline{2{\cdot}48 \text{ m/s}^2}$$

Example 1.28 A motor car has a wheelbase of 2·8 m, its centre of mass is 0·6 m above ground-level and is equidistant from the axles. The brakes are arranged to apply the same braking torque at each wheel. The coefficient of adhesion between tyres and road may be taken as 0·6. If the brakes are applied on a down grade of 1 vertical in 10 sloping, show that the rear wheels will skid first, and find the retardation when this is about to happen. (U. Lond.)

Solution

From Section 1.29,

$$N_f + N_r = mg\cos\alpha \tag{1}$$

$$F = mg\sin\alpha + mf \tag{2}$$

and

$$F \times 0{\cdot}6 = N_f \times 1{\cdot}4 - N_r \times 1{\cdot}4 \tag{3}$$

From equation (3), it is evident that $N_f > N_r$, so that friction will first become limiting at the rear wheels. Hence, the rear wheels will skid first and thus maximum braking force,

$$F = 2\mu N_r = 1{\cdot}2 N_r$$

Also $\sin\alpha = 0{\cdot}1 \quad \text{and} \quad \cos\alpha = 0{\cdot}995$

Therefore equations (1), (2) and (3) become

$$N_f + N_r = 0{\cdot}995\, mg \tag{4}$$

$$1{\cdot}2 N_r - 0{\cdot}1 mg = mf \tag{5}$$

and

$$0{\cdot}72 N_r + 1{\cdot}4 N_r = 1{\cdot}4 N_f \tag{6}$$

$$\therefore \; N_f = 1{\cdot}515 N_r$$

Then, in equation (4), $2{\cdot}515 N_r = 0{\cdot}995 mg$

$$\therefore \; N_r = 3{\cdot}88 m$$

Therefore from equation (5),

$$1{\cdot}2 \times 3{\cdot}88 m - 0{\cdot}1 m \times 9{\cdot}81 = mf$$

from which $\underline{f = 3{\cdot}67 \text{ m/s}^2}$

Example 1.29 A motor car has brakes which act on all four wheels. The weight of the car is 10 kN and when at rest the loads carried by the rear and front wheels on a level road are 3·5 kN and 6·5 kN respectively. The length between the front and rear axles is 3 m and the height of the centre of mass above the ground is 0·6 m. Find the ratio of the braking forces on the rear and front wheels when all four wheels are on the point of skidding together if the coefficient of friction between the wheels and the road is 0·4. Also find the shortest distance in which the car can be brought to rest from a speed of 56 km/h. (U. Lond.)

Solution

Referring to Fig. 1.19, $a = \dfrac{3{\cdot}5}{10} \times 3 = 1{\cdot}05 \text{ m}$

and $b = \dfrac{6{\cdot}5}{10} \times 3 = 1{\cdot}95 \text{ m}$

From Section 1.29,

$$N_f + N_r = mg \tag{1}$$

$$0{\cdot}4 N_f + 0{\cdot}4 N_r = mf \tag{2}$$

and

$$(0{\cdot}4 N_f + 0{\cdot}4 N_r) \times 0{\cdot}6 = N_f \times 1{\cdot}05 - N_r \times 1{\cdot}95 \tag{3}$$

From equations (1) and (3),

$$N_r = 2{\cdot}65 m$$

and

$$N_f = 7{\cdot}16 m$$

$$\therefore \text{ ratio of braking forces} = \mu N_r / \mu N_f = \underline{0{\cdot}37}$$

From equation (2), $$0{\cdot}4mg = mf$$

$$\therefore f = 0{\cdot}4g = 3{\cdot}92 \text{ m/s}^2$$

$$\therefore s = \frac{v^2}{2f}$$

$$= \frac{(56 \times 1000/3600)^2}{2 \times 3{\cdot}92} = \underline{30{\cdot}9 \text{ m}}$$

Example 1.30 The centre of mass of a motor cycle with rider is 0·36 m in front of the rear axle and 0·6 m above the level ground. The distance between the front and rear axles is 1·2 m. Assuming that the power available is more than adequate, determine the greatest acceleration with which the motor cycle can be driven up an incline of 15°. The coefficient of friction, μ, between the tyres and the incline is 0·7.

Determine whether or not an increase in the above value of μ could permit a greater acceleration to be achieved. (U. Lond.)

Solution

From equations 1.37, 1.38 and 1.39,

$$N_f + N_r = mg \cos 15° \tag{1}$$

$$F = \mu N_r = mg \sin 15° + mf \tag{2}$$

and $$\mu N_r \times 0{\cdot}6 = N_r \times 0{\cdot}36 - N_f \times 0{\cdot}84 \tag{3}$$

If a coefficient of friction of 0·7 is used, equation (3) shows that N_f is negative, i.e. the machine would tip up before the rear wheel slips. Thus no greater acceleration could be achieved by increasing the value of μ.

If the machine is not to tip up, the limiting value of $N_f = 0$. Hence, from equation (3), the greatest useful value of μ is found to be 0·6.

If $N_f = 0$, $N_r = mg \cos 15°$ and so, from equation (2),

$$0{\cdot}6mg \cos 15° = mg \sin 15° + mf$$

from which $$\underline{f = 3{\cdot}15 \text{ m/s}^2}$$

This represents the greatest possible acceleration for all values of μ in excess of 0·6 if the machine is not to tip up.

Problems: vehicle dynamics

51. An electrically driven road vehicle of mass 450 kg has four wheels, of effective diameter 400 mm, radius of gyration 125 mm, and mass 9 kg. The armature of the electric motor has a mass of 65 kg, a radius of gyration of 100 mm, and rotates at 4 times the speed of the road wheels. The rolling resistance to motion of the vehicle is to be assumed constant at 225 N.

Determine the acceleration of the vehicle if it is allowed to run freely down a slope whose inclination to the horizontal is $\sin^{-1} 0{\cdot}2$. (U. Lond.) (*Answer:* 0·91 m/s²)

52. A motor vehicle has a mass of 1 Mg and the moments of inertia of the road wheels and rear axle are together 8·4 kg m². The effective diameter of the road wheels is

620 mm. Find the acceleration on the level at an instant when the engine output torque is 100 N m and the overall speed reduction ratio is 14. Take the air and road resistance as 200 N at this speed, and the transmission efficiency as 88 per cent. (U. Lond.)
(*Answer:* 3·47 m/s^2)

53. A 10-t truck is driven at a steady speed of 40 km/h on a level road, against a tractive resistance of 200 N/t. The wheels are 1·2 m in diameter, the engine runs at 1800 rev/min, and the transmission efficiency may be taken as 0·85. Calculate the engine power and the gear ratio in use, engine to wheels.

With the throttle fully open and the engine giving three times this power, at the same speed, estimate the maximum truck speed in km/h up a 10° slope, and the gear ratio required. Under these conditions, what is the overall efficiency as a weight-lifting machine? (U. Glas.) (*Answer:* 26·15 kW; 10·18; 12·6 km/h; 32·3; 76·1 per cent)

54. A motor car, of total mass 1400 kg, running in top gear at 55 km/h, passes on to a rising gradient of 1 in 20 at that speed. The road and other resistances may be taken as constant at 270 N/t. With conditions unchanged, the speed falls uniformly to 40 km/h in a distance of 360 m. An intermediate gear is then engaged for which the speed ratio between the engine and the road wheels is 9 : 1 and the transmission efficiency is 85 per cent. The engine is then developing a torque of 85 N m.

Determine: (a) the tractive efforts before and after the gear change, given that the road wheels are 750 mm in diameter, (b) the time between the start of retardation and the recovery of the original speed of 55 km/h, and (c) the powers being developed by the engine when the car is retarding and accelerating through 50 km/h, given that the transmission efficiency in top gear is 94 per cent. (I. Mech. E.)
(*Answer:* 851 N; 1·735 kN; 36 s; 12·56 kW; 28·35 kW)

55. A locomotive of mass 90 t pulls a train of 10 coaches, each of mass 30 t, at a speed of 72 km/h on a level track, against frictional resistances totalling 68 N/t.
(a) Find the power of the locomotive and the draw-bar pull on the leading coach.
(b) If the rear four coaches are suddenly released, find the speed of the engine and remaining coaches after 120 s, assuming frictional resistance per tonne and draw-bar pull remain unchanged.
(c) Find the speed of the released coaches after 120 s, if: (i) the resistance remains unchanged, and (ii) the resistance varies directly with the speed. (U. Lond.)
(*Answer:* 530 kW; 20·4 kN; 91·6 km/h; 42·6 km/h; 47·9 km/h)

56. A motor vehicle has a total mass of 3 t of which 180 kg represents the mass of the parts rotating at engine speed. The radius of gyration of these rotating parts is 250 mm. The diameter of the road wheels is 800 mm and the engine crankshaft rotates at five times the speed of the road wheels.

Neglecting the inertia of the wheels and the transmission losses, determine, for a constant engine torque of 400 N m: (a) the total kinetic energy of the vehicle at a speed of 32 km/h, and (b) the acceleration of the vehicle when travelling up a slope of 1 in 10. (U. Lond.) (*Answer:* 188 kJ; 0·432 m/s^2)

57. The total mass of a locomotive is 75 t and two-thirds of this total is carried on the coupled driving wheels. The locomotive has to pull a train of mass 325 t under conditions such that the limiting coefficient of adhesion between wheels and rails is 0·13.

Determine: (a) the starting acceleration on the level, if the initial resistances are assumed to be 100 N/t for the locomotive and 65 N/t for the train, (b) the time to reduce speed from 72 to 48 km/h on a down gradient of 1 in 400 with the power off and brakes on the driving wheels only, if the tractive resistances are 45 N/tonne of total mass, and (c) the power developed when climbing a gradient of 1 in 120 at the maximum possible speed if the resistance, in N/tonne of total mass, is given by $(27 + 0{\cdot}007v^2)$, where v is the speed in km/h. (I, Mech. E.)

(*Answer:* 0·0878 m/s²; 37·1 s; 1509 kW)

58. The engine of a motor car runs at 3420 rev/min when the road speed is 100 km/h. The mass of the car is 1 t. The inertia of the rotating parts of the engine corresponds to 10 kg at a radius of gyration of 150 mm and that of the road wheels to 100 kg at 240 mm. The efficiency of engine and transmission is 0·9 and the wind resistance is 1 kN. The road wheel diameter is 0·75 m.

Estimate the power developed by the engine when the car travels on a level road at 100 km/h with an acceleration of 0·9 m/s². (U. Lond.) (*Answer:* 60·75 kW)

59. A car has a mass of 800 kg, the moment of inertia of the wheels is 6·7 kg m² and that of the engine parts is 0·2 kg m². The effective diameter of the road wheels is 620 mm and the maximum engine torque is 65 N m at an engine speed of 2000 rev/min.

If the resistance to motion in N is given by $R = 135 + 0{\cdot}09v^2$, where v is the speed in km/h, find the road speed and the acceleration of the car when the gear ratio is 13·5 to 1 and the engine is developing full torque at 2000 rev/min. The efficiency of the transmission is 88 per cent. (I. Mech. E.) (*Answer:* 17·3 km/h; 1·935 m/s²)

60. For a typical motor car, the rolling resistance is given by the expression $(180 + 0{\cdot}6V)$ and the air resistance by $0{\cdot}1V^2$, the resistance being in newtons and V the speed in km/h in each case.

Assuming that the power output at engine speed corresponding to 48 km/h is 25 kW with a transmission efficiency of 84 per cent and that the inertia of the vehicle corresponds to a mass of 1·25 Mg, calculate the maximum possible acceleration in m/s² when running on the level under these conditions. (I. Mech. E.)

(*Answer:* 0·908 m/s²)

61. The speed of a car on the level is 65 km/h, the engine indicating 18·5 kW and the mass of the car being 1100 kg. The car will just run down a gradient of 1 in 25 when the clutch is in but with ignition cut off.

Assuming the engine and transmission friction and also the road resistance to be independent of the speed, and the wind resistance to be proportional to the square of the speed, determine the power required to drive the car up a gradient of 1 in 25 at a speed of 55 km/h. (U. Lond.) (*Answer:* 19·7 kW)

62. A motor cycle engine gives a torque of 24 N m at 2000 rev/min. The moment of inertia of each road wheel is 1·2 kg m² and that of the engine parts is 0·1 kg m². The effective diameter of the rear wheel is 620 mm and the total mass of the machine and rider is 180 kg.

If the speed reduction between engine and rear wheel is 9 : 1 and the combined effect of rolling resistance and windage is assumed to amount to 180 N, find the road speed and acceleration of the motor cycle at the above engine speed. (I. Mech. E.)

(*Answer:* 26·1 km/h; 1·78 m/s²)

63. A modern touring car has a mass of 1·8 t unloaded and carries two persons, each having a mass of 70 kg, together with 200 kg of luggage. The maximum output torque from the engine is 200 N m, while the effective windage and rolling resistance at a speed of 100 km/h is 620 N. In third gear the ratio of engine speed to back axle speed is 6·43 : 1, while the effective radius of the road wheels is 350 mm. The total moment of inertia of the rotating parts of the engine and the wheels is equivalent to a flywheel of moment of inertia 25 kg m² at the road wheels.

Determine: (a) the time taken for the vehicle to accelerate from 80 km/h to 110 km/h at full output torque, on the assumption that the windage and rolling resistance may be assumed constant over this range and that transmission losses may be neglected, and (b) the maximum gradient that this vehicle can climb in this gear, at a steady speed of 100 km/h. (U. Lond.) (*Answer:* 6·4 s; $\sin^{-1} 0{\cdot}146$)

64. The engine of a car rotates at 12 times the speed of the road wheels, which are 660 mm in diameter. The mass of the car is 900 kg. The engine flywheel mass is 10 kg and its radius of gyration is 125 mm. The resistance to motion of the car is 270 N. Calculate the engine torque and power required to accelerate the car at 0·6 m/s^2 when travelling on a level road at 6 m/s. The efficiency of power transmission may be taken as 85 per cent and the inertia of the other rotating parts may be neglected. (U. Lond.)
(*Answer:* 29·6 N m; 6·45 kW)

65. A motor vehicle has a mass of 9 t and the engine develops 82 kW at 2200 rev/min. The transmission efficiency is 92 per cent in the top gear of 3·3 to 1 and 85 per cent in the third gear of 8·2 to 1. The performance characteristics are such that the vehicle will just reach a speed of 96 km/h at 2200 rev/min and full throttle when running on the level in still air and at the same engine speed in third gear it will just climb a gradient of 1 in 18.

The combined air and rolling resistance is given by a formula of the form $R = A + BV^2$. Calculate the values of A and B when R is in N and V is in km/h and hence deduce the engine power required for climbing a gradient of 1 in 50 at 48 km/h in top gear. (I. Mech. E.) (*Answer:* 1352; 0·1604; 50·53 kW)

66. A motor car has a mass of 1·15 Mg including the four road wheels which each have an effective diameter of 650 mm, radius of gyration 270 mm, and mass 25 kg. The engine develops 34 kW at 3000 rev/min and the parts rotating at engine speed have a mass of 55 kg with a radius of gyration of 88 mm. The transmission efficiency is 90 per cent and the total road and air resistance at this engine speed in top gear of 4·2 to 1 is 800 N. Calculate the acceleration in m/s^2 under these conditions. (I. Mech. E.)
(*Answer:* 0·357 m/s^2)

67. A motor vehicle of total mass 5·7 Mg starts the ascent of a 1 in 50 gradient at a speed of 32 km/h and maintains a constant tractive effort of 3·5 kN throughout the total length of 600 m. The road resistance amounts to 350 N/t. After the ascent, it has to travel along a straight level stretch of 150 m in length before taking a bend of 45 m radius. The roadway at the curve is unbanked and the coefficient of friction against side slip is 0·4.

Determine the uniform tractive effort that is permissible on the 150 m length if the vehicle is not to reach a speed at the curve in excess of that at which skidding would occur. (I. Mech. E.) (*Answer:* 2·302 kN)

68. A motor vehicle has a total mass of 1·5 t, a wheelbase length of 3 m, and a track width of 1·5 m. The centre of mass is 1·7 m behind the front axle and 0·65 m above road level.

Determine: (a) the time to rest from 48 km/h if there are brakes on all four wheels and these are fully effective on a level road for which the limiting coefficient of friction is 0·35, (b) the normal road reactions on the front and rear wheel pairs when accelerating on a level track at 1·2 m/s^2, and (c) the normal road reactions on the inner and outer wheel pairs when travelling at a steady speed of 12 m/s on a roadway curved to a radius of 60 m and banked at 10°. (I. Mech. E.)
(*Answer:* 3·88 s; 6 kN; 8·73 kN; 7·131 kN; 7·987 kN)

69. (a) Prove that the 'equivalent mass' of a rolling disc of mass M, radius R, and radius of gyration about its axis k, is

$$M\left(1+\frac{k^2}{R^2}\right)$$

(b) A car of *total* mass 800 kg accelerates on a horizontal road. The wheel diameters are 660 mm and the radius of gyration $k = 270$ mm. Each pair of wheels has a mass of 68 kg. The wheelbase is 1·8 m and the co-ordinates of the c.m. of the car are 750 mm from the front wheels and 700 mm above the ground. Determine the maximum acceleration possible if the coefficient of friction between the tyres and the road is 0·5. The drive is through the rear wheels. (U. Lond.) (*Answer:* 2·28 m/s^2)

70. A motor vehicle, mass 1·8 Mg, has to climb a gradient whose sine is 0·4. The drive is via the back wheels, the wheelbase is 3 m, and the centre of mass of the vehicle is 1·2 m behind the front axle and 0·6 m above the ground.

Determine, for ascent at a uniform speed: (a) the minimum coefficient of friction between tyres and road, and (b) the power necessary if the speed is 16 km/h and the transmission efficiency is 60 per cent. (U. Lond.) (*Answer:* 0·895; 52·2 kW)

71. A motor vehicle has a wheelbase of 2·4 m. When standing on a level road its centre of mass is 0·9 m behind the front axle and 0·45 m above the road surface. The coefficient of friction between the tyres and the road is 0·4. The vehicle may be driven (a) by the front wheels, (b) by the rear wheels, or (c) by all four wheels. In case (c) the drive is arranged so that the torque transmitted to the front axle is the same as that transmitted to the rear axle.

Determine, for each of these forms of drive, the maximum gradient that may be climbed at a uniform speed, without wheelspin. (U. Lond.)

(*Answer:* $\sin^{-1}$ 0·227; $\sin^{-1}$ 0·16; $\sin^{-1}$ 0·333)

72. A vehicle having a wheelbase of 3·3 m is driven along a horizontal road by a torque applied to the rear wheels. The centre of mass is 0·75 m above the ground and 1·35 m behind the front axle. The coefficient of friction between the wheels and the ground is 0·3.

Determine: (a) the maximum acceleration of the vehicle if the wheels are not to slip, and (b) the maximum retardation of the vehicle when a braking torque is applied to the rear wheels. (U. Lond.) (*Answer:* 1·29 m/s^2; 1·126 m/s^2)

73. A motor car of mass 1·15 Mg has axles 2·4 m apart. When standing on a level road, the centre of mass of the car is 0·6 m above ground-level, and 0·9 m in front of the rear axle.

Calculate the normal reaction on each wheel when the car is moving down a gradient of 1 in 20, against a wind resistance of 180 N acting parallel to the road and 0·75 m from it, with the engine switched off and the rear wheel brakes applied so as to give a deceleration of 0·6 m/s^2. What must be the coefficient of friction between the rear wheels and the road if the rear wheels are not to skid under these conditions? Neglect the rotational inertia of the wheels and engine. (U. Lond.)

(*Answer:* rear, 3·39 kN, front 2·24 kN; 0·159)

74. The wheelbase of a car is 3·3 m, and the centre of mass is 1·2 m in front of the rear axle and 0·75 m above the ground-level. The coefficient of friction between the wheels and the road is 0·45. Find the maximum acceleration which the car can be given on a level road if it has a rear-wheel drive. Find its maximum deceleration if the braking force on all the wheels is the same and no wheel slip occurs. (U. Lond.)

(*Answer:* 3·13 m/s^2; 4·037 m/s^2)

75. A motor vehicle has a loaded mass of 1·3 Mg with a wheelbase of 2·65 m. The centre of mass is 1·35 m behind the front axle and 0·65 m above the ground.

Calculate the load on the front axle: (a) when the car is running at constant speed, and (b) when the car is decelerating with all four wheels just locked by braking, the limiting road adhesion coefficient being 0·6. (I. Mech. E.) (*Answer:* 6·25 kN; 8·13 kN)

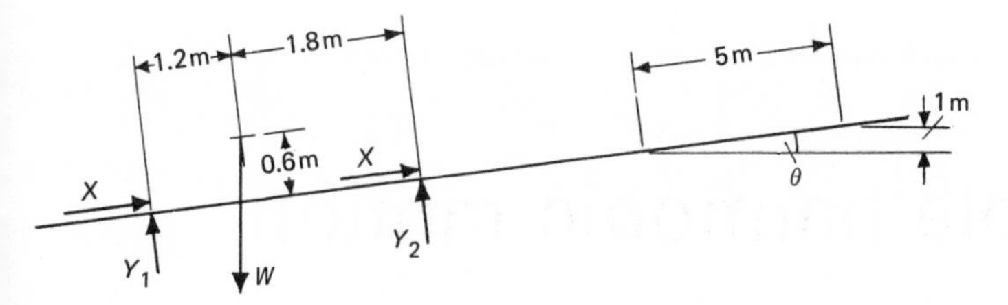

FIG. 1.45

76. The forces acting on a vehicle as it descends an incline are shown in Fig. 1.45. These are its weight W, the normal forces at the contacts of the wheels and the ground, and the tangential forces caused by the application of the brakes. The braking forces are increased equally until slip occurs at the rear wheels. If the friction coefficient is 0·4, find the deceleration then produced and verify that the front wheels have not reached the condition of slip. (I. Mech. E.) (*Answer:* 0·69 m/s^2)

77. A motor cycle engine applies a torque T to the rear wheel.
(a) Assuming there is to be sufficient friction between the rear wheel and the road to prevent slipping, show that the limiting torque, T_0, above which the cycle would tend to tip up, with the body rotating about the rear axle, is given by

$$T_0 = \frac{mgar}{r+h}$$

where m is the mass of the cycle and rider, a is the distance of the centre of mass of the cycle forward of the rear axle, h is the distance of the centre of mass of the cycle above the rear axle, and r is the radius of the wheels.
(b) If the torque applied to the rear wheel is greater than T_0, derive an expression for the coefficient of friction μ which would need to exist between the tyre and the road in order to prevent the cycle from tipping up. (*Answer:* $\mu = a/(r+h)$)

2

Simple harmonic motion

2.1 Simple harmonic motion

A body moves with simple harmonic motion if its acceleration is proportional to its displacement from a fixed point and is always directed towards that point. Let the line OP, of length a, rotate about a fixed point O, with constant angular velocity ω, Fig. 2.1. Then, if time is measured from the position OB, the angle turned through by OP in time t is given by

$$\psi = \omega t$$

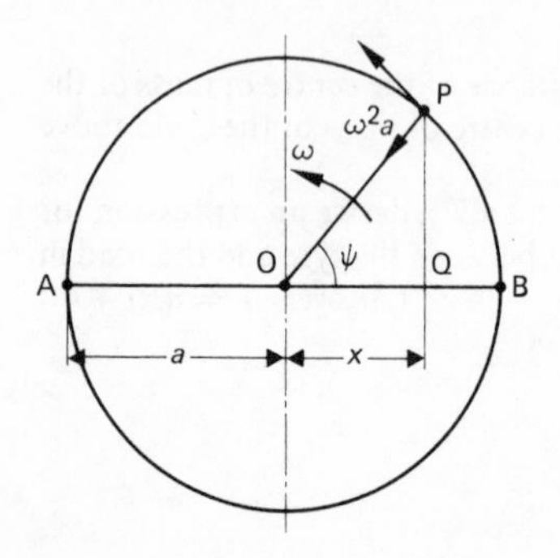

Fig. 2.1

If Q is the projection of P on the diameter AB, the displacement of Q from its mid-position is given by

$$x = a \cos \psi \tag{2.1}$$

The maximum displacement, a, is termed the *amplitude* of the motion. The velocity of Q is the component of the velocity of P parallel to AB,

i.e. $$v = \omega a \sin \psi = \omega \sqrt{(a^2 - x^2)} \tag{2.2}$$

and $$v_{max} = \omega a \quad \text{when} \quad x = 0 \tag{2.3}$$

The acceleration of Q is the component of the acceleration of P parallel to AB,

i.e. $$f = \omega^2 a \cos \psi = \omega^2 x \tag{2.4}$$

and $$f_{max} = \omega^2 a \quad \text{when } x = a \tag{2.5}$$

Thus the acceleration of Q is proportional to its displacement from the fixed point O and is always directed towards O, so that the motion of Q is simple harmonic. ω is called the *circular frequency* of the S.H.M.

These formulae give merely the numerical relationships between displacement, velocity and acceleration without regard to direction.

The periodic time is the time taken for one complete revolution of P,

i.e. $$t_p = 2\pi/\omega \text{ s}$$

but, from equation 2.4, $$\omega^2 = f/x$$

$$\therefore \; t_p = 2\pi\sqrt{\frac{x}{f}} = 2\pi\sqrt{\frac{\text{displacement}}{\text{acceleration}}} \text{ s} \tag{2.6}$$

The unit of frequency is the *hertz* (Hz), which is one cycle per second.

Thus frequency, $$n = \frac{1}{t_p} = \frac{1}{2\pi}\sqrt{\frac{f}{x}} \text{ Hz} \tag{2.7}$$

2.2 Angular simple harmonic motion

The equations of Section 2.1 apply equally well in the case of angular simple harmonic motion. Thus, if the amplitude of the motion is ϕ, Fig. 2.2, the angular velocity of the body at any angular displacement, θ, is given by

$$\Omega = \omega\sqrt{(\phi^2 - \theta^2)} \tag{2.8}$$

where ω is the angular speed of the line generating the S.H.M.

and $$\Omega_{max} = \omega\phi \tag{2.9}$$

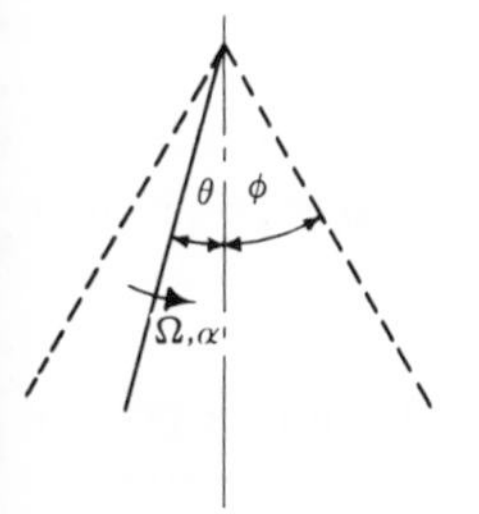

FIG. 2.2

The angular acceleration is given by

$$\alpha = \omega^2\theta \tag{2.10}$$

and $$\alpha_{max} = \omega^2\phi \tag{2.11}$$

The periodic time, $t_p = 2\pi\sqrt{\dfrac{\text{angular displacement}}{\text{angular acceleration}}}$

$$= 2\pi\sqrt{\frac{\theta}{\alpha}} \text{ s} \tag{2.12}$$

and $$n = \frac{1}{2\pi}\sqrt{\frac{\alpha}{\theta}}\,\text{Hz} \tag{2.13}$$

2.3 Linear motion of an elastic system

If a body of mass m, controlled by an elastic system, is given a displacement x, then

$$\text{restoring force} = Sx$$

where S is the stiffness of the system, i.e. the restoring force per unit displacement.

When released, the acceleration of the body is given by

$$\text{force} = \text{mass} \times \text{acceleration}$$

i.e. $$Sx = mf$$

i.e. $$\frac{x}{f} = \frac{m}{S} = \text{constant}$$

$$\therefore\ t_p = 2\pi\sqrt{\frac{m}{S}}\,\text{s} \qquad \text{from equation 2.6}$$

$$= 2\pi\sqrt{\frac{\delta}{g}}\,\text{s} \tag{2.14}$$

where δ is the static deflection under the load mg,

or $$n = \frac{1}{2\pi}\sqrt{\frac{g}{\delta}}\,\text{Hz}$$

$$\simeq \frac{1}{2\sqrt{\delta}}\,\text{Hz} \tag{2.15}$$

where δ is in metres.

2.4 Angular motion of an elastic system

If a body of moment of inertia I, controlled by an elastic system, is given an angular displacement θ, then

$$\text{restoring torque} = q\theta$$

where q is the torsional stiffness of the system, i.e. the restoring torque per unit angular displacement.

When released, the angular acceleration of the body is given by

$$\text{torque} = \text{moment of inertia} \times \text{angular acceleration}$$

i.e. $$q\theta = I\alpha$$

i.e. $$\frac{\theta}{\alpha} = \frac{I}{q} = \text{constant}$$

$$\therefore\ t_p = 2\pi\sqrt{\frac{I}{q}}\,\text{s}$$

or $$n = \frac{1}{2\pi}\sqrt{\frac{q}{I}}\,\text{Hz} \tag{2.16}$$

2.5 Differential equation of motion

If a body of mass m is acted upon by a restoring force S per unit displacement from the equilibrium position, the equation of motion is

$$m\frac{d^2x}{dt^2} = -Sx$$

The negative sign arises because the restoring force Sx is opposite in direction to the displacement x.

This may be written

$$\frac{d^2x}{dt^2} + \omega^2 x = 0 \quad \text{where} \quad \omega^2 = S/m \tag{2.17}$$

The solution is*

$$x = A\cos\omega t + B\sin\omega t \tag{2.18}$$

This equation represents an oscillatory motion of periodic time

$$t_p = 2\pi/\omega$$

and the constants A and B are determined by the initial conditions of the motion.

Thus if $\quad x = a \quad$ when $t = 0$ and $\dfrac{dx}{dt} = 0$ when $t = 0$

then

$$A = a \quad \text{and} \quad B = 0$$

so that

$$x = a\cos\omega t \tag{2.19}$$

In Fig. 2.1, these conditions correspond to the motion commencing when P coincides with B, so that $\psi = \omega t$.

Differentiating equation 2.19 twice,

$$v = -\omega a\sin\omega t = \omega a\cos\left(\omega t + \frac{\pi}{2}\right) \tag{2.20}$$

and

$$f = -\omega^2 a\cos\omega t = \omega^2 a\cos(\omega t + \pi) \tag{2.21}$$

Equations 2.20 and 2.21 correspond to equations 2.2 and 2.4 except that the correct signs are automatically obtained by differentiation.

Equation 2.20 shows that the velocity leads the displacement by 90° and equation 2.21 shows that the acceleration leads the displacement by 180°.

The corresponding equations for angular motion are

$$I\frac{d^2\theta}{dt^2} = -q\theta$$

i.e.

$$\frac{d^2\theta}{dt^2} + \omega^2\theta = 0 \quad \text{where } \omega^2 = q/I \tag{2.22}$$

$$\therefore \theta = A\cos\omega t + B\sin\omega t \tag{2.23}$$

If $\quad \theta = \phi \quad$ when $t = 0$ and $\dfrac{d\theta}{dt} = 0$ when $t = 0$

* See the Appendix.

then $$\theta = \phi \cos \omega t \tag{2.24}$$
$$\Omega = -\omega\phi \sin \omega t \tag{2.25}$$
and $$\alpha = -\omega^2 \phi \cos \omega t \tag{2.26}$$

2.6 Energy method

For a system oscillating freely about a position of static equilibrium, the total energy, U, remains constant,

i.e. $$U = \text{P.E.} + \text{K.E.} + \text{S.E.} = \text{constant}$$

The circular frequency, ω, of the oscillation may then be obtained by equating the total energy of the system at an extreme position to that at the mid-position. (See Examples 2.2, 2.8 and 2.9.)

Since U is a constant,
$$\frac{\mathrm{d}U}{\mathrm{d}t} = 0$$

Thus an alternative method is to obtain an expression for the total energy of the system at a typical point in the oscillation, differentiate with respect to time, and equate to zero. This leads to the equation of motion of the system, from which the frequency of oscillation can be deduced. (See Example 2.8.)

2.7 Simple pendulum

If the pendulum is given a *small* angular displacement θ, Fig. 2.3,

$$\text{restoring moment about O} = mgl \sin \theta$$
$$\simeq mgl\,\theta \quad \text{since } \theta \text{ is small}$$
$$\therefore \; mgl\theta = I_O \alpha = ml^2 \alpha$$
$$\therefore \; \frac{\theta}{\alpha} = \frac{l}{g}$$
$$\therefore \; t_p = 2\pi \sqrt{\frac{l}{g}} \text{ s} \tag{2.27}$$

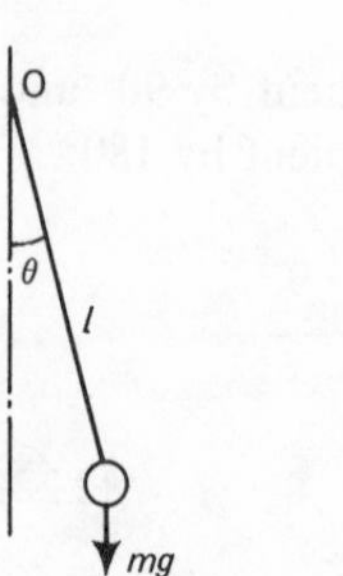

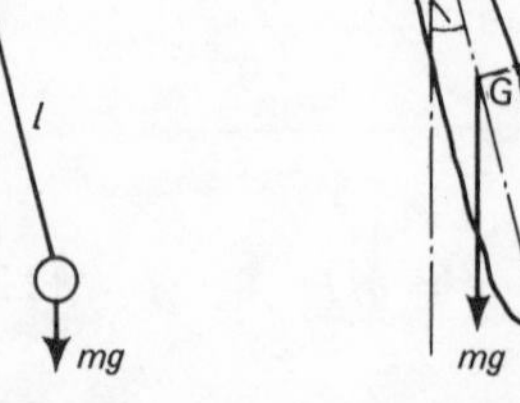

FIG. 2.3 FIG. 2.4

2.8 Compound pendulum

Let the radius of gyration about the centre of mass G be k and the distance of the point of suspension from G be h, Fig. 2.4.

If the pendulum is given a *small* angular displacement θ,

$$\text{restoring moment about O} = mgh\sin\theta \simeq mgh\theta$$

$$\therefore\ mgh\theta = I_O\alpha = m(k^2+h^2)\alpha$$

$$\therefore\ \frac{\theta}{\alpha} = \frac{I_O}{mgh} = \frac{k^2+h^2}{gh}$$

$$\therefore\ t_p = 2\pi\sqrt{\frac{k^2+h^2}{gh}}\ \text{s} \qquad (2.28)$$

Examples

Example 2.1 A spring of stiffness 2 kN/m is suspended vertically and two equal masses of 4 kg each are attached to the lower end. One of these masses is suddenly removed and the system oscillates.

Determine: (a) the amplitude and frequency of the vibration, (b) the velocity and acceleration of the mass when passing through the half amplitude position, and (c) the energy of the vibration in joules. (I. Mech. E.)

Solution

(a) Static deflection under a load of 8 kg $= \dfrac{8\times 9{\cdot}81}{2\times 10^3} = 0{\cdot}039\,24$ m

Static deflection under a load of 4 kg $= \dfrac{4\times 9{\cdot}81}{2\times 10^3} = 0{\cdot}019\,62$ m

Therefore, when one mass is removed, the remaining mass will oscillate about the static deflection position of that mass with an amplitude of $\underline{0{\cdot}019\,62\text{ m.}}$

The frequency, $n = \dfrac{1}{2\sqrt{\delta}}$ from equation 2.15

$$= \frac{1}{2\sqrt{0{\cdot}019\,62}} = \underline{3{\cdot}57\text{ Hz}}$$

(b) $\omega = \sqrt{(g/\delta)}$ from equation 2.14

$$= \sqrt{(9{\cdot}81/0{\cdot}019\,62)} = 22{\cdot}35\text{ rad/s}$$

$v = \omega\sqrt{(a^2 - x^2)}$ from equation 2.2

$$= 22{\cdot}35\sqrt{(0{\cdot}019\,62^2 - 0{\cdot}009\,81^2)} = \underline{0{\cdot}38\text{ m/s}}$$

$f = \omega^2 x$ from equation 2.4

$$= 22{\cdot}35^2 \times 0{\cdot}009\,81 = \underline{4{\cdot}905\text{ m/s}^2}$$

(c) Velocity of mass in mid-position $= \omega a$

$$= 22{\cdot}35 \times 0{\cdot}019\,62 = 0{\cdot}4385\text{ m/s}$$

$$\therefore\ \text{K.E. in mid-position} = \tfrac{1}{2}mv^2 = \tfrac{1}{2}\times 4\times 0{\cdot}4385^2 = \underline{0{\cdot}3844\text{ J}}$$

Example 2.2 Derive an expression for the periodic time of vibration of a mass M attached to the free end of a close-coiled helical spring of stiffness S and mass m, allowing for the mass of the spring.

A spring of stiffness 200 N/m has a mass of 0·75 kg. A mass of 5 kg is attached to the free end and set in motion. Find the time of oscillation: (a) neglecting the mass of the spring, and (b) allowing for the mass of the spring.

Solution

Let a be the amplitude of the vibration. Fig. 2.5, and ω be the angular velocity of the line generating the S.H.M.

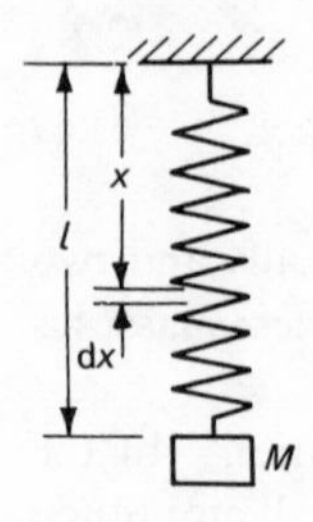

FIG. 2.5

Then velocity of M in mid-position $= \omega a$

$$\therefore \text{ velocity of element of spring in mid-position} = \frac{x}{l}\omega a$$

$$\therefore \text{ K.E. of element in mid-position} = \frac{m(\mathrm{d}x/l)}{2}\left(\frac{x}{l}\omega a\right)^2$$

$$\therefore \text{ total K.E. of spring in mid-position} = \frac{m}{2}\frac{\omega^2 a^2}{l^3}\int_0^l x^2\,\mathrm{d}x = \frac{m}{6}\omega^2 a^2$$

$$\therefore \text{ total K.E. of system in mid-position} = \frac{M}{2}\omega^2 a^2 + \frac{m}{6}\omega^2 a^2$$

$$= \frac{\omega^2 a^2}{2}\left(M + \frac{m}{3}\right)$$

Taking the lowest position as datum,

$$\text{P.E. of system in mid-position} = Mg + \frac{mg}{2}$$

since the centre of mass of the spring is displaced only $a/2$.

At mid-position, mean spring tension $= Mg + \dfrac{mg}{2}$

At lowest position, mean spring tension $= Mg + \dfrac{mg}{2} + Sa$

$\therefore$ work done in stretching spring to lowest position

$$= \frac{\left(Mg + \frac{mg}{2}\right) + \left(Mg + \frac{mg}{2} + Sa\right)}{2} \times a$$

$$= Mga + \frac{mga}{2} + \tfrac{1}{2}Sa^2$$

$$= \text{energy stored in spring at lowest position}$$

Equating the kinetic energy and potential energy of the system in the mid-position to the energy stored in the spring at the lowest position, since the total energy of the system remains constant,

$$\frac{\omega^2 a^2}{2}\left(M+\frac{m}{3}\right)+Mga+\frac{mga}{2}=Mga+\frac{mga}{2}+\tfrac{1}{2}Sa^2$$

i.e.

$$\omega^2 = S\bigg/\left(M+\frac{m}{3}\right)$$

$$\therefore\ t_p=\frac{2\pi}{\omega}=2\pi\sqrt{\left(M+\frac{m}{3}\right)\bigg/S}$$

The mass of the spring may therefore be allowed for by adding one-third of its mass to that of the concentrated load.

(a) Neglecting the mass of the spring,

$$n=\frac{1}{2\pi}\sqrt{\frac{S}{M}}=\frac{1}{2\pi}\sqrt{\frac{200}{5}}=\underline{1{\cdot}005\text{ Hz}}$$

(b) Allowing for the mass of the spring,

$$n=\frac{1}{2\pi}\sqrt{S\bigg/\left(M+\frac{m}{3}\right)}=\frac{1}{2\pi}\sqrt{200\bigg/\left(5+\frac{0{\cdot}75}{3}\right)}=\underline{0{\cdot}982\text{ Hz}}$$

Example 2.3 A valve of piston type has a mass of 7 kg and is driven with S.H.M. that can be referred to a vector of length 60 mm rotating at 24 rad/s. The valve works in oil which creates a viscous resistance proportional to the velocity at the rate of 10 N/(m/s). Events controlled by the valve occur when it is 30 mm from its mid-position and moving towards it and, again, when it is 20 mm beyond its mid-position on the same stroke. Determine the time between these events. Evaluate the viscous resisting force at the first event and the force in the driving rod at the second. (I. Mech. E.)

Solution

$$\omega = 24\text{ rad/s} \quad\text{and}\quad a = 60\text{ mm}$$

In Fig. 2.6, let Q_1 and Q_2 represent the two positions of the valve.

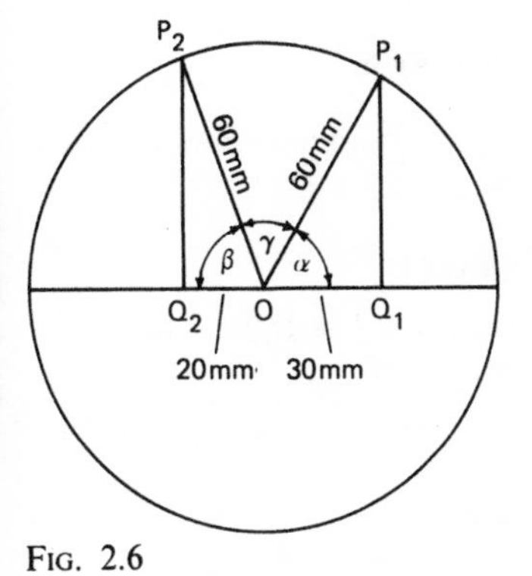

FIG. 2.6

Then $\alpha = \cos^{-1}(30/60) = 60°$

and $\beta = \cos^{-1}(20/60) = 70{\cdot}53°$

$\therefore \gamma = 49{\cdot}47° = 0{\cdot}863$ rad

$\therefore$ time taken to travel from Q_1 to $Q_2 = 0{\cdot}863/24 = \underline{0{\cdot}036 \text{ s}}$

At Q_1, $v = \omega \sqrt{(a^2 - x^2)}$ from equation 2.2

$= 24\sqrt{(0{\cdot}06^2 - 0{\cdot}03^2)} = 1{\cdot}247$ m/s

$\therefore$ viscous force $= 1{\cdot}247 \times 10 = \underline{12{\cdot}47 \text{ N}}$

At Q_2, $v = 24\sqrt{(0{\cdot}06^2 - 0{\cdot}02^2)} = 1{\cdot}358$ m/s

$\therefore$ viscous force $= 1{\cdot}358 \times 10 = 13{\cdot}58$ N

$f = \omega^2 x$ from equation 2.4

$= 24^2 \times 0{\cdot}02 = 11{\cdot}52 \text{ m/s}^2$

$\therefore$ retarding force $= 7 \times 11{\cdot}52 = 80{\cdot}64$ N

$\therefore$ net force in driving rod $= 80{\cdot}64 - 13{\cdot}58 = \underline{67{\cdot}06 \text{ N}}$

Example 2.4 The uniform thin rod, AB, shown in Fig. 2.7 has a mass of 1 kg and carries a concentrated mass of 2·5 kg at B. The rod is hinged at A and is maintained in the horizontal position by a spring of stiffness 1·8 kN/m at C. Find the frequency of oscillation, neglecting the effect of the mass of the spring.

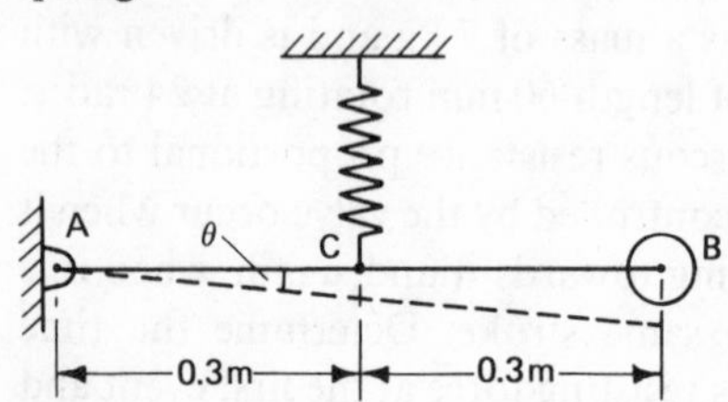

FIG. 2.7

Solution

Moment of inertia of system about A,

$$I_A = 1 \times \frac{0{\cdot}6^2}{3} + 2{\cdot}5 \times 0{\cdot}6^2 = 1{\cdot}02 \text{ kg m}^2$$

Let the rod be given a *small* angular displacement θ rad and then released.

Extension of spring $= 0{\cdot}3\theta$ m

$\therefore$ spring force $= 0{\cdot}3\theta \times 1{\cdot}8 \times 10^3 = 540\theta$ N

$\therefore$ restoring moment about A $= 540\theta \times 0{\cdot}3 = 162\theta$ N m

$\therefore 162\theta = 1{\cdot}02\alpha$

$$\therefore \frac{\alpha}{\theta} = \frac{162}{1{\cdot}02} = 159$$

$$n = \frac{1}{2\pi}\sqrt{\frac{\alpha}{\theta}} \qquad \text{from equation 2.13}$$

$$= \frac{1}{2\pi}\sqrt{159} = \underline{2{\cdot}01\ \text{Hz}}$$

Alternative solution

The given system may be replaced by an equivalent mass concentrated at the lower end of the spring. Then, for equivalence

K.E. of system at any instant = K.E. of equivalent mass

Let the instantaneous angular velocity of AB be Ω. Then

$$\text{K.E. of system} = \tfrac{1}{2}I_A\Omega^2 = \tfrac{1}{2} \times 1{\cdot}02\,\Omega^2$$

$$\text{K.E. of equivalent mass, } m_e = \tfrac{1}{2}m_e v^2$$

$$= \tfrac{1}{2}m_e(0{\cdot}3\,\Omega)^2 = \tfrac{1}{2} \times 0{\cdot}09\,m_e\Omega^2$$

Hence

$$0{\cdot}09\,m_e = 1{\cdot}02$$

$$\therefore\ m_e = 11{\cdot}333\ \text{kg}$$

The problem is then reduced to the standard case considered in Section 2.3, so that

$$n = \frac{1}{2\pi}\sqrt{(S/m)} = \frac{1}{2\pi}\sqrt{\left(\frac{1{\cdot}8 \times 10^3}{11{\cdot}333}\right)} = \underline{2{\cdot}01\ \text{Hz}}$$

Example 2.5 A uniform slender rod 1·2 m long is fitted with a transverse pair of knife-edges, so that it can swing in a vertical plane as a compound pendulum, the position of the knife-edges being variable.

Find the time of swing of the rod: (a) if the knife-edges are 50 mm from one end of the rod, and (b) if they are placed so that the time of swing is a minimum.

In case (a) find also the maximum angular velocity and the maximum angular acceleration of the rod supposing it to swing through 3° on either side of the vertical. (U. Lond.)

Solution

(a)

$$t_p = 2\pi\sqrt{(k^2 + h^2)/gh} \qquad \text{from equation 2.28}$$

$$= 2\pi\sqrt{\left(\frac{1{\cdot}2^2}{12} + 0{\cdot}55^2\right)\Big/(9{\cdot}81 \times 0{\cdot}55)} = \underline{1{\cdot}76\ \text{s}}$$

(b) For the time of swing to be a minimum,

$$\frac{d}{dh}\left(\frac{k^2 + h^2}{h}\right) = 0$$

i.e.

$$h = k$$

$$\therefore\ \text{minimum } t_p = 2\pi\sqrt{(2k/g)} = 2\pi\sqrt{\left(2 \times \frac{1{\cdot}2}{\sqrt{12}}\right)\Big/9{\cdot}81} = \underline{1{\cdot}67\ \text{s}}$$

In case (a), $\omega = 2\pi/1{\cdot}76 = 3{\cdot}565$ rad/s

$\therefore$ maximum angular velocity $= \omega\phi$ from equation 2.9

$$= 3{\cdot}565 \times \frac{3 \times \pi}{180} = \underline{0{\cdot}1868 \text{ rad/s}}$$

Maximum angular acceleration $= \omega^2\phi$ from equation 2.11

$$= 3{\cdot}565^2 \times \frac{3 \times \pi}{180} = \underline{0{\cdot}6666 \text{ rad/s}^2}$$

Example 2.6 Obtain an expression for the moment of inertia of a body as obtained by the trifilar suspension method.

A connecting rod of mass 5·5 kg is placed on a horizontal platform which is suspended by three equal wires, each 1·25 m long, from a rigid support. The wires are equally spaced round the circumference of a circle of 125 mm radius. When the centre of mass of the connecting rod coincides with the axis of the circle, the platform makes 10 angular oscillations in 30 s. Determine its moment of inertia about an axis through its centre of mass.

The unladen platform has a mass of 1·5 kg and makes 10 angular oscillations in 35 seconds.

Solution

Let the platform be displaced through a *small* angle θ and let the corresponding angular displacement of the wires be ϕ, Fig. 2.8.

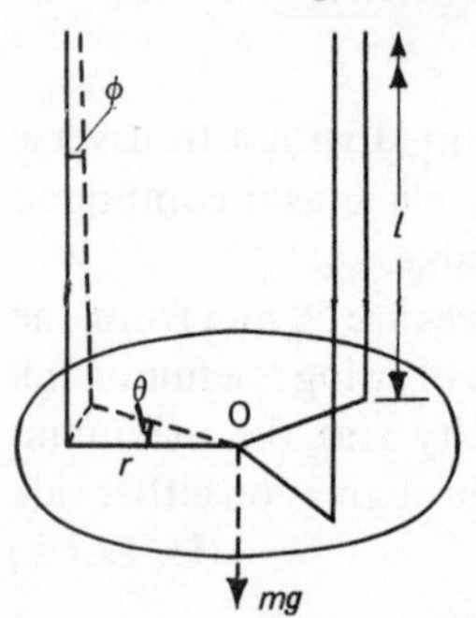

FIG. 2.8

Then
$$r\theta = l\phi$$

i.e.
$$\phi = \frac{r}{l}\theta$$

$$\text{Tension in each wire} = \frac{mg}{3}$$

$$\text{Component perpendicular to } r = \frac{mg}{3}\sin\phi \simeq \frac{mg}{3}\phi = \frac{mg}{3}\frac{r}{l}\theta$$

$$\therefore \text{ total restoring couple} = 3\left(\frac{mg}{3}\frac{r}{l}\theta \times r\right) = \frac{mgr^2\theta}{l}$$

$$\therefore \frac{mgr^2\theta}{l} = I_O\alpha$$

$$\therefore \frac{\theta}{\alpha} = \frac{I_O l}{mgr^2}$$

$$\therefore\ t_p = 2\pi\sqrt{\frac{I_O l}{mgr^2}}$$

or

$$I_O = \frac{t_p^2}{4\pi^2}\frac{mgr^2}{l}$$

For connecting rod and platform,

$$I_O = \frac{3{\cdot}0^2}{4\pi^2}\frac{(1{\cdot}5+5{\cdot}5)\times 9{\cdot}81\times 0{\cdot}125^2}{1{\cdot}25} = 0{\cdot}1955\ \text{kg m}^2$$

For platform alone, $I_O = \frac{3{\cdot}5^2}{4\pi^2}\frac{1{\cdot}5\times 9{\cdot}81\times 0{\cdot}125^2}{1{\cdot}25} = 0{\cdot}057\ \text{kg m}^2$

Therefore, for the connecting rod alone,

$$I_O = 0{\cdot}1955 - 0{\cdot}057 = 0{\cdot}1385\ \text{kg m}^2$$

Example 2.7 A uniform rod of mass $2m$ and length $4r$ is arranged to pivot freely at its centre O as shown in Fig. 2.9. To one end of the rod is attached a uniform disc of mass m and radius r. At the centre of the lower half of the rod is attached a spring of stiffness S, where $Sr = 5\,mg$. The system may oscillate in a vertical plane. Determine an expression for the frequency of *small* oscillations about the vertical position shown. (U. Lond. *modified*)

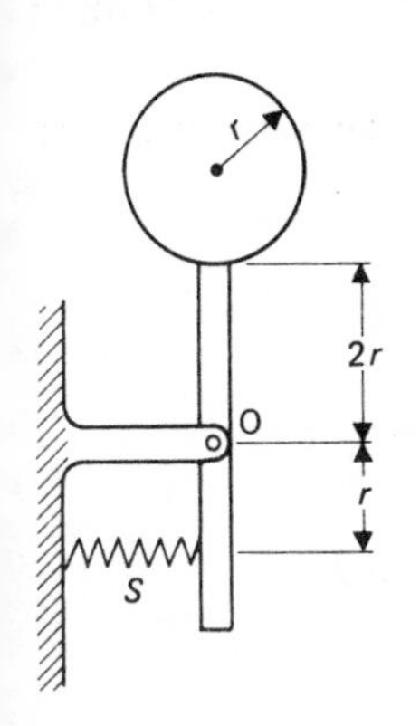

Fig. 2.9

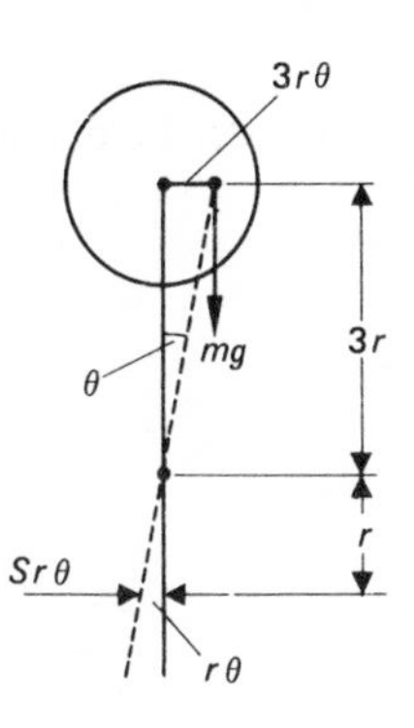

Fig. 2.10

Solution

Let the rod be displaced through a small angle θ, Fig. 2.10. Then

$$\text{spring force} = S\times r\theta$$

$$\therefore\ \text{restoring moment about O} = (Sr\theta\times r) - (mg\times 3r\theta)$$

$$= \frac{5\,mg}{r}r^2\theta - 3\,mgr\theta = 2\,mgr\theta$$

$$\text{Moment of inertia of rod about O} = \frac{(2m)(4r)^2}{12} = \frac{8}{3}mr^2$$

$$\text{Moment of inertia of disc about O} = \frac{mr^2}{2} + m(3r)^2 = \frac{19}{2}mr^2$$

Therefore the total moment of inertia is given by

$$I_O = \frac{8}{3}mr^2 + \frac{19}{2}mr^2 = \frac{73}{6}mr^2$$

When released, restoring moment $= I_O\alpha$

i.e.
$$2\,mgr\theta = \frac{73}{6}mr^2\alpha$$

$$\therefore \frac{\alpha}{\theta} = \frac{12\,g}{73\,r}$$

$$\therefore n = \frac{1}{2\pi}\sqrt{\frac{12\,g}{73\,r}} = \frac{1}{\pi}\sqrt{\frac{3\,g}{73\,r}}\ \text{Hz}$$

Example 2.8 A uniform cylinder of mass m and radius r rolls without slip on the inside of a cylindrical surface of radius R, as shown in Fig. 2.11. Show that the kinetic energy of the cylinder is given by $\frac{3}{4}m\,\Omega^2(R-r)^2$.

Hence, if θ is small and the axis of the cylinder is horizontal, show that the cylinder moves with simple harmonic motion and determine its natural frequency. (U. Lond.)

Solution

Let the angular velocity of the cylinder be ω and the angular velocity of the radius OQ be Ω, Fig. 2.12.

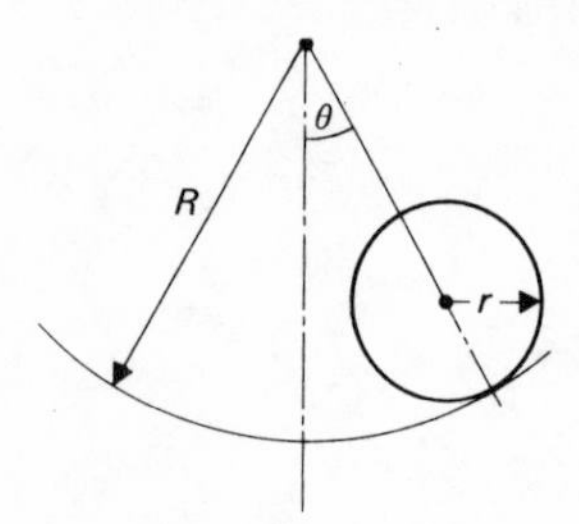

FIG. 2.11

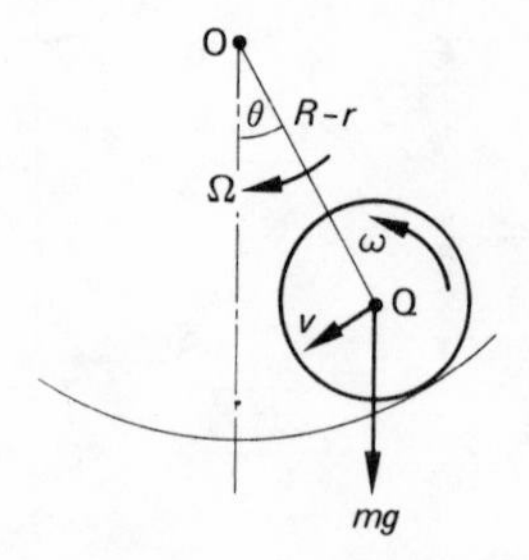

FIG. 2.12

Then the linear velocity of the centre of the cylinder is given by

$$v = \omega r = \Omega(R-r)$$

from which
$$\omega = \Omega\left(\frac{R-r}{r}\right)$$

$$\text{K.E. of cylinder} = \tfrac{1}{2}mv^2 + \tfrac{1}{2}I\omega^2$$

$$= \tfrac{1}{2}m\Omega^2(R-r)^2 + \tfrac{1}{2}\frac{mr^2}{2}\Omega^2\left(\frac{R-r}{r}\right)^2$$

$$= \tfrac{3}{4}m\Omega^2(R-r)^2$$

Taking the lowest position of the cylinder as datum,

$$\text{P.E. of cylinder} = mg(R-r)(1-\cos\theta)$$

Therefore the total energy is given by

$$U = mg(R-r)(1-\cos\theta) + \tfrac{3}{4}m\Omega^2(R-r)^2 = \text{constant}$$

$$\therefore \quad \frac{dU}{dt} = -mg(R-r)\sin\theta\,\frac{d\theta}{dt} + \tfrac{3}{4}m(R-r)^2\,2\Omega\frac{d\Omega}{dt} = 0$$

i.e.

$$-g\sin\theta\,\Omega + \tfrac{3}{4}(R-r)2\Omega\alpha = 0$$

where α is the angular acceleration of OQ.

$$\therefore \quad \alpha = \frac{2g\sin\theta}{3(R-r)} \simeq \frac{2g\theta}{3(R-r)} \quad \text{when } \theta \text{ is small}$$

$$\therefore \quad \frac{\alpha}{\theta} = \frac{2g}{3(R-r)}$$

This ratio is a constant and the cylinder is always accelerating towards its mid-position, so that the motion of the cylinder is simple harmonic.

Thus

$$n = \frac{1}{2\pi}\sqrt{\frac{2g}{3(R-r)}} = \frac{1}{\pi}\sqrt{\frac{g}{6(R-r)}}\ \text{Hz}$$

Example 2.9 A uniform cylinder of mass m is constrained to roll up and down an incline by a spring of stiffness S which wraps round the cylinder as shown in Fig. 2.13.

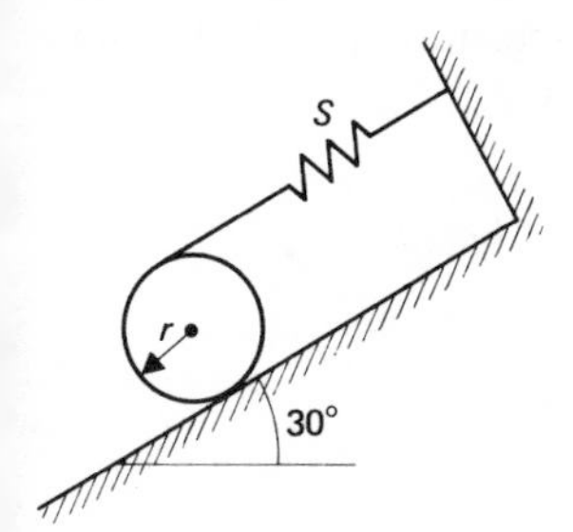

FIG. 2.13

(a) If the cylinder is released from a position where the spring is unstretched, what is the frequency of oscillation of the cylinder if there is no slipping?
(b) What is the maximum friction force developed? (U. Lond.)

Solution

Let the equilibrium position be at a distance a below the initial position, measured along the slope, Fig. 2.14(a). Then the stretch of the spring is $2a$ and the spring force is $2Sa$.

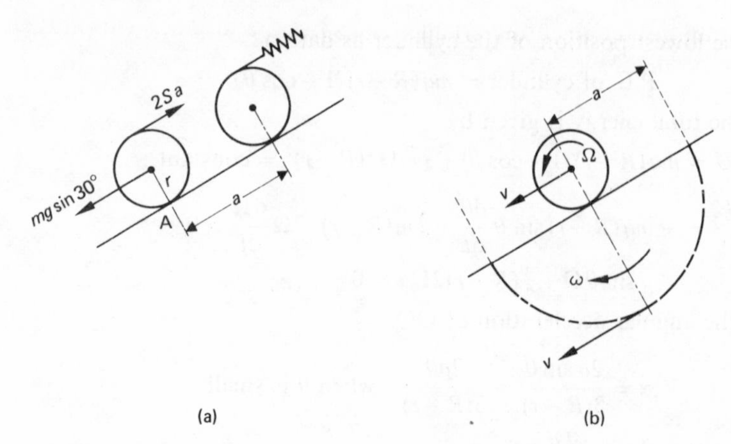

Fig. 2.14

Taking moments about A,

$$2Sa \times 2r = mg \sin 30° \times r$$

from which

$$a = mg/8S$$

When the cylinder is released from the initial position, it will oscillate about the equilibrium position with an amplitude a, Fig. 2.14(b). In the mid-position, the velocity of the centre of the cylinder is given by

$$v = \omega a = \Omega r$$

where ω is the circular frequency of the oscillation and Ω is the angular velocity of the cylinder.

In moving from the highest position to the mid-position, the total energy remains constant.

In the highest position,

$$\text{P.E. relative to mid-position} = mga \sin 30° = \tfrac{1}{2} mga$$

$$\text{K.E.} = 0$$

and

$$\text{S.E.} = 0$$

In the mid-position,

$$\text{P.E.} = 0$$

$$\text{K.E.} = \tfrac{1}{2} mv^2 + \tfrac{1}{2} I\Omega^2 = \tfrac{1}{2} m(\omega a)^2 + \tfrac{1}{2}\frac{mr^2}{2}\left(\frac{\omega a}{r}\right)^2 = \tfrac{3}{4} m\omega^2 a^2$$

and

$$\text{S.E.} = \tfrac{1}{2} S(2a)^2 = 2Sa^2$$

$$\therefore \tfrac{1}{2} mga = \tfrac{3}{4} m\omega^2 a^2 + 2Sa^2$$

from which

$$\omega^2 = 8S/3m \quad \text{since } a = mg/8S$$

Hence

$$n = \frac{\omega}{2\pi} = \frac{1}{2\pi}\sqrt{\frac{8S}{3m}} = \frac{1}{\pi}\sqrt{\frac{2S}{3m}}$$

Consider an instant when the cylinder is displaced a distance x from the equilibrium position, Fig. 2.15.

$$\text{Stretch of spring} = 2(a+x)$$
$$\therefore \text{ spring force} = 2S(a+x)$$

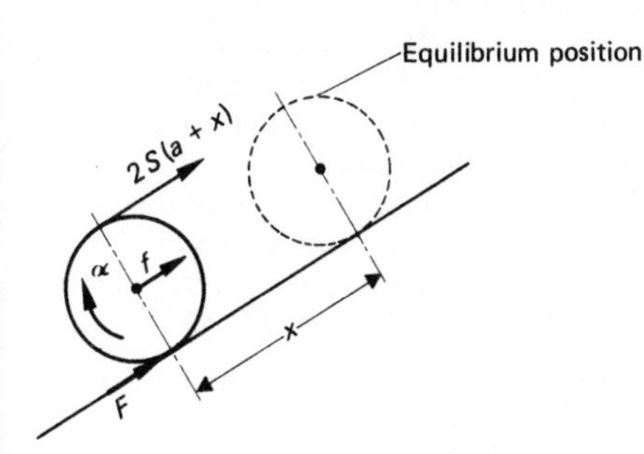

FIG. 2.15

The angular acceleration of the cylinder is clockwise and so the frictional force, F, is up the plane, to prevent slipping.

For the linear motion,

$$2S(a+x)+F = mf \tag{1}$$

For the angular motion,

$$\{2S(a+x)-F\}r = I\alpha = \frac{mr^2}{2}\frac{f}{r}$$

i.e.

$$2S(a+x)-F = mf/2 \tag{2}$$

From equations (1) and (2),

$$F = \frac{mf}{4} = \frac{m}{4}\omega^2 x$$

F is a maximum when $x = a$,

i.e.

$$F_{max} = \frac{m}{4}\omega^2 a = \frac{m}{4}\frac{8S}{3m}\frac{mg}{8S} = \underline{\frac{mg}{12}}$$

Problems

1. A mass of 25 kg is suspended from a spring which has a stiffness of 14 kN/m of extension. It vibrates freely with an amplitude of 12 mm. Find the periodic time, the velocity and acceleration when displaced 8 mm from the equilibrium position and determine the time interval in passing from this position to the position of maximum displacement. (I. Mech. E.) (*Answer:* 0·266 s; 0·212 m/s; 4·48 m/s^2; 0·0356 s)

2. On a packaging machine mechanism, a crosshead moves in a straight guide with a simple harmonic motion. At distances of 125 mm and 200 mm from its mean position, the crosshead has velocities of 6 and 3 m/s respectively.

Determine: (a) the amplitude of the motion, (b) the maximum velocity, and (c) the periodic time. If the crosshead has a mass of 0·2 kg, what is the maximum inertia force? (I. Mech. E.) (*Answer:* 219·5 mm; 7·3 m/s; 0·187 s; 48·6 N)

3. A small plunger receives simple harmonic motion from a crank driving through a long connecting rod. The crank length can be set to give a long stroke of 450 mm or a short stroke of 150 mm. The change of speed is such that the product of crank speed and crank length is constant. The plunger has a mass of 25 kg and has to overcome a maximum resistance of 4·5 kN which, on the long stroke, occurs at one-third of the stroke length from the beginning and, on the short stroke, at one-sixth of the stroke length from the end. The speed on the long stroke is 150 rev/min. Compare, for the positions of maximum resistance, the crank angles and the driving forces in the connecting rod. (I. Mech. E.) (*Answer*: 70·53°, 4·963 kN; 131·82°, 1·72 kN)

4. A mass m when gradually applied to a spring of elastic force S per unit deflection produces a static deflection of δ. Show that, if disturbed, the resulting motion will be simple harmonic and that the frequency n, if the effect of the mass of the spring is neglected, is as given by the formula $2\pi n = \sqrt{(g/\delta)}$.

A horizontal shaft, supported in bearings at the ends, deflects at the centre by 0·005 mm per 100 N of load applied there. When a wheel of mass 300 kg is centrally fitted the system responds to an external disturbance and free vertical vibrations of amplitude 0·25 mm are set up. Calculate the values of the frequency, the maximum velocity and the maximum acceleration for this vibration. (I. Mech. E.)

(*Answer:* 41·1 Hz; 0·0645 m/s; 16·67 m/s²)

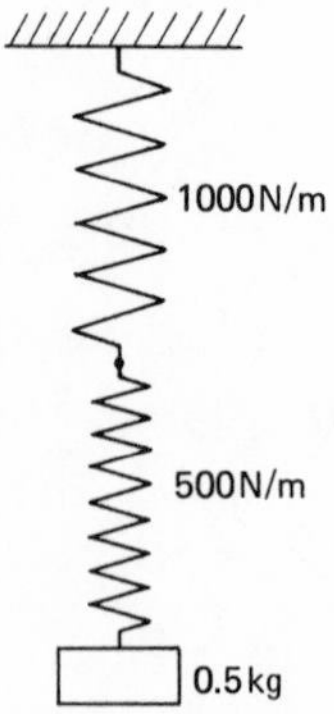

5. In the system shown in Fig. 2.16, the top spring has a stiffness of 1000 N/m and the bottom spring has a stiffness of 500 N/m. The suspended mass is 0·5 kg. Find the natural frequency of vibration. (*Answer:* 4·11 Hz)

1000 N/m
500 N/m
0.5 kg

Fig. 2.16

6. A helical spring has both ends securely fixed, one vertically above the other, and a mass is attached to the spring at some intermediate point. Show that the frequency of the vibrations is a minimum when the load point is midway between the fixed ends.

A helical spring has a stiffness of 3·5 kN/m when one end is fixed and the load is applied to the free end. Determine the minimum value of the frequency when both ends are fixed and a mass of 18 kg is applied to the spring. (U. Lond.) (*Answer:* 4·44 Hz)

7. A torsional pendulum consists of a wire 0·5 m long, 10 mm diameter, fixed at its upper end and attached at its lower end to a heavy disc having a moment of inertia of 0·06 kg m². The modulus of rigidity of the wire is 44 GN/m². Find the frequency of torsional oscillation of the disc. If the maximum displacement to one side of the rest position is 5°, find the maximum angular velocity and acceleration of the disc.

(*Answer:* 6·04 Hz; 3·31 rad/s; 125·5 rad/s²)

8. A solid metal cylinder 450 mm diameter is suspended with its axis vertical by means of a wire coaxial with the cylinder and rigidly attached to it. The wire develops an elastic resistance to twisting of 22 N m per radian of twist. Find the necessary mass of the cylinder in order that when given a small angular displacement about its axis, it shall make 40 vibrations/minute. (I. Mech. E.) (*Answer:* 49·5 kg)

9. A ship is pitching 10° above and 10° below the horizontal. Assuming the motion to be simple harmonic having a period of 12 s, find the maximum angular velocity and angular acceleration of the ship during pitching.

(*Answer:* 0·0914 rad/s; 0·047 75 rad/s^2)

10. A flywheel is suspended by resting the inside of the rim on a horizontal knife-edge so that the wheel can swing in a vertical plane. The flywheel has a mass of 350 kg. The knife-edge is parallel to and 350 mm from the axis of the wheel. The time for making one small oscillation is 1·77 s. Assuming that the centre of mass is in the axis of the wheel, find the radius of gyration about this axis.

Also find the torque to increase the speed of the wheel at a uniform rate from 240 to 250 rev/min in 0·75 s when the wheel is revolving about its axis. (U. Lond.)

(*Answer:* 387·2 mm; 73·25 N m)

11. Prove the formula for the time of swing of a compound pendulum.

Two straight light rods AB and CD intersect at right angles at E. AE = 350 mm, BE = 250 mm, CE = 625 mm and DE = 225 mm. Masses are fixed to the extremities of the rods, 3 kg at A, 2·5 kg at B, 4 kg at C and 3·5 kg at D. The system is set swinging in a vertical plane, as a compound pendulum, about a horizontal axis through E perpendicular to the plane of the rods. Find the time of swing.

If the rods swing through 5° on either side of the equilibrium position, find the maximum kinetic energy and the maximum angular acceleration of the system. (U. Lond.)

(*Answer:* 2·277 s; 0·0656 J; 0·6638 rad/s^2)

12. A solid disc flywheel has a diameter of 600 mm and is mounted with its axis horizontal on frictionless centres. A cylindrical mass of 5 kg, 200 mm in diameter, is bolted to the disc so that its axis is parallel to that of the wheel and 200 mm away from it. The whole system is found to swing with a time period of 3·18 s. What is the mass of the flywheel? (U. Lond.)

(*Answer:* 50·8 kg)

13. A body of uniform thickness has a small hole drilled at distance h from the centre of mass G, from which it is suspended to swing in its plane as a compound pendulum as shown in Fig. 2.17. Show that for oscillations of small amplitude, the periodic time is the same as for a simple pendulum of length $h + k^2/h$, where k is the radius of gyration about an axis through G parallel to that through O.

If the body is a disc of diameter 400 mm and with distance h = 150 mm, find the periodic time and the position of an alternative point O_1 for which the time of swing would be the same. (I. Mech. E.)

(*Answer:* 1·068 s; h_1 = 133 mm)

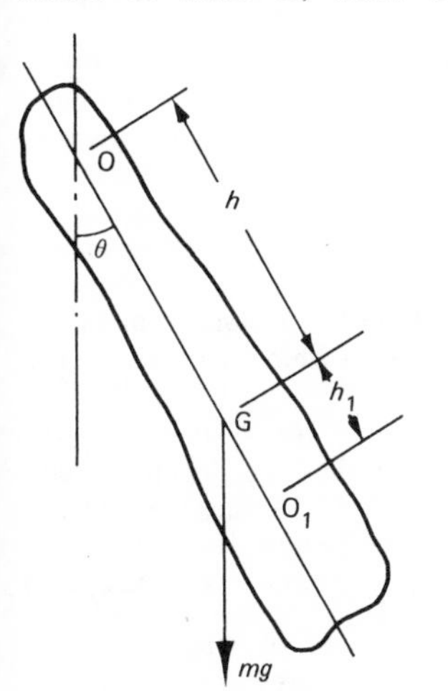

FIG. 2.17

14. A large double helical gear has a mass of 2 t and is suspended from knife-edges so that it is free to swing in a vertical plane at right angles to the gear axis. If the point of suspension is 700 mm from the gear axis and the period of a complete oscillation is 2·25 s, determine: (a) the moment of inertia of the gear about its axis, (b) the minimum possible period of oscillation if the point of suspension may be moved, and (c) the distance of the point of suspension from the gear axis under these conditions. (U. Lond.)

(*Answer:* 781 kg m^2; 2·245 s; 625 mm)

15. A compound pendulum, whose mass is 6·5 kg, is observed to make 34 double swings per minute.

A concentrated bob, of mass 9 kg, is now fixed firmly to the pendulum at a distance of 1 m from the knife-edge, so that the static equilibrium position of the pendulum is unaltered. The pendulum, with bob attached, is now observed to make 30·5 double swings per minute.

Calculate the distance from the knife-edge to the centre of mass of the original pendulum and the radius of gyration of the original pendulum about an axis through its centre of mass parallel to the knife-edge. (U. Lond.) (*Answer:* 283·6 mm; 373 mm)

16. A gear wheel has a diameter of 5 m and a mass of 9 t. It is suspended from knife-edges so that it is free to swing in a vertical plane at right angles to the axis of the gear. The point of suspension is 100 mm inside the pitch circle. The period of one complete oscillation is 4·29 s.

Determine: (a) the moment of inertia of the gear about its axis of symmetry, stating the units carefully and justifying any compound pendulum equation used, and (b) the tangential force that must be applied at the pitch radius of this gear to give it an angular acceleration of 0·25 rad/s^2. (U. Lond.) (*Answer:* 46 950 kg m^2; 4·695 kN)

17. The moment of inertia of a connecting rod was determined by means of bifilar suspension, as shown in Fig. 2.18. The length of the equal wires was 1·1 m and their distance apart was 0·3 m. The mass of the rod was 16 kg and the time for a complete small angular oscillation was 3·7 s.

Determine the moment of inertia of the rod about the axis AG. (U. Lond.)

(*Answer:* 1·113 kg m^2)

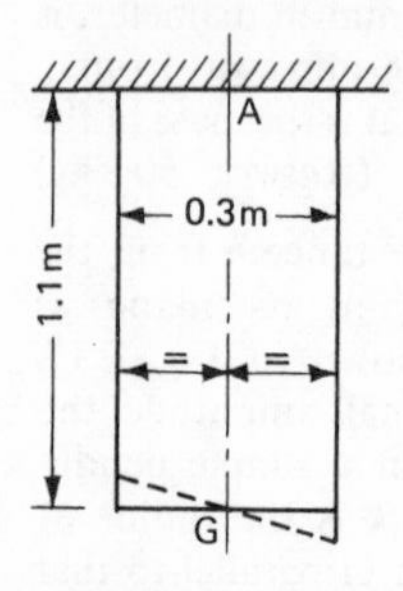

FIG. 2.18

18. Describe the trifilar method of determining experimentally the moment of inertia of a body about a given axis and deduce the expression from which the value of the moment of inertia is calculated.

In a particular case, the points of attachment of the three parallel suspension wires form an equilateral triangle inscribed in a circle of 75 mm radius. The length of each wire is 0·75 m and the period of oscillation is observed to be 2·5 s. Deduce the moment of inertia of the body. The body has a mass of 4·5 kg. (U. Lond.)

(*Answer:* 0·0524 kg m^2)

19. A 125 mm diameter disc is suspended by three wires of length 550 mm symmetrically attached to the rim. Two identical bellcrank levers are placed on the disc as shown in Fig. 2.19(a). With this arrangement the disc completes 100 oscillations about a vertical axis in 109 s.

The levers are then placed on the disc as shown in Fig. 2.19(b), and in this condition the disc requires 106 s to complete 100 oscillations.

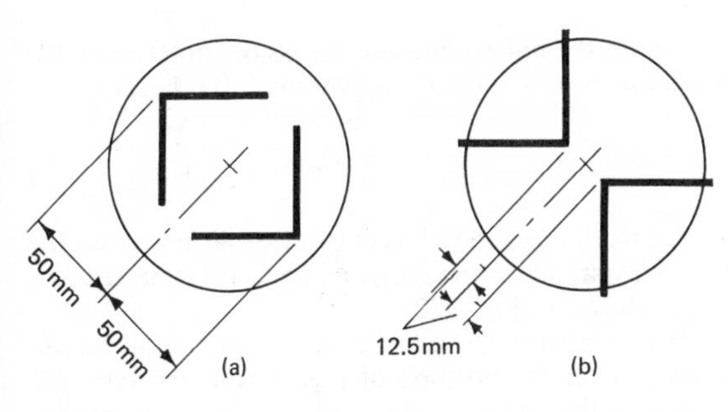

FIG. 2.19

The disc has a mass of 0·88 kg and each lever has a mass of 0·65 kg.

From the results of these experiments, determine the position of the centre of mass of a lever and the radius of gyration of a lever about its centre of mass. (U. Lond.)

(*Answer:* 17·2 mm from apex; 33·5 mm)

20. A uniform disc of mass m and radius $a/2$ is freely supported at its centre by a light, rigid rod of length $2a$ which is freely pinned to a fixed support as shown in Fig. 2.20. Two springs, each of stiffness S, are attached to the centre of the rod and to fixed supports. The system may be set in motion to oscillate as a pendulum in two conditions: (a) with a pin P locking the disc and rod together, thus causing them to rotate together, and (b) without the locking pin.

Show that the ratio of the frequency of *small* oscillations for condition (b) to that of condition (a) is $(33/32)^{1/2}$.

Given that m is 2 kg, S is 400 N/m and a is 0·1 m, find the values of the two frequencies. (U. Lond.) (*Answer:* 1·913 Hz; 1·943 Hz)

21. The system shown in Fig. 2.21 consists of a pulley of mass m, radius r and moment of inertia about its central axis I, supported by a cord and spring of stiffness S, and a body of mass $2m$ freely attached to the centre of the pulley.

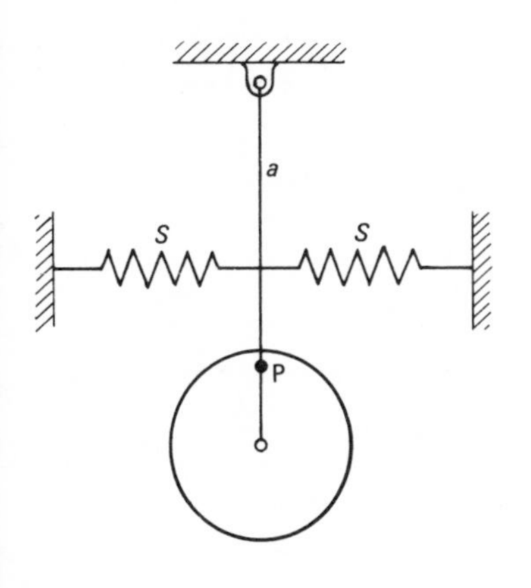

FIG. 2.20

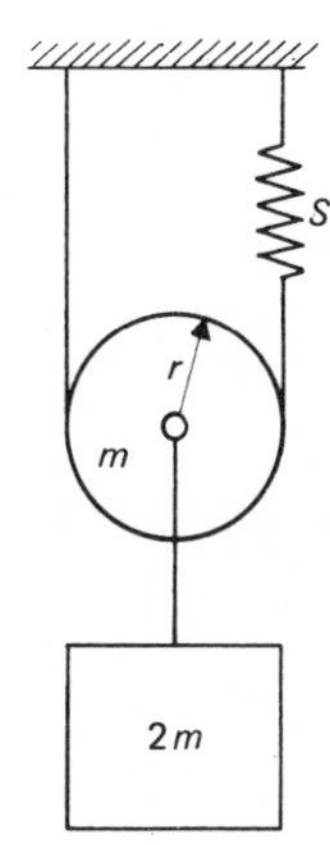

FIG. 2.21

Given that the cord does not slip on the pulley, find an expression, in terms of the given quantities, for the natural frequency of vertical oscillations. (U. Lond.)

$$\left(\textit{Answer: } \frac{r}{\pi}\sqrt{\left(\frac{S}{3mr^2+I}\right)}\right)$$

22. A 200 mm diameter cylinder of mass 2 kg and radius of gyration 80 mm about the cylinder axis rests on a surface inclined at 30° to the horizontal and is connected to a spring, S, of stiffness 1200 N/m, as shown in Fig. 2.22.
(a) If the cylinder rolls on the surface without slipping, find the frequency of the motion after the cylinder has been displaced from the position of equilibrium and released.
(b) If the coefficient of friction between the cylinder and surface is 0·2, find the greatest amplitude of the motion of the centre of the cylinder if the cylinder is not to slip on the surface. (U. Lond.) (*Answer:* 3·04 Hz; 7·27 mm)

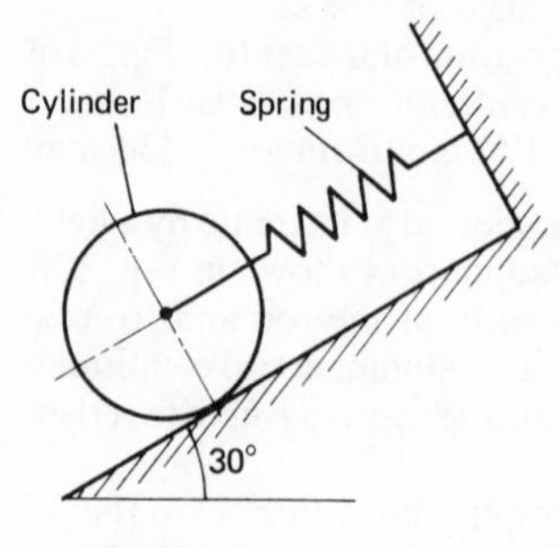

FIG. 2.22

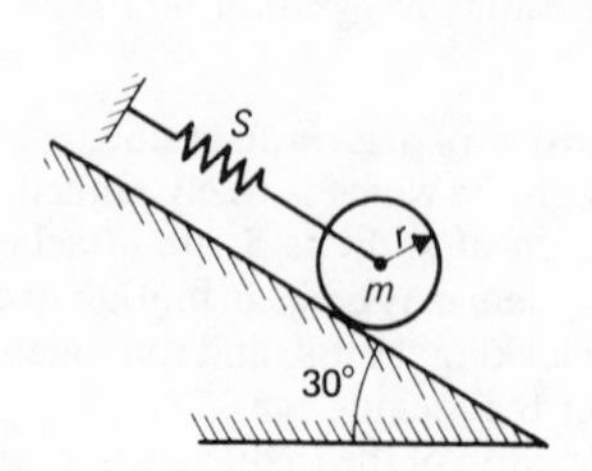

FIG. 2.23

23. A cylinder of mass m and radius r is constrained by the spring of stiffness S to roll up and down the inclined plane as shown in Fig. 2.23.
(a) What is the frequency of oscillation of the system when there is no slipping?
(b) If the cylinder is released from a position where the spring is unstretched, how large must the coefficient of friction be to prevent slipping? (U. Lond.)

$$\left(\textit{Answer: } \frac{1}{\pi}\sqrt{\left(\frac{S}{6m}\right)};\ 0{\cdot}1925\right)$$

3

Velocity and acceleration

3.1 Velocities in mechanisms: relative velocity diagrams

If the ends A and B of a rigid link, Fig. 3.1, are moving with velocities v_a and v_b respectively, the velocity of A relative to B is given by *ba* in the relative velocity diagram and is denoted by v_{ab}, Fig. 3.2.

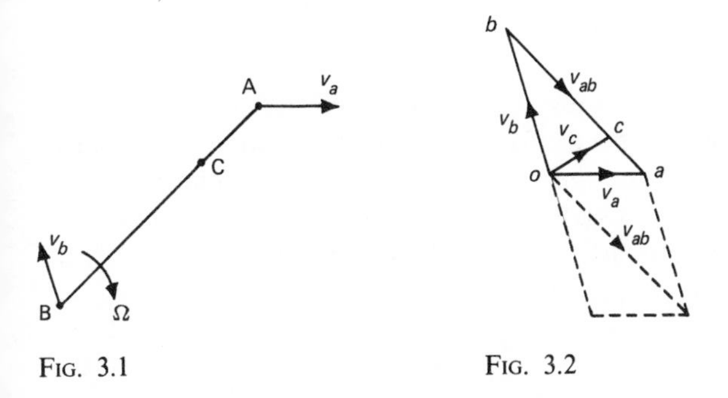

FIG. 3.1

FIG. 3.2

v_{ab} may also be obtained by imparting a velocity equal and opposite to v_b to the link, thereby bringing B to rest, in which case the velocity of A relative to B is as shown dotted. The link may then be regarded as rotating about the point B and the angular velocity of the link is given by

$$\Omega = \frac{v_{ab}}{\mathrm{AB}} \tag{3.1}$$

It is evident that v_{ab} must be perpendicular to AB, otherwise the link would have to extend or contract along its length. Thus, if the magnitude and direction of v_b are known, but only the direction of v_a is known, then the magnitude of v_a may be determined by drawing from *b* a line perpendicular to AB to intersect the line of action of A at *a*. Thus, in a mechanism consisting of a number of rigid links, it is only necessary to know the velocity of one point in magnitude and direction in order to find the velocities of all other points for a given configuration.

The velocity of any point C on AB may be obtained by dividing *ab* so that $bc:ba = \mathrm{BC}:\mathrm{BA}$. The line *oc* then represents the velocity of C.

3.2 Velocity diagram for a block sliding on a rotating link

Let Ω be the angular velocity of the link about the fixed point O, Fig. 3.3, and v_a be the velocity of the block, assumed known in magnitude and direction. If A′ is the point on the link coincident with the block, then the velocity of A′ relative to O is perpendicular to OA′ and the velocity of A relative to A′ is parallel to OA′. Therefore, if a line perpendicular to OA′ is drawn from the point o, Fig. 3.4, and a line parallel to OA′ is drawn from the point a, the intersection gives the point a'. The velocity of sliding of A relative to A′ is then given by $a'a$.

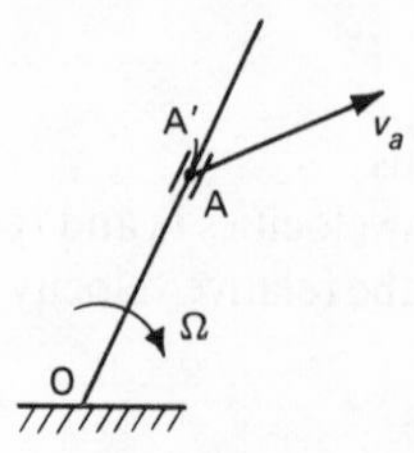

FIG. 3.3

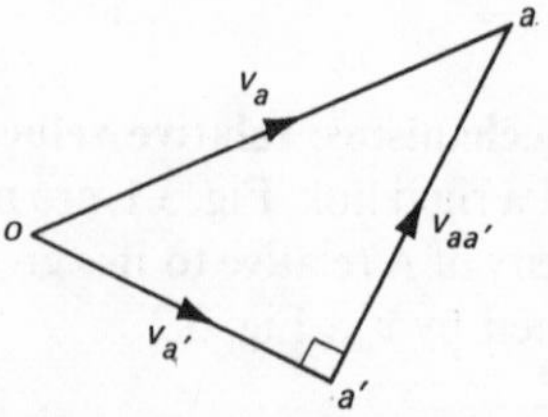

FIG. 3.4

3.3 Velocities in mechanisms: instantaneous centre method

If AI and BI are drawn perpendicular to the lines of action of v_a and v_b respectively, Fig. 3.5, then the link may be considered as rotating instantaneously about the point I. This is termed the *instantaneous* or *virtual centre of rotation* and moves continuously as the link itself moves. Thus, if Ω is the angular velocity of the link about I, then

$$\frac{v_a}{\text{IA}} = \frac{v_b}{\text{IB}} = \Omega \tag{3.2}$$

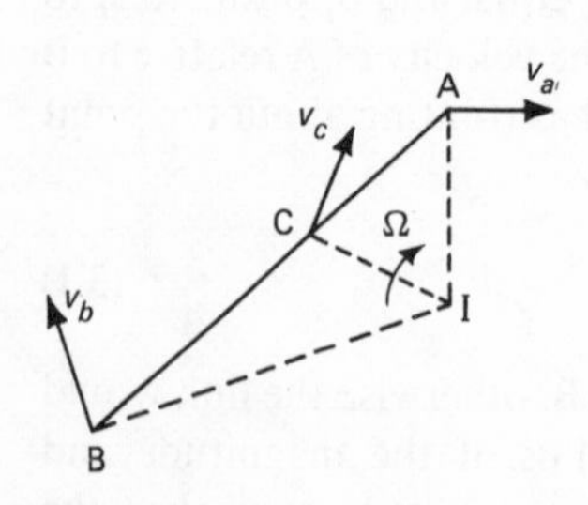

FIG. 3.5

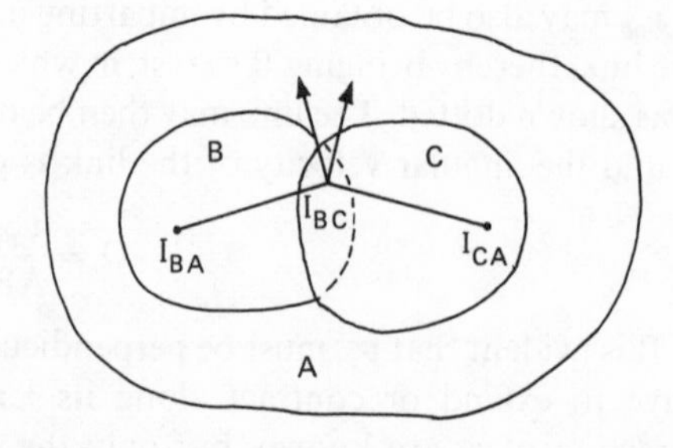

FIG. 3.6

or, since triangle AIB is similar to triangle aob, Fig. 3.2,

$$\frac{oa}{\text{IA}} = \frac{ob}{\text{IB}} = \frac{ab}{\text{AB}}$$

i.e.

$$\frac{v_a}{\text{IA}} = \frac{v_b}{\text{IB}} = \Omega$$

The velocity of any point C on AB is proportional to the distance of that point from I and is perpendicular to IC, so that

$$v_c = \Omega \times \mathrm{IC} \quad \text{or} \quad \frac{v_c}{\mathrm{IC}} = \frac{v_a}{\mathrm{IA}}$$

3.4 Three-centres-in-line theorem

Let A, B and C be the three bodies having plane motion relative to each other, Fig. 3.6. Let I_{BA} and I_{CA} be the instantaneous centres of B and C relative to A, and let I_{BC} be that of B relative to C. Assume that I_{BC} is not on the line joining I_{BA} and I_{CA}. I_{BC} is a point fixed to either B or C because it is the instantaneous centre about which one body moves relative to the other. If I_{BC} is considered as a point on B, then it is moving relative to A about I_{BA} in a direction perpendicular to the line $I_{BA}I_{BC}$. If I_{BC} is considered as a point on C, it is moving relative to A about I_{CA} in a direction perpendicular to the line $I_{BC}I_{CA}$. For these two conditions to be satisfied, I_{BC} must be on the line joining I_{BA} and I_{CA}, i.e. the three instantaneous centres lie in a straight line.

3.5 Rubbing velocity at a pin joint

If r is the radius of the pin at the joint shown in Fig. 3.7 and Ω_1 and Ω_2 are the angular velocities of the two links, then

$$\text{linear rubbing velocity at surface of pin} = (\Omega_1 \pm \Omega_2)r \qquad (3.3)$$

The angular velocities are added when the links rotate in opposite directions and subtracted when rotating in the same direction.

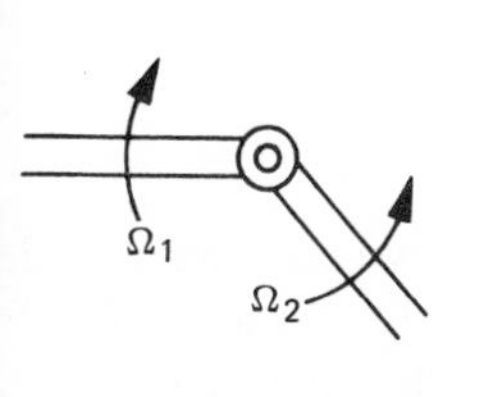

FIG. 3.7

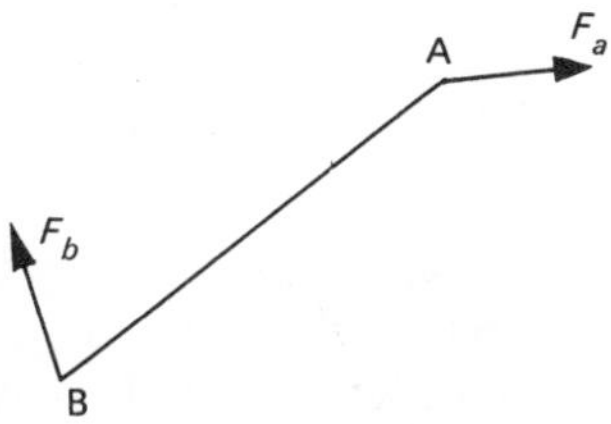

FIG. 3.8

3.6 Forces in mechanisms

Let F_a and F_b be the external forces acting at the ends A and B respectively of a rigid link, Fig. 3.8, and let v_a and v_b be the velocities of these points *in the directions of the lines of action* of F_a and F_b. Then, if A is the driving end and B is the driven end,

input work per unit time = output work per unit time

i.e.
$$F_a v_a = F_b v_b^* \qquad (3.4)$$

neglecting any work done against friction or in changing the K.E. of the link.

* In problems involving static forces, the ratio v_a/v_b represents the ratio of the virtual velocities and displacements (see Example 3.3).

If friction forces are present, then the efficiency of the transmission is given by

$$\eta = \frac{F_b v_b}{F_a v_a} \tag{3.5}$$

3.7 Accelerations in mechanisms

Consider a link AB, Fig. 3.9, rotating with angular velocity and acceleration Ω and α respectively. Let f_a and f_b be the accelerations of A and B respectively. Then, if $\boldsymbol{oa}$ and $\boldsymbol{ob}$ are drawn from the point o, Fig. 3.10, to represent f_a and f_b, the acceleration of A relative to B is given by $\boldsymbol{ba}$ in the relative acceleration diagram. This relative acceleration has two components: (a) a centripetal acceleration, Ω^2 AB or v_{ab}^2/AB, acting in the direction of A to B and (b) a tangential acceleration, αAB, perpendicular to AB. These are represented respectively by $\boldsymbol{ba_1}$ and $\boldsymbol{a_1a}$ in Fig. 3.10.

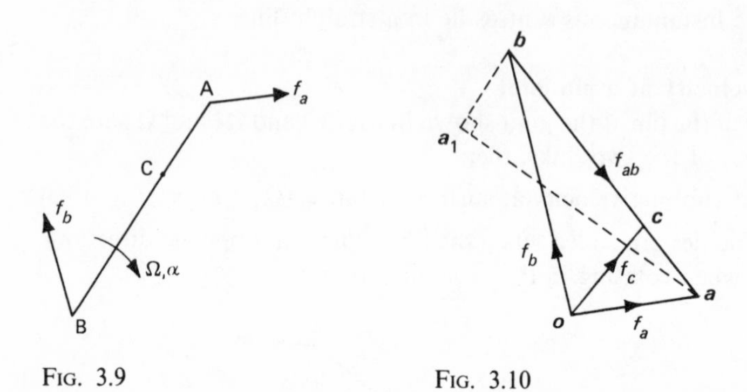

FIG. 3.9 FIG. 3.10

Normally, only one acceleration, such as f_b, will be known completely and the other, f_a, in direction only. If f_b is drawn as before, the centripetal acceleration, $\boldsymbol{ba_1}$, can be set off from $\boldsymbol{b}$, parallel to AB. The intersection of a line through o, parallel to f_a, and a line through $\boldsymbol{a_1}$, perpendicular to AB, will then give the required point $\boldsymbol{a}$.

$\boldsymbol{b}$ is called the acceleration image of AB, and to find the acceleration of any point C, a point c is taken on $\boldsymbol{ac}$ such that $\boldsymbol{ac}:\boldsymbol{ab}$ = AC:AB. The acceleration of C is then represented by oc.

The accelerations of points on other links connected to AB can be determined in a similar manner.

3.8 Crank and connecting rod: graphical constructions for velocity and acceleration

Graphical construction for velocity Let ω be the angular velocity of the crank OC, Fig. 3.11, and Ω be the angular velocity of the connecting rod, PC.

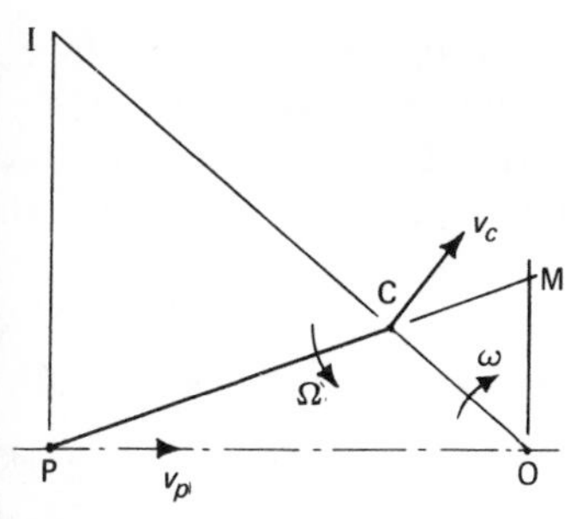

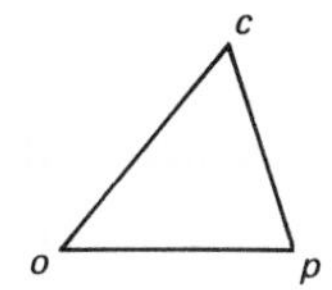

FIG. 3.11

Then, since I is the instantaneous centre for PC,

$$\frac{v_c}{IC} = \frac{v_p}{IP} = \frac{v_{pc}}{PC} = \Omega$$

If PC is produced to intersect a vertical through O at M, then triangles PIC and OCM are similar.

Hence $$v_p = v_c \times \frac{IP}{IC} = v_c \times \frac{OM}{OC} = \omega \times OM \tag{3.6}$$

Also $$v_{pc} = v_c \times \frac{PC}{IC} = v_c \times \frac{CM}{OC} = \omega \times CM \tag{3.7}$$

and $$\Omega = \frac{v_{pc}}{PC} = \omega \times \frac{CM}{PC} \tag{3.8}$$

The triangle OCM is similar to the velocity diagram shown by triangle *ocp*.

Graphical construction for acceleration—Klein's construction* Draw a circle with PC as diameter, Fig. 3.12, and produce PC to cut the vertical

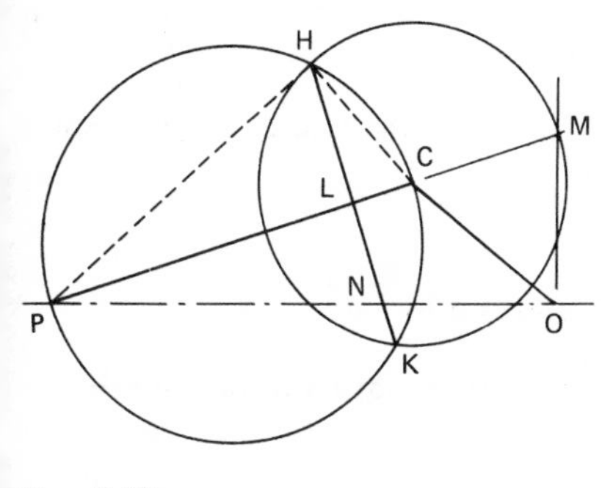

FIG. 3.12

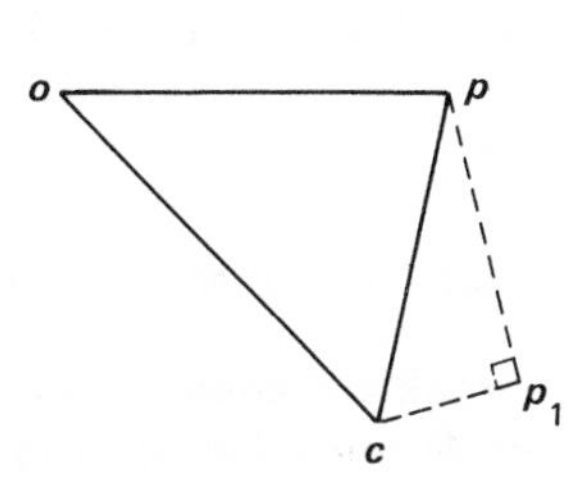

FIG. 3.13

* This construction can only be used when the crank has a uniform angular velocity.

through O at M. With centre C and radius CM, describe another circle, intersecting the first at H and K. Draw the common chord HK, intersecting OP at N and PC at L. Then the quadrilateral OCLN represents the acceleration diagram to the same scale that OC represents the centripetal acceleration of C.

Thus
$$f_p = \omega^2 \mathrm{ON} \tag{3.9}$$

Also
$$\text{centripetal acceleration of P relative to C} = \omega^2 \mathrm{LC} \tag{3.10}$$
$$\text{tangential acceleration of P relative to C} = \omega^2 \mathrm{LN} \tag{3.11}$$

and
$$\text{angular acceleration of PC, } \alpha = \omega^2 \frac{\mathrm{LN}}{\mathrm{PC}} \tag{3.12}$$

Proof Figure 3.13 shows the acceleration diagram for the mechanism. oc represents the centripetal acceleration of C relative to O ($= \omega^2$OC) and cp_1 represents the centripetal acceleration of P relative to C ($= v_{cp}^2/$CP). The tangential acceleration of P relative to C is perpendicular to PC and the acceleration of P relative to O is horizontal, thus giving the point p.

From equation 3.7,
$$v_{cp} = \omega \mathrm{CM} \tag{1}$$

Also
$$oc = \omega^2 \mathrm{OC}$$

and
$$cp_1 = \frac{v_{cp}^2}{\mathrm{CP}} = \frac{(\omega \mathrm{CM})^2}{\mathrm{CP}}$$

$$\therefore \frac{cp_1}{oc} = \frac{\omega^2 \mathrm{CM}^2}{\omega^2 \mathrm{OC} \times \mathrm{CP}} = \frac{\mathrm{CM}^2}{\mathrm{OC} \times \mathrm{CP}} \tag{2}$$

If P and C are joined to H, triangles CLH and CHP are similar,

$$\therefore \frac{\mathrm{CL}}{\mathrm{CH}} = \frac{\mathrm{CH}}{\mathrm{CP}}$$

or
$$\mathrm{CL} = \frac{\mathrm{CH}^2}{\mathrm{CP}} = \frac{\mathrm{CM}^2}{\mathrm{CP}}$$

$$\therefore \frac{\mathrm{CL}}{\mathrm{OC}} = \frac{\mathrm{CM}^2}{\mathrm{OC} \times \mathrm{CP}} \tag{3}$$

Therefore, from equations (2) and (3), OCLN is similar to $\boldsymbol{ocp_1p}$ and, since

$$f_c = \omega^2 \mathrm{OC}$$

then
$$f_p = \omega^2 \mathrm{ON}$$

The direction of the acceleration of P is given by $\overrightarrow{\mathrm{NO}}$.

3.9 Crank and connecting rod: analytical determination of piston velocity and acceleration for uniform angular velocity*

Let x be the displacement of the piston from the inner (or top) dead centre position and θ be the corresponding angle of rotation of the crank, Fig. 3.14.

* Approximate expressions are derived in this section; for exact expressions, see p. 133.

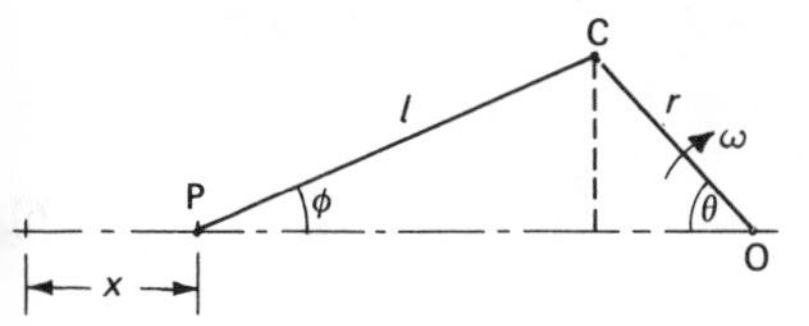

FIG. 3.14

Then $$x = (r+l) - (r\cos\theta + l\cos\phi)$$

Also $$r\sin\theta = l\sin\phi$$

$$\therefore\ \sin\phi = \frac{r}{l}\sin\theta = \frac{\sin\theta}{n} \quad \text{where } n = l/r$$

$$\cos\phi = \sqrt{\left[1-\left(\frac{\sin\theta}{n}\right)^2\right]} \simeq 1-\frac{\sin^2\theta}{2n^2} \quad \text{since } 1/n \text{ is small}$$

$$\therefore\ x = r(1-\cos\theta) + l\frac{\sin^2 2\theta}{2n^2}$$

$$= r\left(1-\cos\theta+\frac{\sin^2\theta}{2n}\right)$$

$$\therefore\ v_p = \frac{\mathrm{d}x}{\mathrm{d}t} = r\left(\sin\theta+\frac{\sin 2\theta}{2n}\right)\frac{\mathrm{d}\theta}{\mathrm{d}t}$$

$$= \omega r\left(\sin\theta+\frac{\sin 2\theta}{2n}\right) \tag{3.13}$$

$$\therefore\ f_p = \frac{\mathrm{d}^2x}{\mathrm{d}t^2} = \omega r\left(\cos\theta+\frac{\cos 2\theta}{n}\right)\frac{\mathrm{d}\theta}{\mathrm{d}t}$$

$$= \omega^2 r\left(\cos\theta+\frac{\cos 2\theta}{n}\right) \tag{3.14}$$

The sign of f_p changes from positive to negative when $\cos\theta + (\cos 2\theta)/n = 0$; this occurs when the crank and connecting rod are approximately perpendicular.

For an infinitely long connecting rod, the motion of P becomes simple harmonic, i.e. $f_p = \omega^2 r\cos\theta$.

Let Ω be the angular velocity and α be the angular acceleration of the connecting rod.

$$\sin\phi = \frac{\sin\theta}{n}$$

$$\therefore\ \cos\phi\frac{\mathrm{d}\phi}{\mathrm{d}t} = \frac{\cos\theta}{n}\frac{\mathrm{d}\theta}{\mathrm{d}t}$$

i.e.
$$\Omega = \frac{d\phi}{dt} = \frac{\cos\theta}{\cos\phi}\frac{\omega}{n} \simeq \omega\frac{\cos\theta}{n} \tag{3.15}$$

$$\alpha = \frac{d\Omega}{dt} \simeq -\omega^2\frac{\sin\theta}{n} \tag{3.16}$$

3.10 Forces in crank and connecting rod

Let F be the force on the piston and X be the corresponding force at the crank pin, perpendicular to the crank, Fig. 3.15.

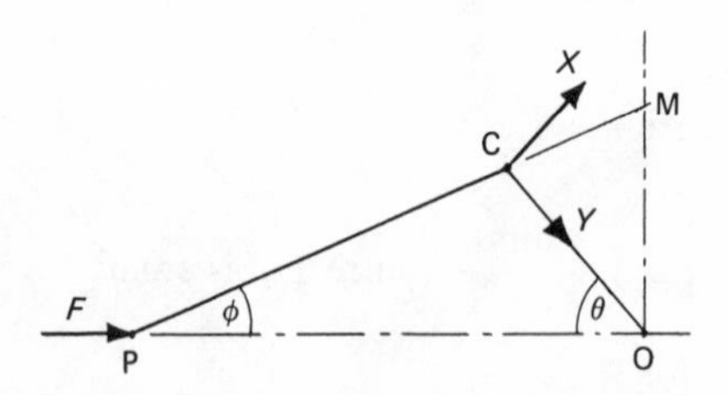

FIG. 3.15

Then work done/s by F = work done/s by X

neglecting the inertia and gravity effects of the connecting rod,

i.e.
$$F \times v_p = X \times v_c$$

$$\therefore\ X = F \times \frac{v_p}{v_c}$$

$$= F \times \frac{\text{OM}}{\text{OC}} \quad \text{from equation 3.6} \tag{3.17}$$

$\therefore$ crankshaft torque, or *crank effort*,

$$T = X \times \text{OC} = F \times \text{OM} \tag{3.18}$$

If p is the gas pressure in the cylinder and a is the area of the piston, then

force on the piston due to gas pressure = pa

In a double acting steam engine,

$$\text{force on piston} = p_1a_1 - p_2a_2$$

where p_1 and p_2 are the steam pressures on the two sides of the piston, and a_1 and a_2 are the respective areas.

If R is the mass of the reciprocating parts (i.e. of the piston and gudgeon pin and possibly including an allowance for the mass of the connecting rod), the force required to accelerate or decelerate these parts is given by Rf_p. This force must be subtracted from the gas force during the accelerating period and added to it during the retardation period.

Thus
$$F = pa - R\omega^2 r\left(\cos\theta + \frac{\cos 2\theta}{n}\right) \tag{3.19}$$
the sign of the inertia force changing at the point of maximum velocity.

In vertical engines, the dead weight Rg assists the piston effort on the downstroke and opposes it on the upstroke.

The force in the connecting rod, Q, and the sidethrust on the cylinder, S, can be obtained from the triangle of forces, Fig. 3.16.

Thus
$$Q = F\sec\phi \tag{3.20}$$
and
$$S = F\tan\phi \tag{3.21}$$

The components of Q perpendicular to and along CO give the tangential and radial forces, X and Y respectively, acting on the crankshaft.

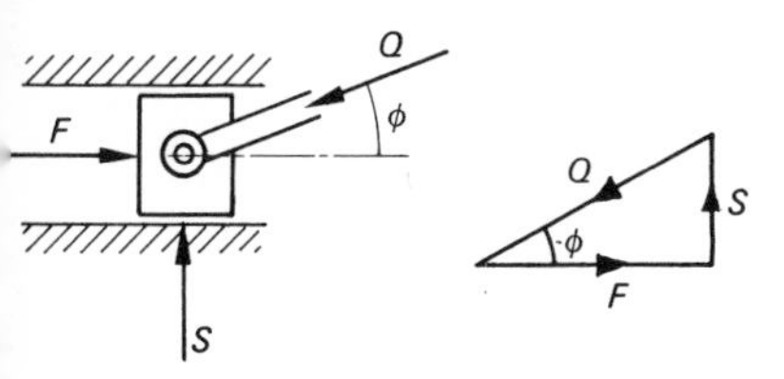

FIG. 3.16

Examples: velocity diagrams

Example 3.1 An aircraft, C, is flying in a vertical plane containing two tracking stations, A and B, Fig. 3.17, which are 15 km apart. At a certain instant, θ is 60°, θ is $-0{\cdot}025$ rad/s, ϕ is 150° and $\dot{\phi}$ is $-0{\cdot}02$ rad/s. Determine the magnitude and direction of the velocity of the aircraft.

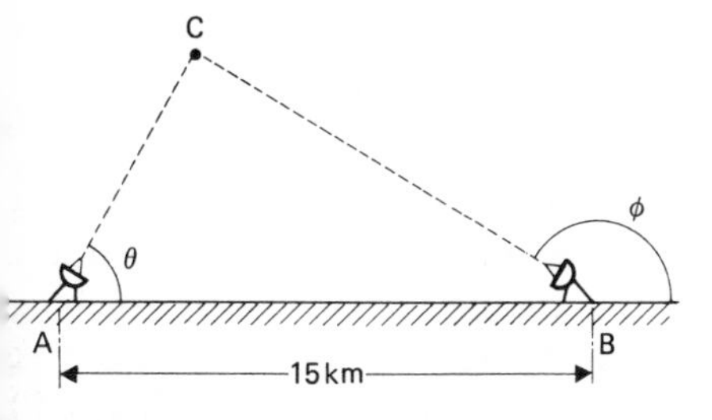

FIG. 3.17

Solution

The equivalent mechanism is shown in Fig. 3.18, where C_1 and C_2 are the points on imaginary links rotating about A and B respectively.

Since ABC is a right-angled triangle, AC = 7·5 km and BC = 12·99 km.

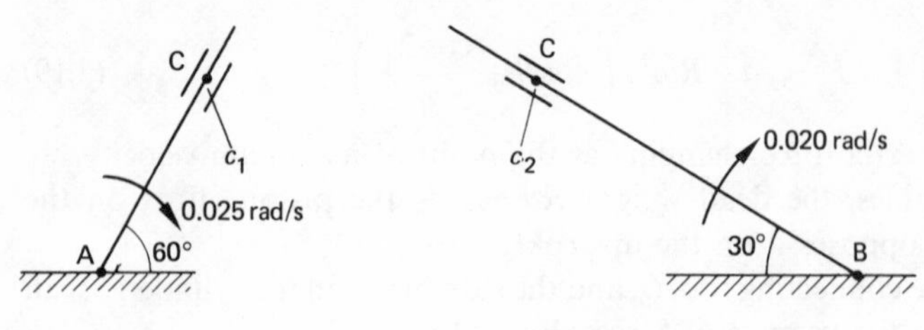

FIG. 3.18

Hence velocity of $C_1 = 0{\cdot}025 \times 7{\cdot}5 = 0{\cdot}1875$ km/s

and velocity of $C_2 = 0{\cdot}020 \times 12{\cdot}99 = 0{\cdot}2598$ km/s

These velocities are set out in Fig. 3.19 in which lines drawn perpendicular to ac_1 and bc_2 represent the sliding velocities of C relative to C_1 and C relative to C_2 respectively. Then

$$v_c = \sqrt{(0{\cdot}1875^2 + 0{\cdot}2598^2)} = 0{\cdot}3204 \text{ km/s} = \underline{1154 \text{ km/hr}}$$

$$\tan\alpha = 0{\cdot}2598/0{\cdot}1875$$

$$\therefore \alpha = 54{\cdot}2°$$

i.e. the aircraft is climbing at $\underline{24{\cdot}2°}$ to the horizontal.

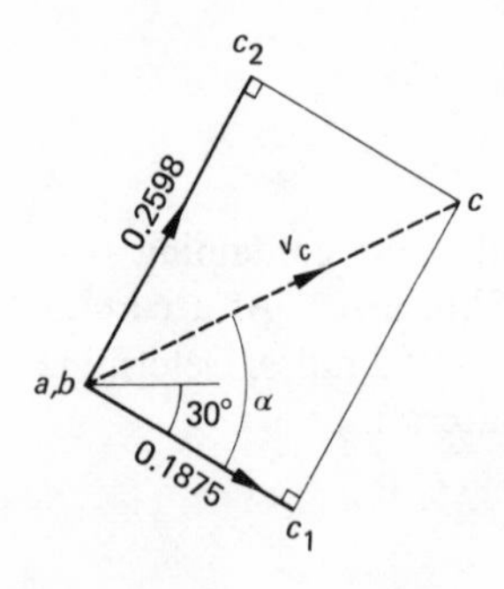

FIG. 3.19

Example 3.2 In the mechanism shown in Fig. 3.20, the crank AB revolves anticlockwise at 16 rad/s and the point D is constrained to move vertically downwards at 2 m/s. Draw the velocity diagram and determine: (a) the rubbing speed at E if the pin diameter is 30 mm, and (b) the velocity of F. The link CE = 450 mm and is horizontal in the position shown. (U. Lond.)

Solution

$$v_b = \omega r = 16 \times 0{\cdot}09 = 1{\cdot}44 \text{ m/s}$$

The velocity diagram for the mechanism is shown in Fig. 3.21. The vector *ab* is drawn perpendicular to AB from the fixed point *a* to represent the absolute velocity of B and *ad* is drawn vertically downwards to represent the absolute

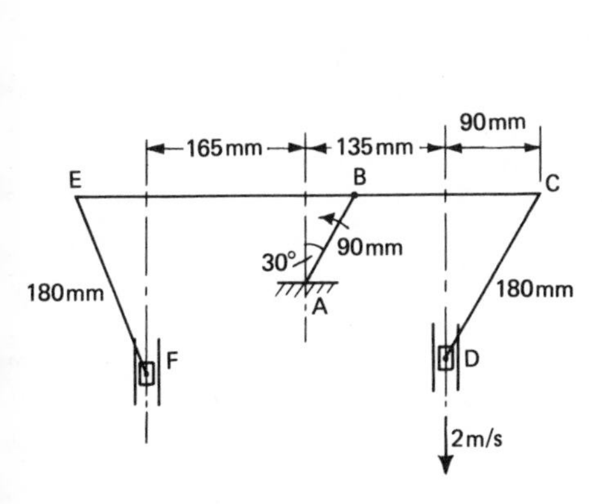

FIG. 3.20

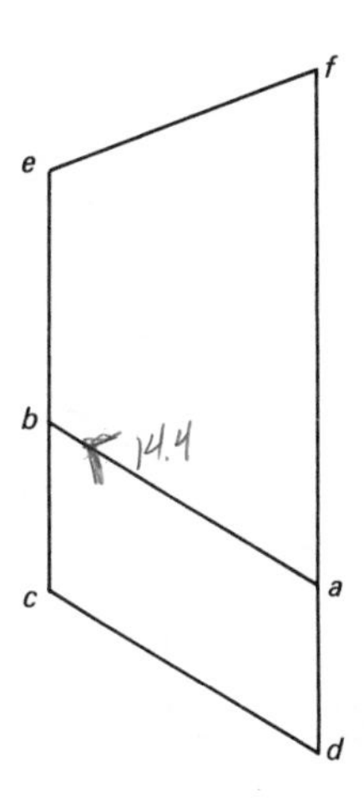

FIG. 3.21

velocity of d (2 m/s). The velocity of B relative to C is perpendicular to BC and the velocity of C relative to D is perpendicular to CD; hence the intersection of lines through b and d, perpendicular to BC and DC respectively, gives the point c. The line cb is produced to e, such that $cb:be = \text{CB}:\text{BE}$. The velocity of F relative to E is perpendicular to EF and the velocity of F relative to A is vertical and thus the point f is obtained.

From the diagram, $v_{fe} = fe = 3{\cdot}7$ m/s

and $v_{be} = be = 3{\cdot}0$ m/s

Thus $$\Omega_{fe} = \frac{v_{fe}}{\text{FE}} = \frac{3{\cdot}7}{0{\cdot}18} = 20{\cdot}55 \text{ rad/s} \quad \text{(anticlockwise)}$$

and $$\Omega_{be} = \frac{v_{be}}{\text{BE}} = \frac{3{\cdot}0}{0{\cdot}27} = 11{\cdot}11 \text{ rad/s} \quad \text{(clockwise)}$$

$$\therefore \text{ rubbing speed at E} = (\Omega_{fe} + \Omega_{be}) \times r$$
$$= (20{\cdot}55 + 11{\cdot}11) \times 0{\cdot}03 = \underline{0{\cdot}95 \text{ m/s}}$$

$$\text{Velocity of F, } v_{fa} = \underline{6{\cdot}25 \text{ m/s}}$$

Example 3.3 In Fig. 3.22, two equal links AC and BD, 120 mm long, are pivoted at A and B, which are 150 mm apart. The links are moved symmetrically by being attached to two gear wheels in mesh. DE and CE each measure 210 mm, and the length CF is 600 mm.

Draw the velocity diagram for the mechanism when the angle between BD and the horizontal line through AB is 30°, the vector representing the velocity of D being 50 mm long, BD rotating clockwise.

If a tension spring is connected between C and D, find the force it must exert to balance a vertical load of 200 N applied at F. (U. Lond.)

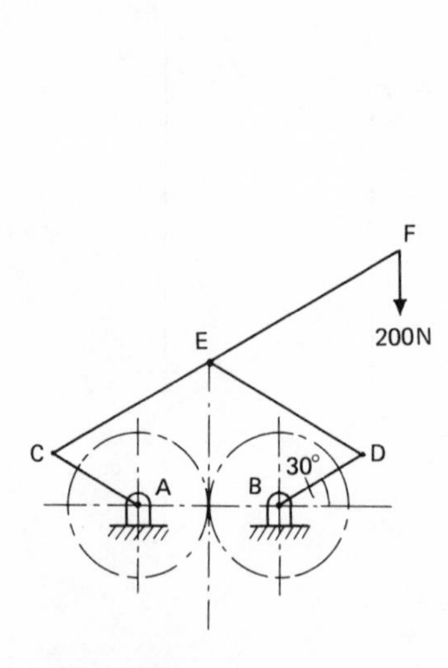

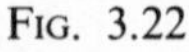
FIG. 3.22

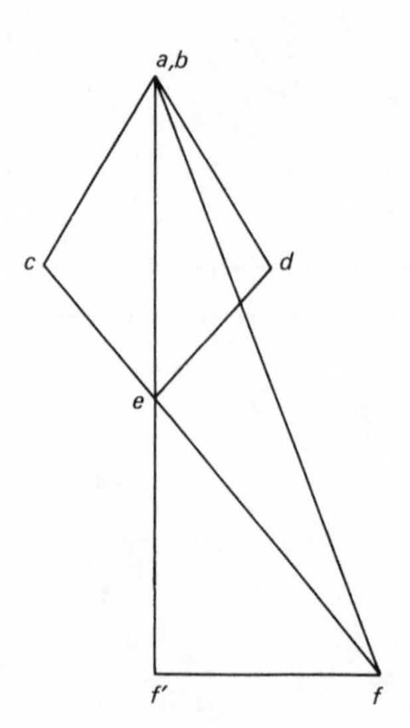

FIG. 3.23

Solution

The absolute velocities of C and D are represented by ac and bd respectively in the velocity diagram, Fig. 3.23. The velocity of e relative to C and D is perpendicular to CE and DE respectively, thus giving the point e. ce is then extended to f such that $ce:cf = \mathrm{CE:CF}$.

The tension, T, in the spring CD is obtained by the principle of virtual work, i.e. if the cranks are given an indefinitely small displacement from the given position, then

$$\text{work done by spring CD} = \text{work done by load at F}$$

i.e.
$$T \times \delta_\mathrm{T} = 200 \times \delta_\mathrm{F}$$

where δ_F is the virtual displacement at F and δ_T is the stretch of the spring. But the ratio $\delta_\mathrm{T}/\delta_\mathrm{F}$ is equal to the velocity ratio $v_\mathrm{T}/v_\mathrm{F}$ since each of these displacements takes place in the same time.

$$\text{Vertical velocity of F, } v_\mathrm{F} = af' = 160$$

$$\text{Velocity of C relative to D, } v_\mathrm{T} = cd = 50$$

$$\therefore\ T = 200 \times 160/50 = \underline{640\ \mathrm{N}}$$

Example 3.4 In the quick return mechanism shown in Fig. 3.24, the distance between the fixed centres CB = 90 mm, crank BA = 180 mm, CD = 180 mm, and the rod DE = 360 mm with its centre of mass, G, 90 mm from D. If AB rotates at 120 rev/min clockwise, find graphically for the given position the linear velocity of G and the angular velocity of DE. If the rod DE has a mass of 7 kg, and the radius of gyration about G is 120 mm, find its kinetic energy.

(I. Mech. E.)

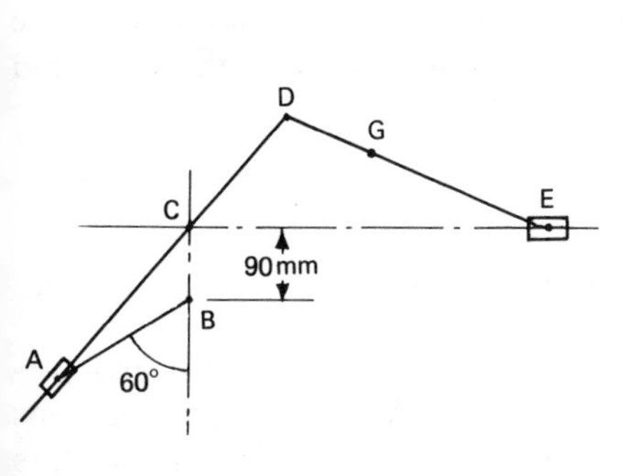

FIG. 3.24

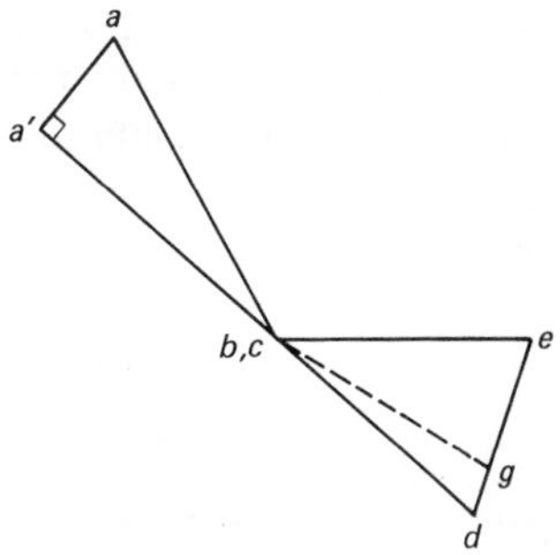

FIG. 3.25

Solution

$$v_a = \frac{2\pi}{60} \times 120 \times 0{\cdot}18 = 2{\cdot}26 \text{ m/s}$$

In Fig. 3.25, *ba* represents the absolute velocity of A. If A′ is the point on the link CD which at the given instant is coincident with A, then the velocity of A′ relative to C (and B) is perpendicular to ACD. Also the velocity of A′ relative to A is parallel to ACD, hence the point a' in the velocity diagram is obtained. $a'c$ is then produced to d such that $a'c : cd$ = A′C : CD and the diagram is completed by drawing through d a line perpendicular to DE and a line through c, parallel to CE, to give the point e. The point g divides ed in the same proportion as G divides ED in the configuration diagram. The absolute velocity of G is then represented by the line cg, shown dotted.

$$v_g = \underline{1{\cdot}56 \text{ m/s}}$$

$$\omega_{de} = \frac{de}{\text{DE}} = \frac{1{\cdot}136}{0{\cdot}36} = \underline{3{\cdot}16 \text{ rad/s}}$$

$$\text{K.E. of DE} = \tfrac{1}{2}mv^2 + \tfrac{1}{2}I\omega^2$$
$$= \tfrac{1}{2} \times 7 \times 1{\cdot}56^2 + \tfrac{1}{2} \times 7 \times 0{\cdot}12^2 \times 3{\cdot}16^2 = \underline{9 \text{ J}}$$

Example 3.5 In the oscillating cylinder mechanism shown in Fig. 3.26, the crank OA is 50 mm long while the piston rod AB is 150 mm long. The crank OA rotates uniformly about O at 300 rev/min.

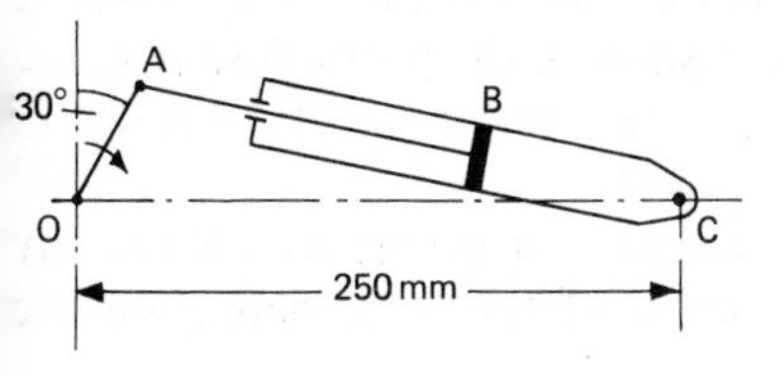

FIG. 3.26

Determine, for the position shown: (a) the velocity of the piston B relative to the cylinder walls, (b) the velocity of the piston B relative to the frame, and (c) the angular velocity of the piston rod AB.

Solution

$$v_a = \frac{2\pi}{60} \times 300 \times 0{\cdot}05 = 1{\cdot}571 \text{ m/s}$$

The velocity of A can be resolved into components parallel and perpendicular to AB; these are shown in Fig. 3.27 by Aa_1 and Aa_2 respectively. Aa_1 represents the velocity along its length of all points on AB and is therefore the velocity of sliding of the piston relative to the cylinder,

i.e. velocity of sliding $= Aa_1 = \underline{1{\cdot}475 \text{ m/s}}$

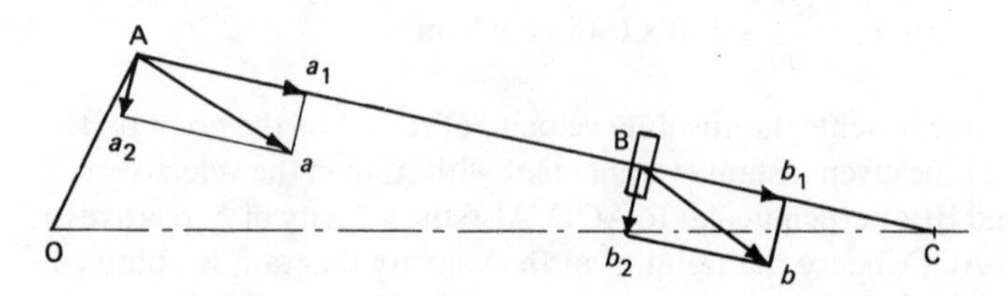

FIG. 3.27

Aa_2 represents the velocity of A perpendicular to the cylinder axis so that

$$\text{angular velocity of the piston rod} = \Omega = \frac{Aa_2}{AC} = \frac{0{\cdot}515}{0{\cdot}23} = \underline{2{\cdot}24 \text{ rad/s}}$$

$$\text{Velocity of piston parallel to its axis} = Bb_1 = Aa_1 = 1{\cdot}475 \text{ m/s}$$

$$\text{Velocity of piston perpendicular to its axis} = Bb_2 = \Omega \times BC$$

$$= 2{\cdot}24 \times 0{\cdot}0792$$

$$= 0{\cdot}1775 \text{ m/s}$$

$$\therefore \text{ resultant velocity of piston} = Bb = \sqrt{(1{\cdot}475^2 + 0{\cdot}1775^2)} = \underline{1{\cdot}486 \text{ m/s}}$$

Example 3.6 In the mechanism shown in Fig. 3.28, the crank DC is turning clockwise at 50 rev/min. When the mechanism is in the position shown, the clockwise torque applied to DC resists a vertical force of 50 N at F. AB = 60 mm, BC = 180 mm and CD = 30 mm.

Determine, for the given position of the mechanism: (a) the vertical velocity of EF, and (b) the magnitude and direction of the force acting on the mechanism at point D.

Neglect the effects of friction. (U. Lond.)

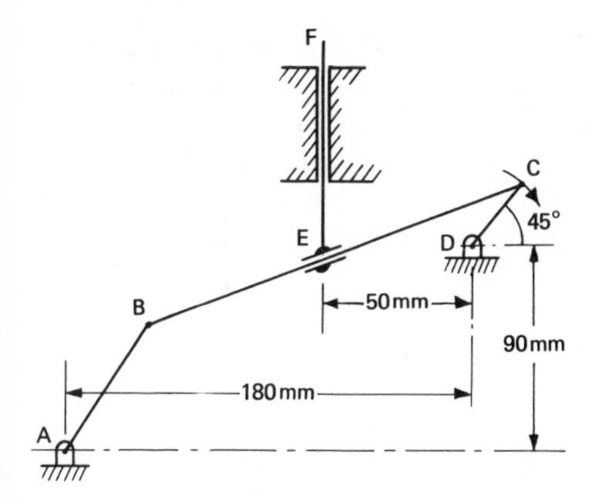

FIG. 3.28

Solution

$$v_c = \frac{2\pi}{60} \times 50 \times 0{\cdot}03 = 0{\cdot}157 \text{ m/s}$$

In Fig. 3.29, *ac* represents the absolute velocity of C. The velocity of B relative to C is perpendicular to BC and relative to A it is perpendicular to BA, thus giving the point *b*. If E′ is the point on BC, coincident with E at the given instant, then *ce′* : *e′b* = CE′ : E′B.

The velocity of swivel E relative to link point E′ is parallel to BC, and relative to A and D it is parallel to EF. Hence the point *e* is obtained.

From the diagram,

$$\text{vertical velocity of EF} = ea = \underline{0{\cdot}122 \text{ m/s}}$$

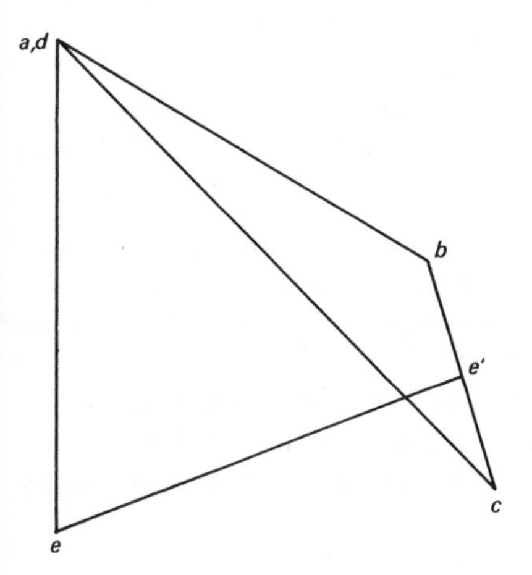

FIG. 3.29

The force P applied to BC at E, Fig. 3.30 (a), must be perpendicular to BC in the absence of friction. The vertical component of this force is 50 N, the horizontal component being supplied by the guides for EF.

The other forces acting on BC are the reactions at B and C. Since AB oscillates freely, the reaction R_b lies along AB and the remaining reaction R_c is then obtained from the triangle of forces, Fig. 3.30 (b), the three forces P, R_b and R_c being concurrent.

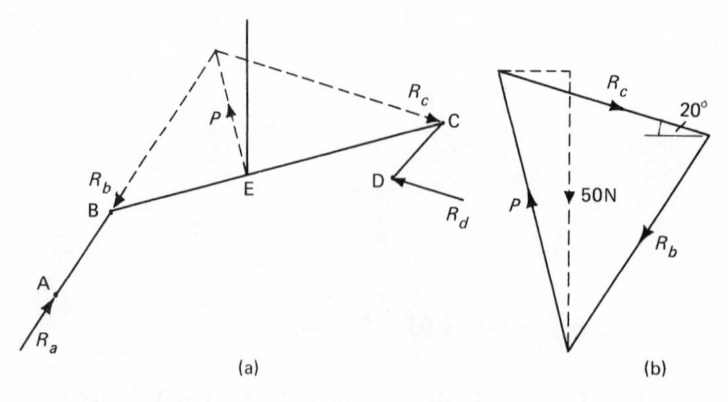

FIG. 3.30

From the diagram, $R_c = 42{\cdot}5$ N and inclination to horizontal = 20°. The reaction at D is equal and opposite to R_c, the couple produced by these two forces being the driving torque on the crank DC.

Example 3.7 In the mechanism shown in Fig. 3.31, the links OA and OB are free to turn about the same fixed axis at O. The link CD carries sliders at C and D, which are free to turn on pins at these points, and is connected to link OB by a pin joint at B. Link OB is 100 mm long, link CD is 400 mm long and the length BC is 300 mm. Link OA is rotating clockwise at 4 rad/s.

Use the instantaneous centre method to determine: (a) the velocity of the slider at C, and (b) the angular velocity of the link CD. (U. Lond.)

Solution

The mechanism is drawn for the given configuration by describing a circle of radius OB = 100 mm and setting out the centre lines for the movements of D and C. The position of B is then found, by trial, such that DB = 100 mm and BC = 300 mm.

The point B is moving perpendicular to OB and the point C is moving horizontally, so that the instantaneous centre for DBC is at I, Fig. 3.32.

If D′ is the point on link OA which is coincident with the block D then, by measurement, OD′ = 0·15 m.

The velocity of D′, perpendicular to OA, is given by

$$v_{D'} = 4 \times 0{\cdot}15 = 0{\cdot}6 \text{ m/s}$$

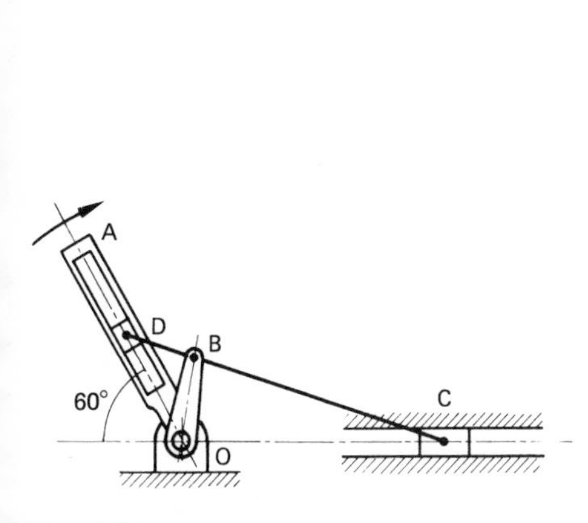

Fig. 3.31

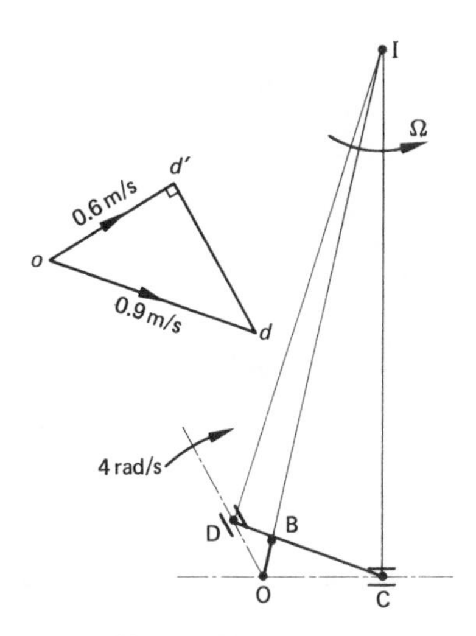

Fig. 3.32

The absolute velocity of D′, perpendicular to ID, is then represented by *od* in the velocity triangle *od′d*, since D and D′ must have the same velocity perpendicular to OA.

Thus

$$v_D = 0{\cdot}9 \text{ m/s}$$

Using the instantaneous centre for DBC,

$$v_C = v_D \times \frac{IC}{ID} = 0{\cdot}9 \times \frac{1{\cdot}34}{1{\cdot}26} = \underline{0{\cdot}96 \text{ m/s}}$$

$$\Omega_{CD} = \frac{v_D}{ID} = \frac{0{\cdot}9}{1{\cdot}26} = \underline{0{\cdot}714 \text{ rad/s}}$$

Example 3.8 Figure 3.33 shows the mechanism of a variable-stroke pump. The drive is taken from the crank OA to the pin B on the connecting rod CBD. The end C of the connecting rod carries a slider which is free to move along the curved link EFG. The stroke of the pump may be varied by rotating the link EFG about the fixed pivot G, although during operation the position of EFG is fixed.

If the crank OA rotates clockwise at 100 rev/min, find the velocity of the piston D for the position shown in Fig. 3.33, given the following

OA = 75 mm
Radius of curvature of link EFG = 400 mm
Length of connecting rod DC = 400 mm
DB = 300 mm
In the given position, D and A are on the same horizontal line

(U. Lond.)

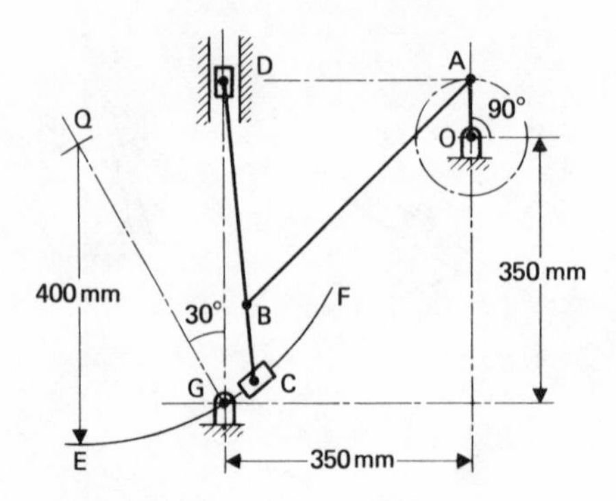

FIG. 3.33

Solution

Instantaneous centre method
The block C, Fig. 3.34, is moving in a direction tangential to the curved link EFG and the block D is moving in a vertical direction, so that the instantaneous centre for DC is at the point I_1. The point B is therefore moving in a direction perpendicular to I_1B and the point A is moving in a horizontal direction, so that the instantaneous centre for AB is at the point I_2.

$$v_a = \frac{2\pi}{60} \times 100 \times 0{\cdot}075 = 0{\cdot}785 \text{ m/s}$$

Then $$v_b = v_a \times \frac{I_2 B}{I_2 A} = 0{\cdot}785 \times \frac{0{\cdot}47}{0{\cdot}64} = 0{\cdot}576 \text{ m/s}$$

$$v_d = v_b \times \frac{I_1 D}{I_1 B} = 0{\cdot}576 \times \frac{0{\cdot}25}{0{\cdot}41} = \underline{0{\cdot}351 \text{ m/s}} \quad \text{(vertically upwards)}$$

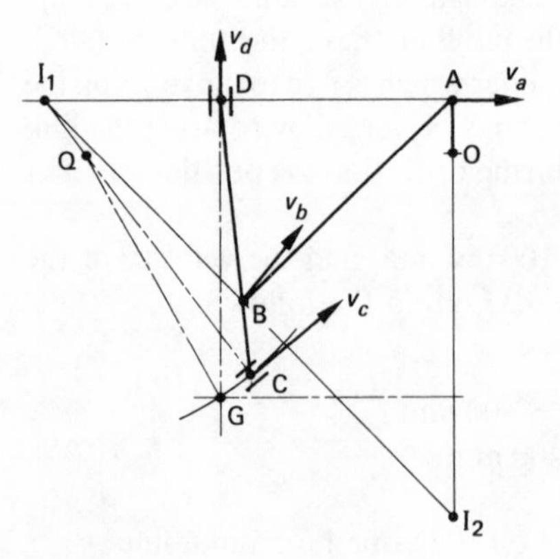

FIG. 3.34

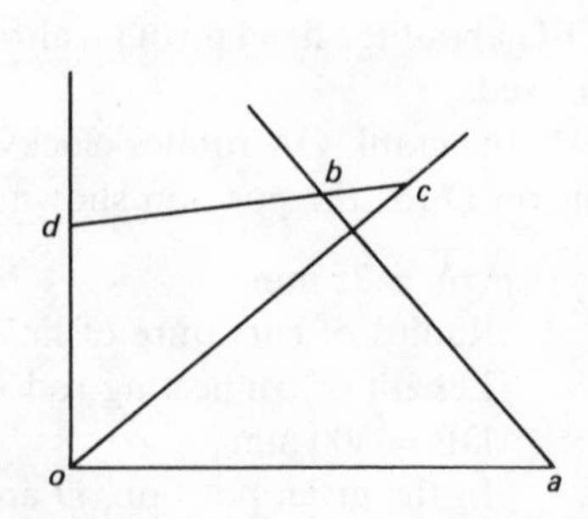

FIG. 3.35

Velocity diagram method

oa represents the absolute velocity of A in the velocity diagram, Fig. 3.35. The velocity of B relative to A is perpendicular to AB, the velocity of C relative to O is tangential to link EFG, and the velocity of D relative to O is vertical. Lines on which *d*, *b* and *c* must lie can then be drawn from *o* and *a*.

The velocities of D, B and C, relative to each other, lie on a line which is perpendicular to DBC and the lengths *db* and *bc* are in the ratio of 3 to 1. The line *dbc* must therefore be positioned, by trial, at the appropriate angle (i.e. perpendicular to CD) until the required ratio of lengths is achieved. Then the velocity of D is represented by the length *od* in the velocity diagram.

Problems: velocity diagrams

1. An aircraft A, flying on a straight path at a constant altitude of 7500 m, is tracked by a radar station, O, in the same vertical plane, as shown in Fig. 3.36. At a certain instant, the angle of elevation $\theta = 60°$ and $\dot{\theta} = -0{\cdot}025$ rad/s. Find the speed of the aircraft. (*Answer:* 900 km/h)

2. The end A of a bar AB, Fig. 3.37, is constrained to move along the vertical path AD and the bar passes through a swivel bearing pivoted at C. When A has a velocity of 0·75 m/s towards D, find the velocity of sliding of the bar through the swivel and the angular velocity of the bar. (*Answer:* 0·375 m/s; 5·63 rad/s)

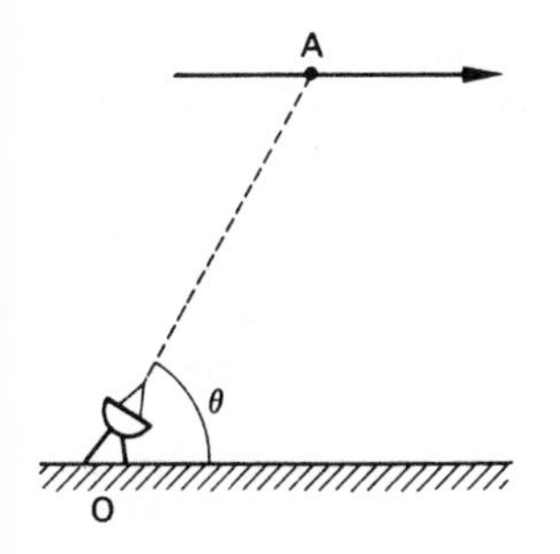

FIG. 3.36

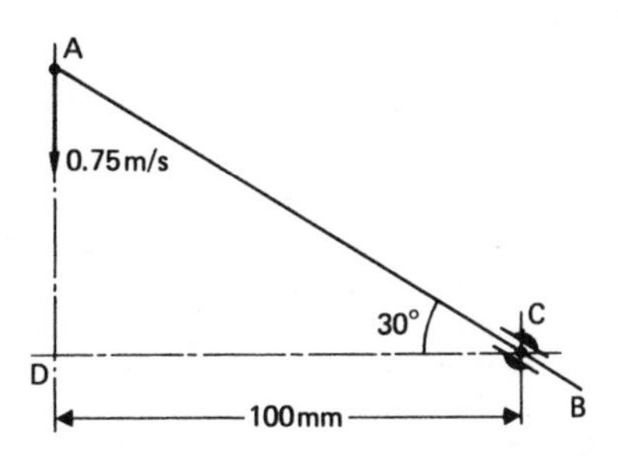

FIG. 3.37

3. In the mechanism shown in Fig. 3.38, the crank OA rotates in a clockwise direction about O at 120 rev/min; the end B of the rigid link AB moves along the straight line XX, while the end D of the link CD moves along the line YY; AB and CD are connected by a pin joint at C. AB = 210 mm, AC = 80 mm and CD = 240 mm.

For an angle EOA of 75°, draw the velocity diagram.

If a force of 100 N acting along YY resists the motion of D, determine the forces, in magnitude and direction, acting on the pins at the ends of the bar AB. The effects of mass and of friction may be neglected. (U. Lond.)

(*Answer:* A, 62 N at 38° to horizontal; B, 59 N at 28° to horizontal)

4. In Fig. 3.39, CDE is a straight bar connected to the fixed vertical bar AB by links BC and AD. AD is 550 mm long, CD is 100 mm and DE is 250 mm. In the position indicated, AD is inclined to the vertical at an angle of 30° and CDE is inclined to the

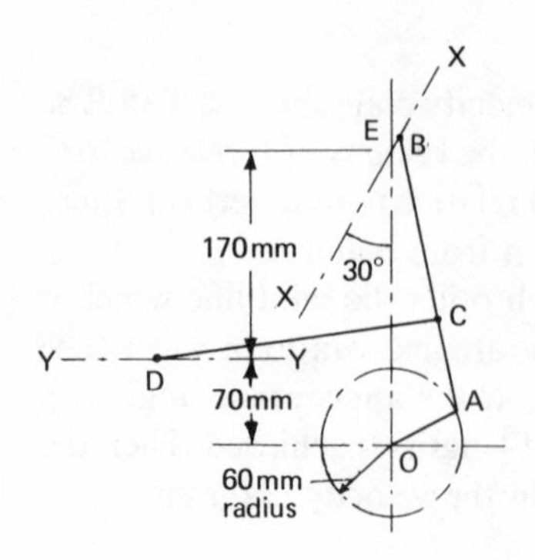

FIG. 3.38

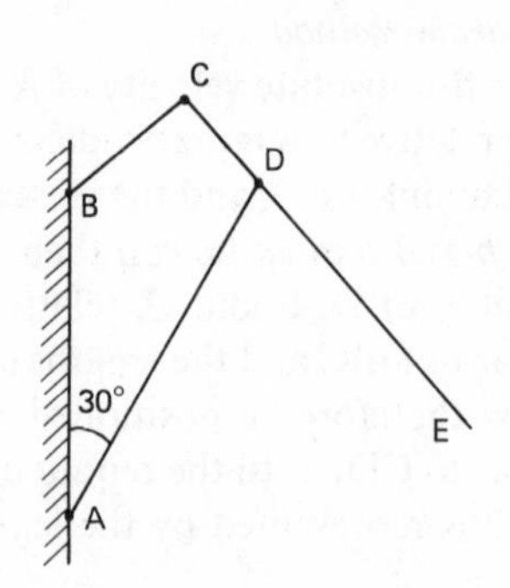

FIG. 3.39

vertical at an angle of 45°. Find the lengths of AB and BC so that in small movements from the given position the direction of motion of E will be horizontal.

If a horizontal force of 5 N is applied at E, what resisting moment must be applied to the bar AD at A in order to prevent movement? Neglect the mass of the bars. (U. Lond.) (*Answer:* 353 mm; 282 mm; 3·75 N m)

5. A four-bar kinematic chain is represented by a quadrilateral ABCD in which AD is fixed and is 400 mm long. The crank AB, 75 mm long, rotates in a clockwise direction at 120 rev/min and drives the link CD, 125 mm long, by means of the connecting link BC, 350 mm long.

Determine the angle through which CD oscillates and find the angular velocities of the links BC and CD in one of the positions when BC is perpendicular to AB. (U. Lond.) (*Answer:* 93°; 1·1075 rad/s; 8·13 rad/s)

6. A straight rod PQ, 180 mm long, forms part of a mechanism. The end P of the rod is constrained to move in a straight vertical path with simple harmonic motion, making 5 complete oscillations per second. The travel of P between extreme positions is 60 mm. The rod PQ slides in a small block pivoted at a fixed point O. O is situated 60 mm to the right, and 60 mm below the mean position of the point P.

Determine the velocity of Q at an instant when P is 15 mm below the centre of the line of stroke, and is moving upwards. (U. Lond.) (*Answer:* 1·03 m/s)

7. In the slotted lever mechanism shown in Fig. 3.40, the crank OA rotates at 90 rev/min in an anticlockwise direction about the fixed centre O, swinging the lever DEQ about the fixed centre Q.

Find the magnitude and direction of the angular velocity of the lever DEQ and the velocity of the slider C when $\theta = 45°$. (U. Lond.)
(*Answer:* 1·66 rad/s anticlockwise; 312 m/s)

8. In the valve gear shown in Fig. 3.41, O_1, O_2 and O_3 are fixed centres. O_1 is the crankshaft centre, and O_2 and O_3 are located from the limit of stroke of the crosshead pin A. The valve at V slides along a surface parallel to the engine centre line.

The lengths of the members are given in mm as follows: O_1C, 300; O_2D, 900; O_3F, 750; AB, 675; BC, 1000; DE, 350; BE, 175; EF, 450; FG, 100; VG, 850.

Draw the velocity diagram when the crank angle is 45° as shown. State the velocity of the valve V when the crank rotates at 90 rev/min in the direction given. (U. Lond.)
(*Answer:* 0·54 m/s)

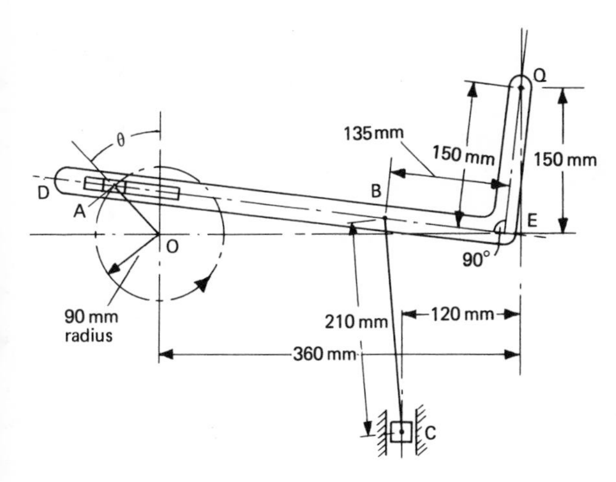

FIG. 3.40

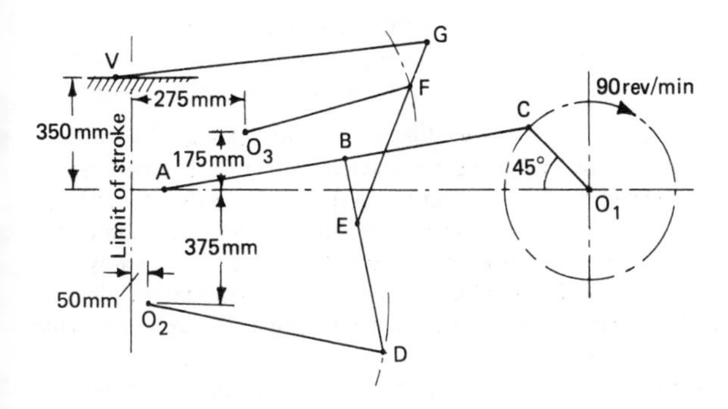

FIG. 3.41

9. In the linkwork shown in Fig. 3.42, P and Q are fixed axes. APE rotates about P and along it moves a sliding block attached to the end of BQD which rotates about Q. A and B are connected by the links AC and BC. AC = 70 mm, PE = 140 mm, QD = 110 mm, BC = 120 mm, and QB = PA = 60 mm.

If the velocity of E is 80 mm/s, find the velocity of C in magnitude and direction when the angle EPQ is 60°. (U. Lond.) (*Answer:* 46 mm/s)

10. In the mechanism shown in Fig. 3.43, the crank AB drives the bent link CDE by means of the sliding block at B. AB = 120 mm, CD = 90 mm, DE = 450 mm, and EF = 450 mm.

When the crank is horizontal, as shown, and is rotating at 60 rev/min anticlockwise, find: (a) the velocity of the slider F, and (b) the angular velocity of the link CDE. (U. Lond.) (*Answer:* 0·75 m/s; 1·78 rad/s)

11. Define 'instantaneous centre' and show that, if two rigid bodies have plane motion relative to a third, to which they are connected, the three instantaneous centres lie on a straight line.

Reproduce the centre lines of the mechanism shown in Fig. 3.44. By the use of

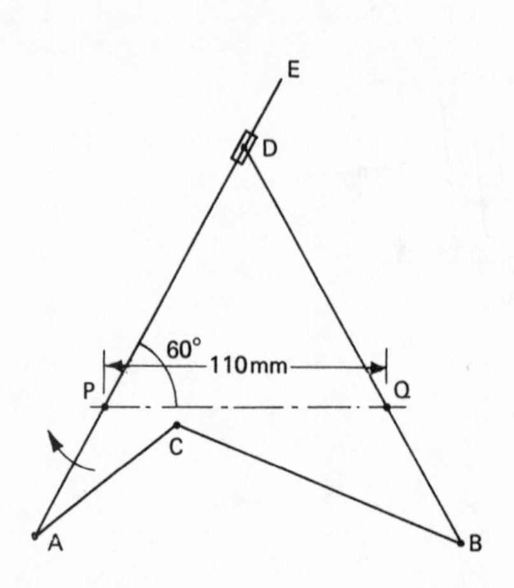

FIG. 3.42

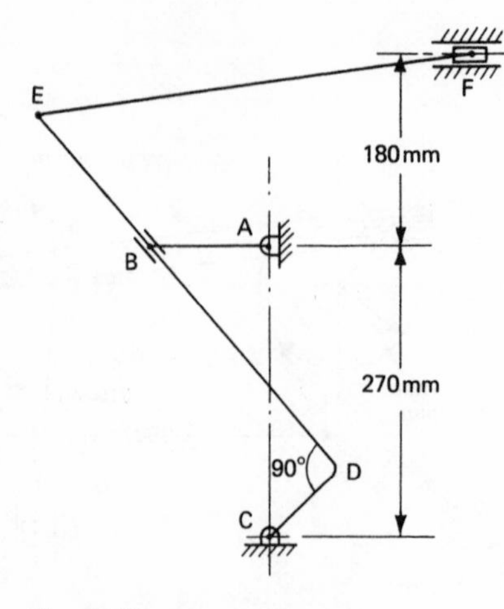

FIG. 3.43

instantaneous centres, or otherwise, determine, for the given configuration and bell-crank speed, the angular velocity of the link EC and the sliding velocity of the block F. (U. Lond.) (*Answer:* 3·88 rad/s; 0·15 m/s)

12. In the mechanism shown in Fig. 3.45, the double crank EOA rotates uniformly at 200 rev/min, in an anticlockwise direction, about the fixed centre O; the bell-crank lever BCD is rocked about the fixed centre C by the link AB, and the movement of the rigid bar DFG is controlled by the pin joint at D and the link EF. When the angle θ is 60°, there is a force P of 500 N, acting on G, at 30° to the horizontal, as shown.

Find: (a) the linear velocity of G and the angular velocity of DG in magnitude and direction, and (b) the magnitude and direction of the force acting at D, and the torque acting on AE.

OA = 200 mm, CD = 150 mm, OE = 75 mm, DG = EF = 450 mm, AB = 600 mm, DF = 350 mm and BC = 350 mm (U. Lond.)

(*Answer:* 2·34 m/s; 4·6 rad/s anticlockwise; 350 N at 30° to vertical; 15 N m)

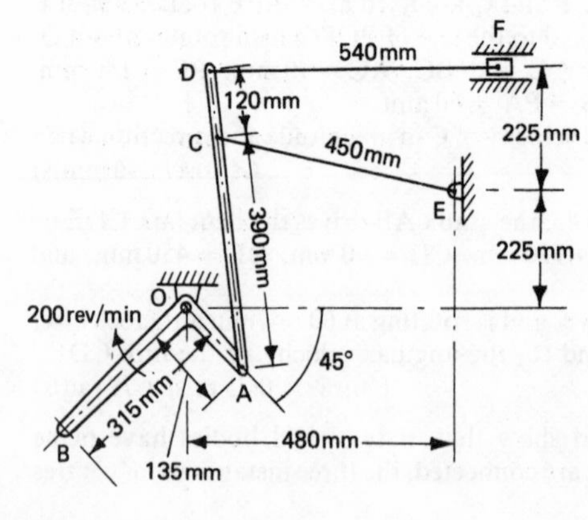

FIG. 3.44

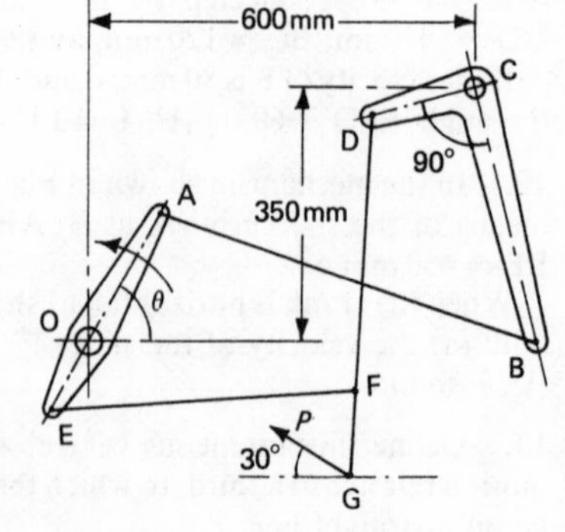

FIG. 3.45

13. The diagram of a linkage is given in Fig. 3.46. Find the velocity of the slider D and the angular velocity of DC when the crank O_1A is in the given position and the speed of rotation is 90 rev/min in the direction of the arrow. $O_1A = 24$ mm, $O_2B = 60$ mm, CD = 96 mm, AB = 72 mm and CB = 48 mm.

If a force of 500 N is applied to the slider in the direction of its motion, what will be the torque produced at O_1? (U. Lond.)

(*Answer:* 0·046 m/s; 2·18 rad/s; 2·52 N m)

14. In the mechanism shown in Fig. 3.47, the crank OA rotates uniformly in a clockwise direction at 500 rev/min; the rod AB is constrained to move in contact with the surface of the fixed block D, which has an effective radius of 60 mm about the fixed centre Q. OA = 48 mm, AB = 168 mm and BC = 96 mm.

Find the linear speed of C, the velocity of sliding of AB on D, and the angular velocity of AB when the angle θ is 30°. (U. Lond.)

(*Answer:* 1·08 m/s; 1·9 m/s; 26·2 rad/s)

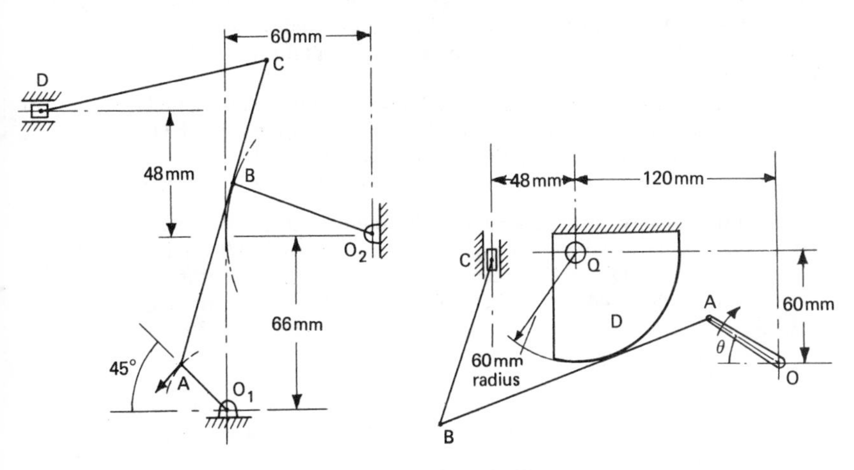

FIG. 3.46

FIG. 3.47

15. In the rudder mechanism shown in Fig. 3.48, the pinion A meshes with a semicircular gear with centre B, the ratio of the pitch circle diameters being 11 to 3. The coupling rod CD is connected to the arm DE, the latter being connected rigidly at right angles to EF. When EF lies along the centre line BG, ED and the gear diameter HBJ are parallel. BE = 2·5 m, BC = 0·25 m and ED = 0·425 m.

Find the angular velocity ratio between the pinion and arm EF, when the latter makes an angle of 35° with EG. If the normal force in EF is 80 kN acting at 0·9 from E and the efficiency of the mechanism is 60 per cent, find the couple exerted by the pinion. (U. Lond.) (*Answer:* 12·7 : 1; 100 kN m)

16. In the four-bar mechanism ABCD, Fig. 3.49, the fixed link AD is 600 mm long. In the position shown, AB is inclined at 60° to AD. The lengths of the links AB, BC and CD are 200, 400 and 300 mm respectively. A force P of 250 N acting parallel to AD and applied at a point E, half-way along AB, is sufficient to overcome a resisting force R, acting along a line inclined at 30° to AD and applied at a point F, half-way along CD.

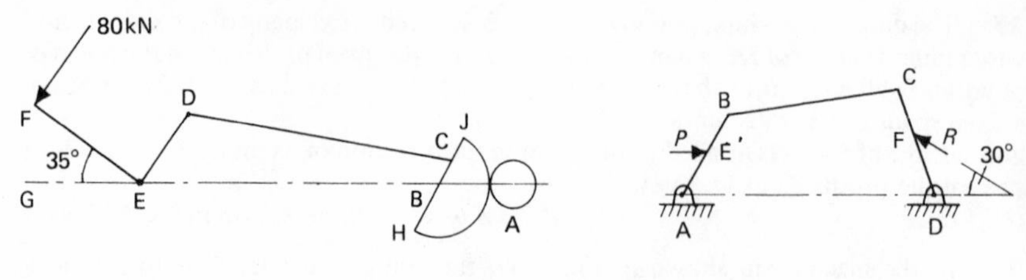

FIG. 3.48

FIG. 3.49

Find the magnitude of the force R and of the force at each of the joints A, B, C and D. (U. Lond.) (*Answer:* 500 N; 111 N; 150 N; 150 N; 410 N)

17. Figure 3.50 shows a crank OA, 72 mm long which rotates anticlockwise about O at 150 rev/min. A straight bar DBC is pivoted at B, which is 150 mm vertically below O, and the length BC = 75 mm. The portion BD of this bar slides in a trunnion fitted at A on the crank OA. A slider E slides in horizontal guides 30 mm below B, and is connected to C by a rod CE, 240 mm long. A horizontal force of 150 N opposes the motion of E. Friction is to be neglected.

For the position shown, where angle AOB = 120°, find: (a) the angular velocity of DBC and the linear velocity of E, (b) the driving torque on the crank OA, and (c) the bending moment at B in the bar DBC. (U. Lond.)

(*Answer:* 4·34 rad/s; 0·292 m/s; 2·79 N m; 10·35 N m)

18. Find the velocity of the slider F and the angular velocity of the bell-crank lever CDE in the mechanism shown in Fig. 3.51 when the crank AB rotates clockwise at 250 rev/min. A and D are fixed centres and the relevant lengths are as follows: AB = 100 mm, BC = 440 mm, CD = 200 mm, DE = 120 mm and EF = 400. (C.E.I.) (*Answer:* 1·35 m/s; 11·25 rad/s)

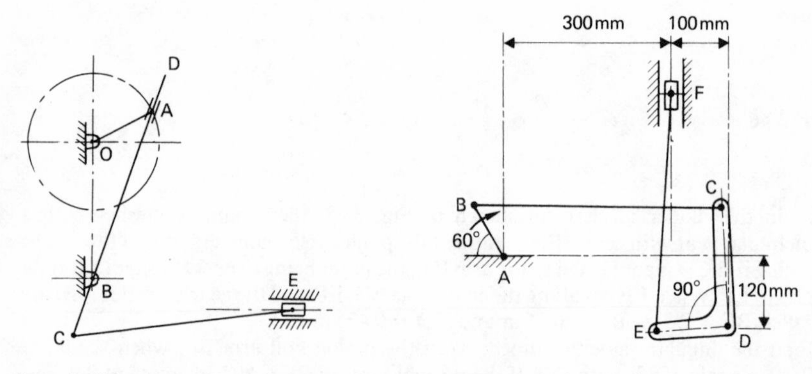

FIG. 3.50

FIG. 3.51

19. The screw clamp shown in Fig. 3.52 consists of a frame, a screw and an L-shaped link. The screw is threaded through a collar which is pin-jointed to the frame at B, and is connected to the L-shaped link through a pin joint at A. The L-shaped link, which is right-angled, pivots in the frame at C and has pinned to it a clamping block at D.

The screw has a lead (advance per revolution) of 4 mm. Assuming no friction, find the torque T required to produce a clamping force of 500 N in the position shown where $\theta = 30°$. (U. Lond.) (*Answer:* 0·45 N m)

20. In the shuttle mechanism shown in Fig. 3.53, the driving crank OA rotates clockwise about O at a constant angular velocity of 600 rev/min, causing the slotted lever to rock about the pivot at C and the sliding block B to oscillate in the vertical guide.

Determine the linear velocity of the block B when the shuttle is in the position shown and calculate the torque on the crank OA required to overcome a force of 5 N resisting the motion of the block, assuming friction in the pivots and slides to be negligible. (U. Lond.) (*Answer:* 0·45 m/s; 0·0358 N m)

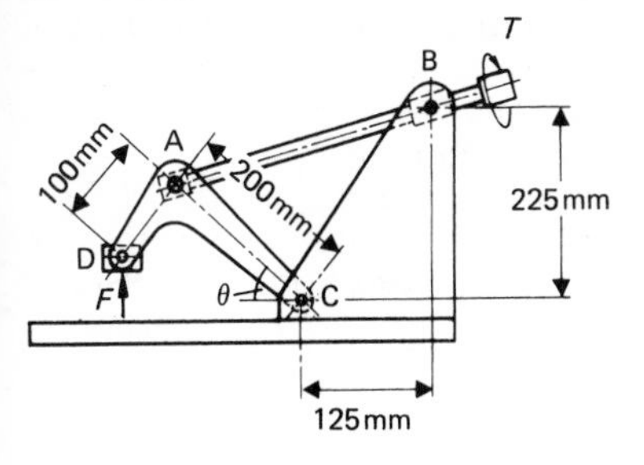

FIG. 3.52

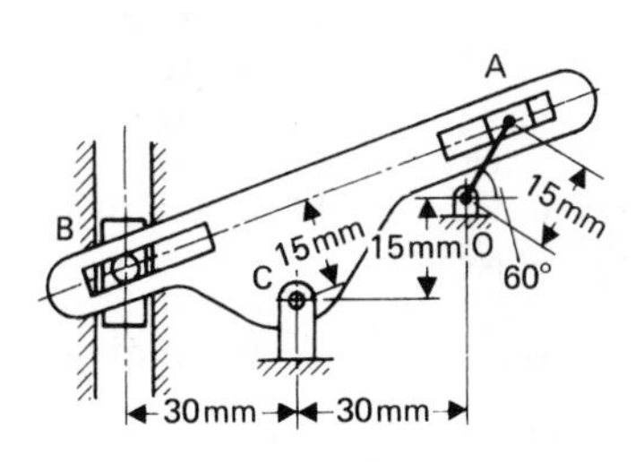

FIG. 3.53

21. In the mechanism shown in Fig. 3.54, the link OX rotates about the fixed centre O in the anticlockwise direction at 5 rad/s. It passes through a slot in block A, which rotates on a pin attached to the gear Y. This gear Y has a pitch circle diameter of 160 mm and rotates about the fixed centre P. The distance between P and the centre of the pin carrying the slotted block A is 50 mm. The gear Z, which has a pitch circle diameter of 80 mm, meshes with gear Y and rotates about the fixed centre Q. The link BD, which is 160 mm long, is pinned to the gear Y at B, the distance of B from P being 60 mm. The link CD, which is also 160 mm long, is pinned to the gear Z at C, the distance of C from Q being 30 mm. The links BD and CD are pinned together at D and also to the end D of the link DE, which is 200 mm long. The end E of this link DE is pinned to a slider, which is constrained to move along a vertical path.

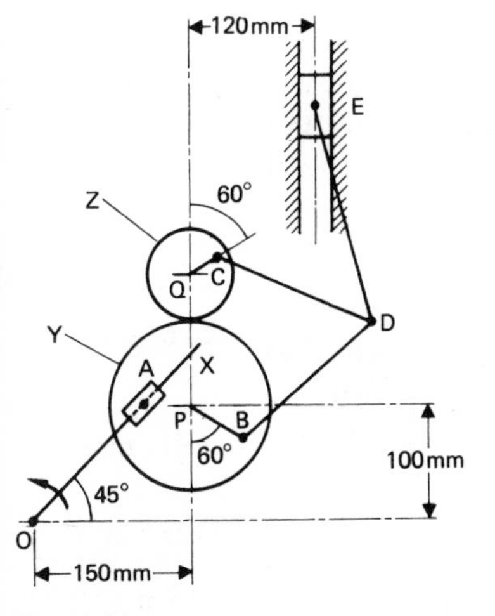

FIG. 3.54

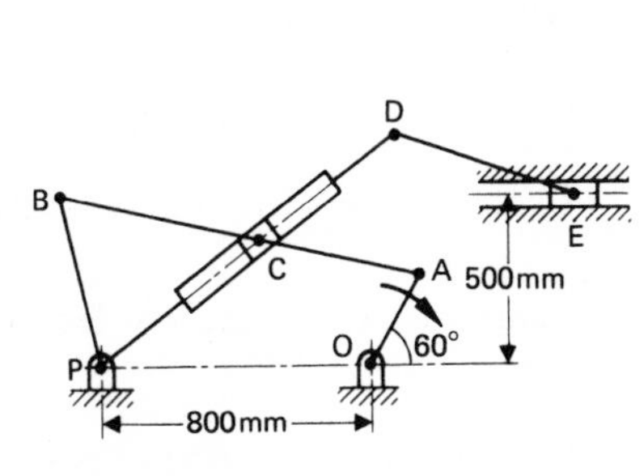

FIG. 3.55

Use the instantaneous centre of rotation method to determine: (a) the velocity of the slider at the end E of the link DE, and (b) the speed of sliding between the link OX and the slotted block A through which it passes. (U. Lond.)

(*Answer:* 0·127 m/s; 0·72 m/s)

22. In the mechanism shown in Fig. 3.55, the member OA, which has a length of 300 mm, rotates about O at 50 rad/s in the clockwise direction. The member PB, which is 500 mm long, rotates about P and is connected to member OA by the member AB, which is 1100 mm long. The slotted member PD rotates about P and is 1100 mm long. A block, which is pinned to the member AB at the point C, slides in the slot in the member PD. The length AC is 500 mm. The member DE is 600 mm long and is pinned at E to a slider, which is constrained to move along a horizontal path.

Use the instantaneous centre of rotation method to determine: (a) the velocity of the slider at E, and (b) the velocity of sliding between the block at C and the member PD. (U. Lond.) (*Answer:* 17 m/s; 9·8 m/s)

Examples: acceleration diagrams

Example 3.9 An engine mechanism is shown in Fig. 3.56; the crank CB = 100 mm and the connecting rod BA = 300 mm with centre of mass, G, 100 mm from B. In the position shown, the crankshaft has a speed of 25π rad/s and an angular acceleration of 400π rad/s^2.

Find the velocity and acceleration of G and the angular velocity and angular acceleration of AB. (I. Mech. E.)

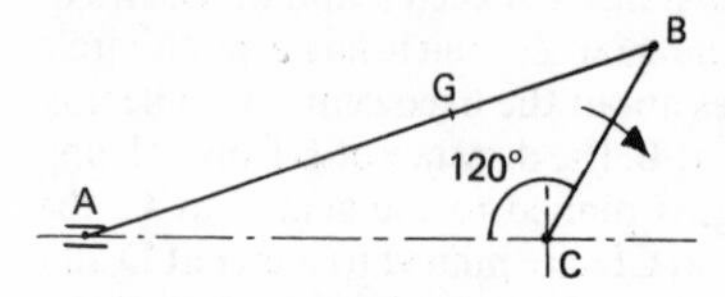

FIG. 3.56

Solution

$$v_b = 25\pi \times 0{\cdot}1 = 7{\cdot}85 \text{ m/s}$$

The velocity diagram, Fig. 3.57, is obtained by drawing *cb* perpendicular to CB to represent 7·85 m/s. *ba* is then perpendicular to BA and *ca* is parallel to CA, thus giving the point *a*. The point *g* divides *ab* in the same ratio as G divides AB in Fig. 3.56.

By measurement, $v_g = gc = \underline{6{\cdot}94 \text{ m/s}}$

and angular velocity of AB $= \dfrac{ab}{\text{AB}} = \dfrac{4{\cdot}1}{0{\cdot}3} = \underline{13{\cdot}65 \text{ rad/s}}$

$$\text{Centripetal acceleration of B relative to C} = \omega^2\text{BC}$$
$$= (25\pi)^2 \times 0{\cdot}1 = 617 \text{ m/s}^2$$
$$\text{Tangential acceleration of B relative to C} = \alpha\text{BC}$$
$$= 400\pi \times 0{\cdot}1 = 125{\cdot}8 \text{ m/s}^2$$
$$\text{Centripetal acceleration of A relative to B} = \frac{ab^2}{\text{AB}} = \frac{4{\cdot}1^2}{0{\cdot}3} = 56 \text{ m/s}^2$$

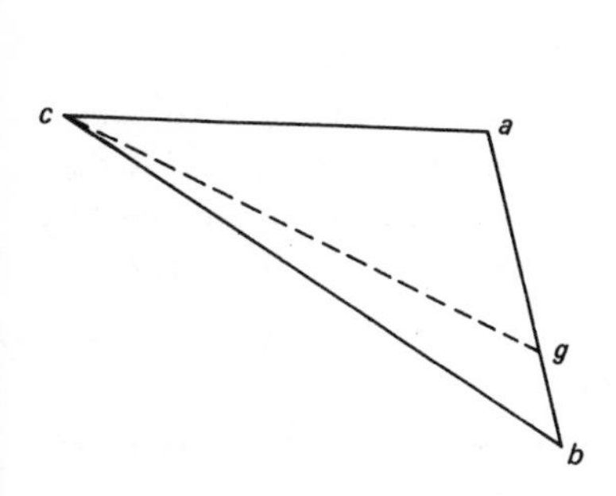

FIG. 3.57

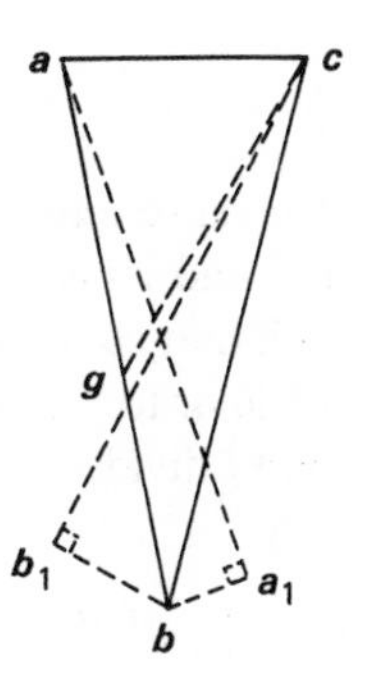

FIG. 3.58

In the acceleration diagram, Fig. 3.58, the centripetal and tangential components of the acceleration of B relative to C, cb_1 and b_1b, are drawn respectively parallel and perpendicular to BC. ba_1 is then drawn to represent the centripetal acceleration of A relative to B and the diagram is completed by drawing a_1a perpendicular to AB and ac parallel to AC, a_1a and ac representing the tangential acceleration of A relative to B and the absolute acceleration of A relative to C respectively. The point g divides ab in the same ratio as G divides AB in Fig. 3.56.

By measurement, $$f_g = gc = \underline{464 \text{ m/s}^2}$$

and $$\alpha_{ab} = \frac{a_1a}{\text{AB}} = \frac{606}{0{\cdot}3} = \underline{2020 \text{ rad/s}^2}$$

Example 3.10 In the mechanism shown in Fig. 3.59, the link AB rotates with a uniform angular velocity of 30 rad/s. Determine the velocity and acceleration of G for the configuration shown.

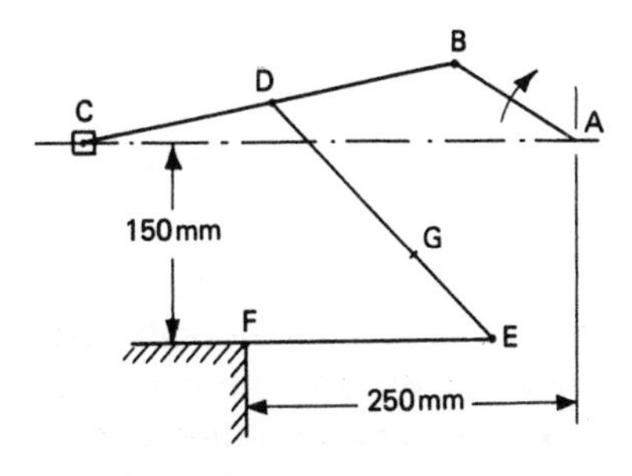

FIG. 3.59

The lengths of the various links are: AB = 100 mm, BC = 300 mm, BD = 150 mm, DE = 250 mm, EF = 200 mm and DG = 167 mm. Angle CAB = 30°. (U. Lond.)

Solution

$$v_b = 30 \times 0{\cdot}1 = 3 \text{ m/s}$$

In Fig. 3.60, *ab* represents the absolute velocity of B. Also, *bc* is perpendicular to BC and *ac* is horizontal, thus giving the point *c*. *d* is then positioned on *bc* such that *bd* : *dc* = BD : DC and the diagram is completed by drawing *de* perpendicular to DE and *fe* perpendicular to FE. *g* then divides *de* in the same ratio as G divides DE.

By measurement, $v_g = ag = \underline{0{\cdot}6 \text{ m/s}}$

Centripetal acceleration of B relative to A $= \omega^2 \text{AB} = 30^2 \times 0{\cdot}1 = 90 \text{ m/s}^2$

$$\text{Centripetal acceleration of C relative to B} = \frac{cb^2}{\text{CB}} = \frac{2{\cdot}625^2}{0{\cdot}3} = 23{\cdot}0 \text{ m/s}^2$$

$$\text{Centripetal acceleration of E relative to D} = \frac{ed^2}{\text{ED}} = \frac{2{\cdot}54^2}{0{\cdot}25} = 25{\cdot}8 \text{ m/s}^2$$

$$\text{Centripetal acceleration of E relative to F} = \frac{ef^2}{\text{EF}} = \frac{0{\cdot}576^2}{0{\cdot}2} = 1{\cdot}66 \text{ m/s}^2$$

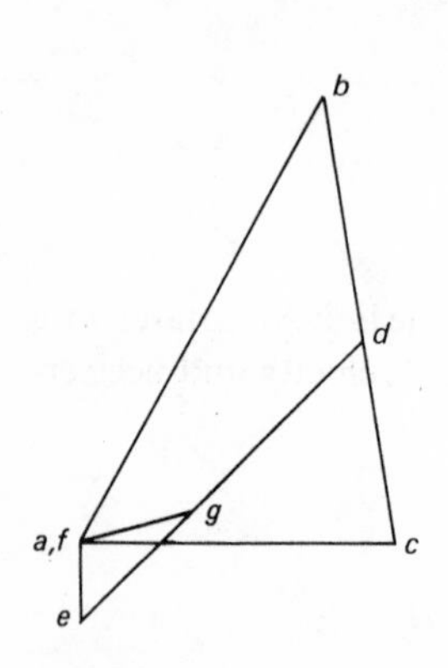

Fig. 3.60

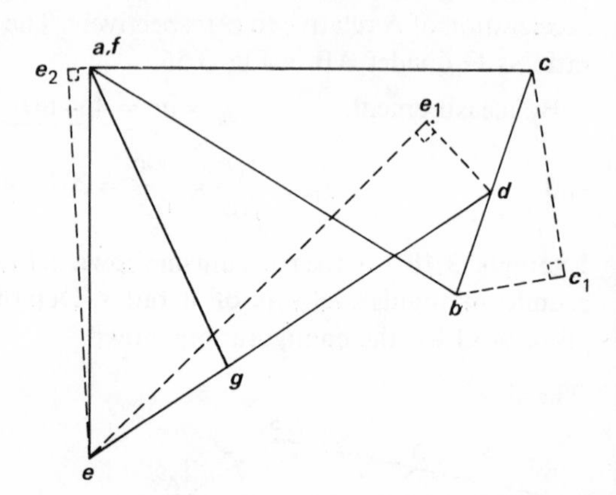

Fig. 3.61

The absolute acceleration of B is represented by ***ab***, Fig. 3.61. The centripetal acceleration of C relative to B (bc_1) is parallel to BC and the tangential acceleration ($c_1 c$) is perpendicular to BC. The absolute acceleration of C is horizontal, so that the point ***c*** is obtained. ***d*** is then positioned on ***bc*** such that ***bd*** : ***dc*** = BD : DC.

The centripetal acceleration of E relative to D (de_1) is parallel to DE and the tangential acceleration ($e_1 e$) is perpendicular to DE. Also, the centripetal acceleration of E relative to F (ae_2) is parallel to EF and the tangential acceleration ($e_2 e$) is perpendicular to EF, so that the intersection of the two tangential acceleration vectors gives the point ***e***. ***g*** then divides ***de*** in the same ratio as G divides DE.

By measurement, $f_g = \boldsymbol{ag} = \underline{66\ \mathrm{m/s^2}}$

Example 3.11 In the mechanism shown in Fig. 3.62, D is constrained to move on a horizontal path. Find, for the given configuration, the velocity and acceleration of D and the angular velocity and acceleration of BD when OC is rotating in an anticlockwise direction at a speed of 180 rev/min, increasing at the rate of 50 rad/s². (U. Lond.)

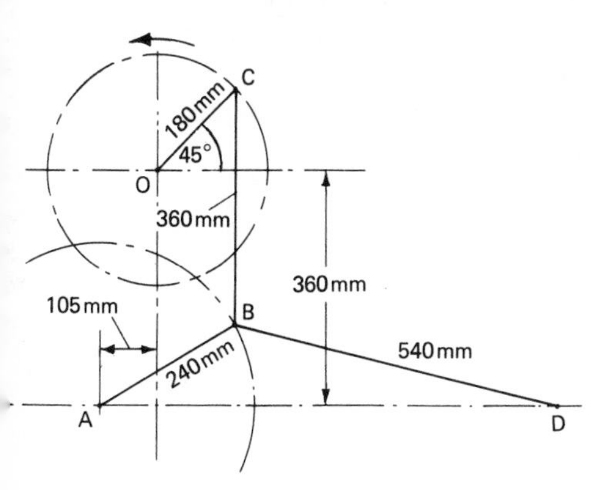

Fig. 3.62

Solution

$$v_c = \left(\frac{2\pi}{60} \times 180\right) \times 0{\cdot}18 = 3{\cdot}39\ \mathrm{m/s}$$

The absolute velocity of C is represented by *oc*, Fig. 3.63. The velocities of B relative to C and A are respectively perpendicular to CB and AB, thus giving the point *b*. The velocity of D relative to B is perpendicular to BD and relative to A it is horizontal, so that point *d* is obtained.

By measurement, $v_d = da = \underline{2{\cdot}055\ \mathrm{m/s}}$

and $$\omega_{bd} = \frac{bd}{\mathrm{BD}} = \frac{2{\cdot}415}{0{\cdot}54} = \underline{4{\cdot}47\ \mathrm{rad/s}}$$

Tangential acceleration of C relative to O $= \alpha\mathrm{OC} = 50 \times 0{\cdot}18 = 9\ \mathrm{m/s^2}$

Centripetal acceleration of C relative to O $= \omega^2\mathrm{OC}$

$$= \left(\frac{2\pi}{60} \times 180\right)^2 \times 0{\cdot}18 = 63{\cdot}9\ \mathrm{m/s^2}$$

Centripetal acceleration of B relative to C $= \dfrac{bc^2}{\mathrm{BC}} = \dfrac{0{\cdot}9^2}{0{\cdot}36} = 2{\cdot}25\ \mathrm{m/s^2}$

$$\text{Centripetal acceleration of B relative to A} = \frac{ba^2}{BA} = \frac{2{\cdot}79^2}{0{\cdot}24} = 32{\cdot}5 \text{ m/s}^2$$

$$\text{Centripetal acceleration of D relative to B} = \frac{db^2}{DB} = \frac{2{\cdot}41^2}{0{\cdot}54} = 10{\cdot}78 \text{ m/s}^2$$

The point c in the acceleration diagram, Fig. 3.64, is obtained by drawing the centripetal and tangential components oc_1 and c_1c respectively parallel and perpendicular to OC. The centripetal components of the accelerations of B relative to C and A (cb_1 and ab_2 respectively) are next drawn and the point b then lies at the intersection of the tangential components, b_1b and b_2b. The diagram is completed by setting off bd_1 to represent the centripetal acceleration of D relative to B and d_1d to represent the tangential acceleration, the acceleration of D relative to A being horizontal.

By measurement, $f_d = da = \underline{13{\cdot}3 \text{ m/s}^2}$

and $\alpha_{bd} = \dfrac{dd_1}{BD} = \dfrac{38{\cdot}5}{0{\cdot}54} = \underline{71{\cdot}3 \text{ rad/s}^2}$

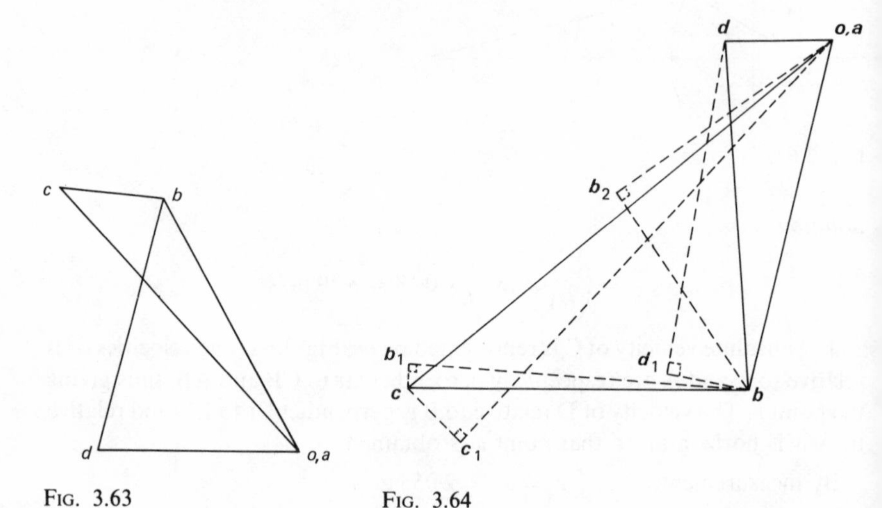

FIG. 3.63

FIG. 3.64

Example 3.12 Figure 3.65 shows a device called 'Peaucellier's inversor', comprising six pin-jointed rigid links. Links OU and OV are 1 m long and the four remaining links are 0·5 m long. The point P is guided so as to move along a straight line through OQ. At the instant when the angle PUQ is 60°, point P is moving towards O with a velocity of 0·5 m/s and an acceleration of 1 m/s².

Draw velocity and acceleration diagrams for the configuration given and use them to determine the velocity and acceleration of point Q. (C.E.I.)

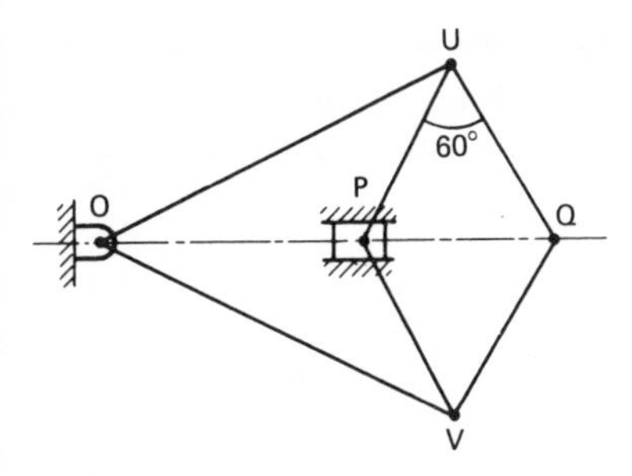

FIG. 3.65

Solution

In the velocity diagram, Fig. 3.66, *op* represents the given velocity of P; *ou*, *pu* and *uq* are perpendicular to OU, PU and UQ respectively.

From the diagram,

$$\text{velocity of Q} = oq = \underline{0{\cdot}88 \text{ m/s}}$$

$$\text{Centripetal acceleration of U relative to P} = \frac{up^2}{\text{UP}} = \frac{0{\cdot}8^2}{0{\cdot}5} = 1{\cdot}28 \text{ m/s}^2$$

$$\text{Centripetal acceleration of U relative to O} = \frac{uo^2}{\text{UO}} = \frac{0{\cdot}44^2}{1} = 0{\cdot}194 \text{ m/s}^2$$

$$\text{Centripetal acceleration of Q relative to U} = \frac{qu^2}{\text{QU}} = \frac{0{\cdot}8^2}{0{\cdot}5} = 1{\cdot}28 \text{ m/s}^2$$

In the acceleration diagram, Fig. 3.67, ***op*** represents the given acceleration of P. $\boldsymbol{pu_1}$ is drawn to represent the centripetal acceleration of U relative to P and $\boldsymbol{ou_2}$ is drawn to represent the centripetal acceleration of U relative to O. The intersection of the tangential accelerations of U relative to P and U relative to O gives the point ***u***.

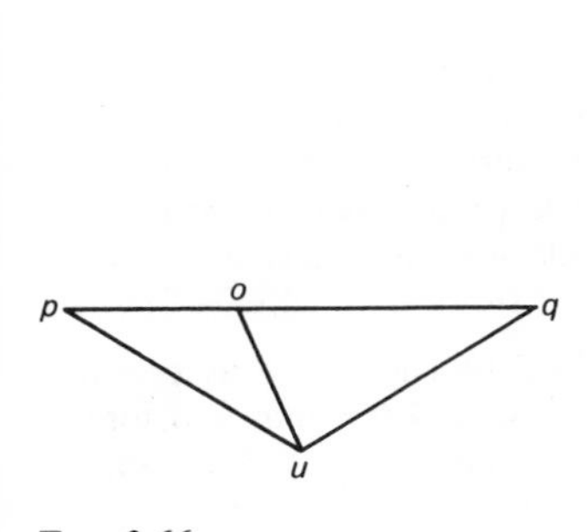

FIG. 3.66

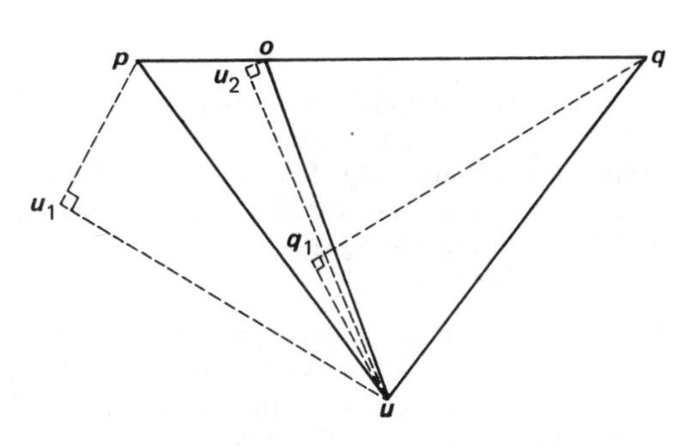

FIG. 3.67

If uq_1 is drawn to represent the centripetal acceleration of Q relative to U, then a perpendicular line drawn through q_1, representing the tangential acceleration, will intersect the horizontal through o at the point q, since the block Q moves horizontally.

From the acceleration diagram,

$$\text{acceleration of Q} = oq = \underline{3{\cdot}2\ \text{m/s}^2}$$

Problems: acceleration diagrams

23. One end A of a link AB of length l shown in Fig. 3.68 moves with uniform velocity v along a horizontal guide towards O, while the end B moves in the vertical guide. From the expressions for the distances of A and B from O, obtain expressions for the angular velocity of the rod, the velocity of B and the acceleration of B, in terms of l, θ and v.

If $l = 625$ mm, $v = 0{\cdot}9$ m/s and the guide block B has a mass of 3 kg, find the force P applied at A in the direction of motion required to overcome the resistance at B due to its mass and inertia for the position in which $\theta = 30°$. Neglect friction.

(*Answer:* $\omega = -v/(l \sin\theta)$; $v_b = -v \cot\theta$; $f_b = -v^2/(l \sin^3\theta)$; $-2{\cdot}91$ N)

24. In the mechanism shown in Fig. 3.69, the crank AB is 75 mm long and rotates uniformly clockwise at 8 rad/s. Given that BD = DC = DE and BC = 300 mm, draw the velocity and acceleration diagrams. State the velocity and acceleration of the pistons at C and E. (U. Lond.)

(*Answer:* v_c, 0·6 m/s; v_e, 0·1875 m/s; f_c, 4·16 m/s²; f_e, 9·06 m/s²)

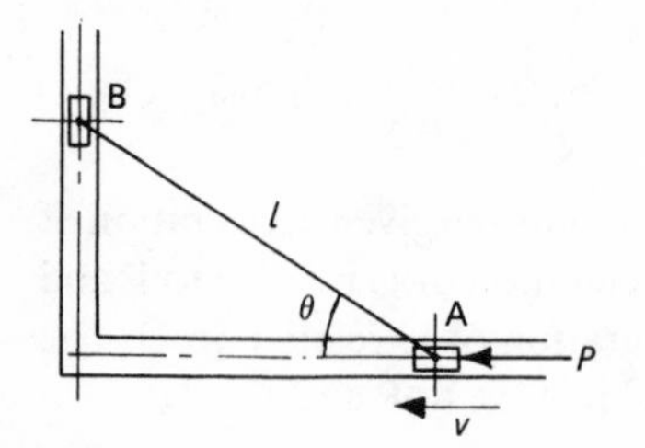

FIG. 3.68

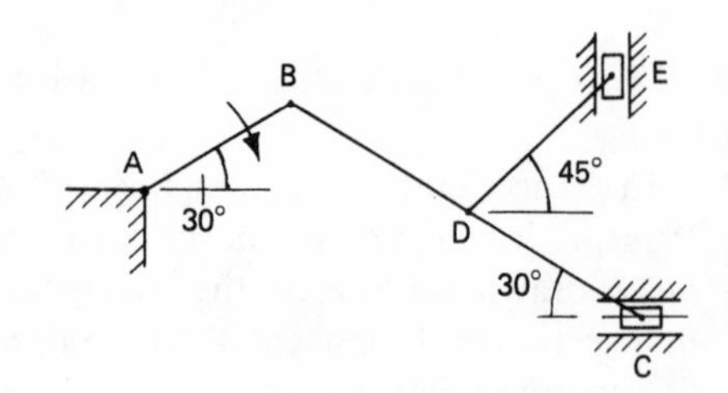

FIG. 3.69

25. Two links OA and QB turn about fixed parallel axes through O and Q. The ends A and B are pin-jointed to a connecting link AB. The lengths of the four links are: OA, 50 mm; AB, 175 mm; BQ, 100 mm; OQ, 200 mm. OA turns at a uniform speed of 150 rev/min. For the position of the mechanism in which A and B are on opposite sides of OQ and the angle AOQ is 30°, find the angular velocity and the angular acceleration of BQ. (I. Mech. E.) (*Answer:* 8·4 rad/s; 24·5 rad/s²)

26. Part of a reversing gear mechanism is shown in Fig. 3.70. The block P reciprocates along the line AB. If the crank OC turns at a uniform speed of 240 rev/min, find, for the given position of the crank: (a) the velocity and acceleration of the block P, and (b) the velocity and acceleration of the point D. (I. Mech. E.)

(*Answer:* 3·72 m/s; 63·9 m/s²; 3·05 m/s; 71·4 m/s²)

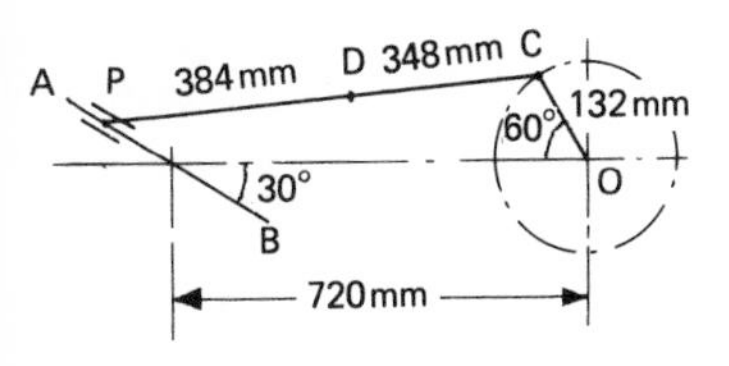

FIG. 3.70

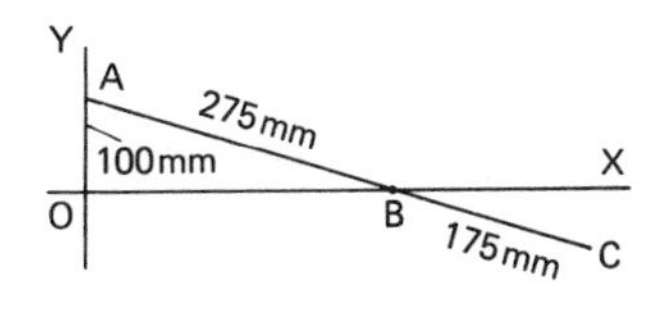

FIG. 3.71

27. In Fig. 3.71, a bar ABC moves with points A and B guided along the OY and OX axes. AB = 275 mm and BC = 175 mm. For the position shown in which A is 100 mm from O, the velocity of A is 0·75 m/s away from O and the acceleration of A is 3·75 m/s^2 towards O.

Find the angular velocity and angular acceleration of the bar and the velocity and acceleration of the point C. Show the directions of these on the diagram. (I. Mech. E.)

(*Answer:* 2·93 rad/s; 11·3 rad/s^2; 0·675 m/s; 2·98 m/s^2)

28. In the mechanism shown in Fig. 3.72, the crank AB rotates in an anticlockwise direction at a uniform speed of 240 rev/min. The crank length AB is 100 mm while the lengths of BC and BD are respectively 400 mm and 500 mm.

Determine, for the position shown, the instantaneous difference in the velocity of C as compared with that of D and the instantaneous difference in the acceleration of C as compared with that of D, along their respective axes. (U. Lond.)

(*Answer:* 0·54 m/s; 35·5 m/s^2)

29. In the mechanism shown in Fig. 3.73, the crank BC, 100 mm long, turns at 300 rev/min about a centre C offset 75 mm from the horizontal line of stroke of A. The connecting rod BA = 300 mm. Find the length of stroke of A, and the times of inward and outward motion. For the position shown with BC turned 30° from the horizontal, find the velocity and acceleration of the point A. (I. Mech. E.)

(*Answer:* 207·5 mm; 0·106 s; 0·094 s; 1·35 m/s; 114 m/s^2)

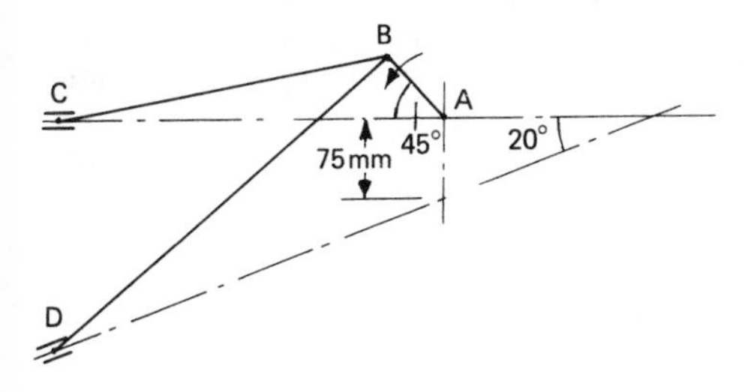

FIG. 3.72

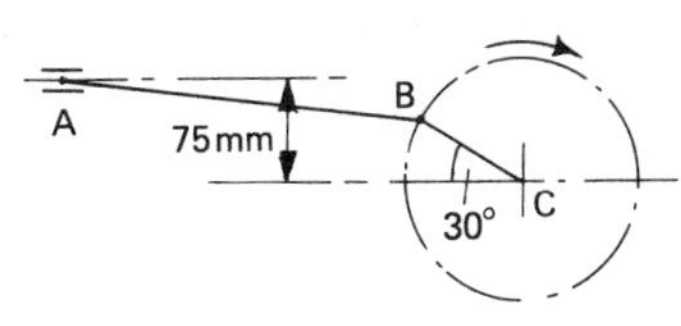

FIG. 3.73

30. Figure 3.74 shows a slider-crank mechanism in which the crank AB is 300 mm long and the connecting rod BC is 1050 mm long. In the position shown, where the crank is inclined at 45° to the vertical, the angular velocity of the crank is 30 rad/s clockwise and it has an angular retardation of 300 rad/s^2.

Find the acceleration of the slider at C and the angular acceleration of the connecting rod. (U. Lond.)

(*Answer:* 250 m/s^2; 119 rad/s^2)

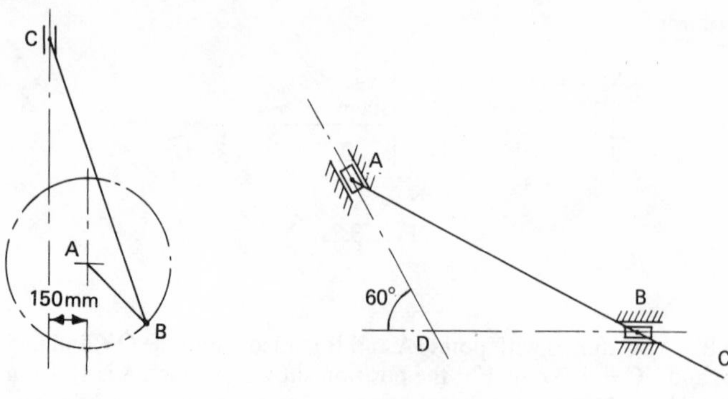

FIG. 3.74 FIG. 3.75

31. In the link ABC in Fig. 3.75, AB = 600 mm and BC = 225 mm. A and B are attached by pin joints to the sliding blocks. If, for the position where BD = 375 mm, A is sliding towards D with a velocity of 6 m/s and a retardation of 150 m/s², find the acceleration of C and the angular acceleration of the link. (U. Lond.)

(*Answer:* 259 m/s²; 294 rad/s²)

32. The connecting rod of an engine is 250 mm long and the stroke is 110 mm. When the crank is 45° past the i.d.c. position, the acceleration of the piston is 500 m/s² and the speed of the engine at this instant is 1000 rev/min. Find the angular acceleration of the crank and that of the connecting rod for this position of the crank. (U. Lond.)

(*Answer:* 1545 rad/s²; 1440 rad/s²)

33. The instantaneous position of a simple slider-crank mechanism OAB is shown in Fig. 3.76. At this instant, the velocity and acceleration of point B are both directed towards the left with magnitudes of 10 m/s and 800 m/s² respectively. OA = 0·1 m and AB = 0·4 m.

Using a graphical construction, determine: (a) the angular velocity of the crank OA, (b) the velocity of point A relative to point B in magnitude and direction, and (c) the absolute tangential component of acceleration of point A.

For the same configuration and velocities of the mechanism, what acceleration of point B is needed in order to reduce the absolute tangential component of acceleration of point A to zero? (C.E.I.)

(*Answer:* 120 rad/s; 8·6 m/s, upwards at 80° to horizontal; 272 m/s²; 1020 m/s²)

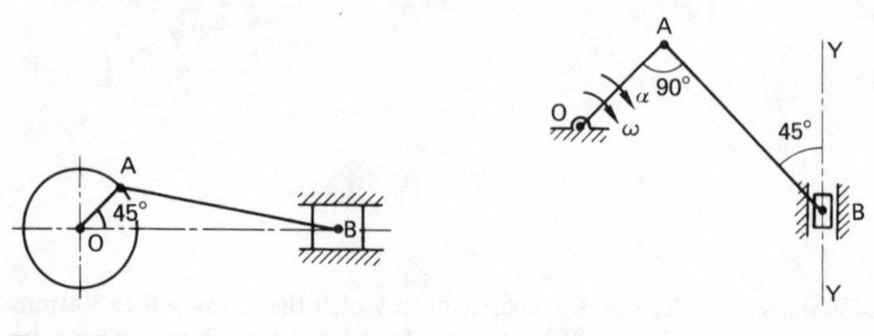

FIG. 3.76 FIG. 3.77

34. In the simple link mechanism shown in Fig. 3.77, the crank OA rotates about the fixed point O and the piston B slides along YY. The lengths of the links are OA = 75 mm and AB = 150 mm.

For the configuration shown, the crank has an angular velocity $\omega = 10$ rad/s and an angular acceleration $\alpha = 30$ rad/s^2, both in the clockwise direction. Sketch the vector diagrams for velocity and acceleration and determine: (a) the velocity of point B, (b) the angular velocity of link AB, and (c) the acceleration of point B. (C.E.I.)

(*Answer:* 1·06 m/s; 5 rad/s; 2·12 m/s^2)

35. The piston, connecting rod and crank of a reciprocating engine are shown in Fig. 3.78. At a particular instant in the cycle, the acceleration of the piston is 4680 m/s^2 (downwards) and the angular velocity of the connecting rod is $-87{\cdot}9$ rad/s (with the clockwise direction taken as positive) when the crank angle θ is 30°. OA = 50 mm and AB = 150 mm.

For this configuration: (a) sketch the velocity diagram and hence *calculate* the piston velocity and angular velocity of the crank, and (b) draw the acceleration diagram to a suitable scale and hence obtain graphically the angular accelerations of the connecting rod and the crank. (C.E.I.)

(*Answer:* 9·7 m/s; 300 rad/s, clockwise; 14 000 rad/s^2; zero)

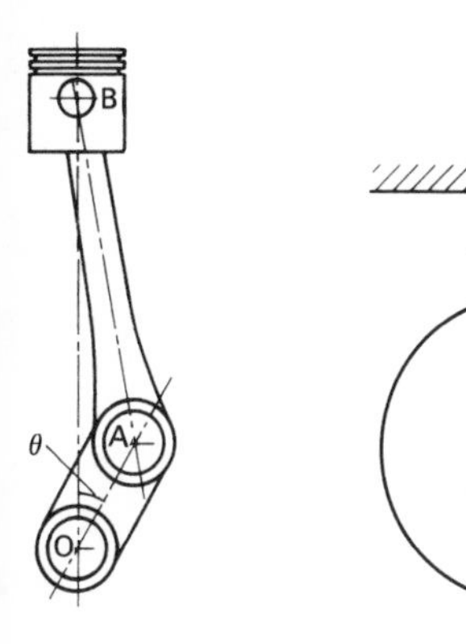

FIG. 3.78

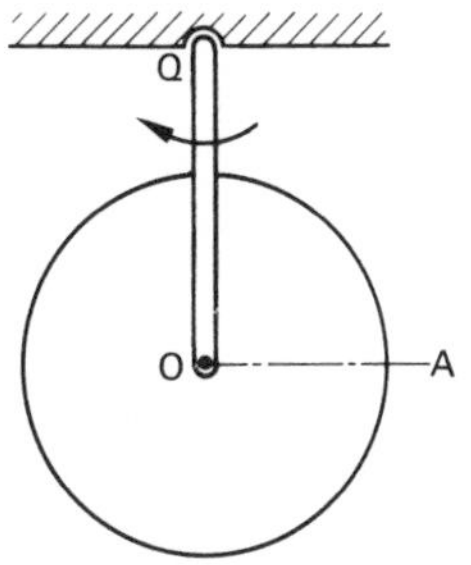

FIG. 3.79

36. The arm OQ, of length 3 m, rotates about a fixed point Q with a constant angular speed of 2 rad/s clockwise, as shown in Fig. 3.79. At the same time, the circular disc, of radius 2 m, rotates about its centre O at an angular speed of 3 rad/s anticlockwise and at an angular acceleration of 6 rad/s^2 clockwise, both *relative* to OQ. The two motions occur in the same plane.

Find the absolute acceleration, in magnitude and direction, of the point A on the periphery of the disc, given that the angle AOQ is 90° at the instant under consideration. (C.E.I.) (*Answer*: 18 m/s^2, to the left)

Examples: crank and connecting rod

Example 3.13 A vertical single-cylinder engine has a cylinder diameter of 250 mm and a stroke of 450 mm. The reciprocating parts have a mass of 180 kg, the connecting rod is 4 cranks long and the speed is 360 rev/min. When the crank has turned through an angle of 45° from the t.d.c., the net pressure on the piston is 1·05 MN/m^2. Calculate the effective turning moment on the crankshaft for this position. (U. Lond.)

Solution

From equation 3.14,

$$f_p = \omega^2 r\left(\cos\theta + \frac{\cos 2\theta}{n}\right)$$

$$= \left(\frac{2\pi}{60} \times 360\right)^2 \times 0{\cdot}225\left(\cos 45^\circ + \frac{\cos 90^\circ}{4}\right) = 226 \text{ m/s}^2$$

Alternatively, from Klein's construction, Fig. 3.80,

$$f_p = \omega^2 \text{ON} = \left(\frac{2\pi}{60} \times 360\right)^2 \times 0{\cdot}159 = 226 \text{ m/s}^2$$

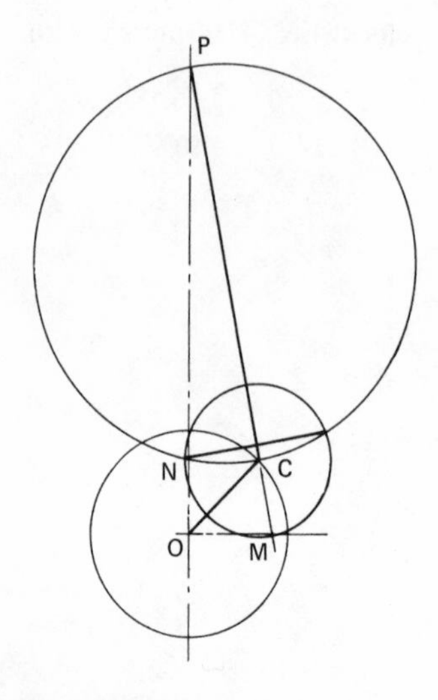

FIG. 3.80

Force to accelerate reciprocating parts $= Rf_p = 180 \times 226 = 40\,680$ N

$$\text{Gas force on piston} = 1{\cdot}05 \times 10^6 \times \frac{\pi}{4} \times 0{\cdot}25^2 = 51\,500 \text{ N}$$

$$\text{Dead weight of piston} = 180 \times 9{\cdot}81 = 1765 \text{ N}$$

$\therefore$ net force on piston,

$$F = 51\,500 + 1765 - 40\,680 = 12\,585 \text{ N}$$

$$\therefore T = F \times \text{OM} \qquad \text{from equation 3.18}$$

$$= 12\,585 \times 0{\cdot}19 = \underline{2390 \text{ N m}}$$

Example 3.14 An air compressor, 200 mm bore and 300 mm stroke, has a connecting rod 625 mm long. The reciprocating parts have a mass of 135 kg and the speed is 250 rev/min. When the crank is at 30° to the i.d.c. and moving inwards, the gauge pressure of the air is 840 kN/m². The crank pin is 60 mm diameter.

Neglecting the effect of the mass of the connecting rod, determine: (a) the rubbing velocity at the surface of the crank pin, (b) the force on the crankshaft bearing, and (c) the torque reaction on the frame. (U. Lond.)

Solution

(a) Referring to Fig. 3.81, $\omega = \frac{2\pi}{60} \times 250 = 26{\cdot}2 \text{ rad/s}$

$$\Omega = \omega \times \frac{\text{CM}}{\text{PC}} \quad \text{from equation 3.8}$$

$$= 26{\cdot}2 \times \frac{0{\cdot}131}{0{\cdot}625} = 5{\cdot}5 \text{ rad/s}$$

$$\therefore \text{ rubbing speed at crank pin} = (\Omega + \omega) \times \text{radius of pin}$$

$$= (5{\cdot}5 + 26{\cdot}2) \times 0{\cdot}03 = \underline{0{\cdot}95 \text{ m/s}}$$

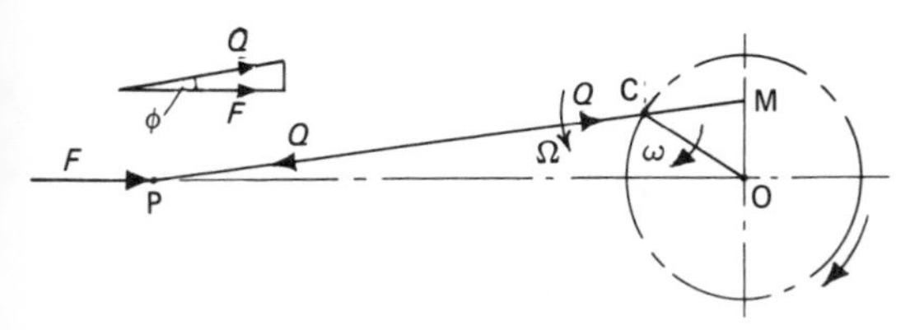

FIG. 3.81

(b) Piston acceleration,

$$f_p = \omega^2 r\left(\cos\theta + \frac{\cos 2\theta}{n}\right)$$

$$= 26{\cdot}2^2 \times 0{\cdot}15\left(\cos 30^\circ + \frac{150}{625}\cos 60^\circ\right) = 111{\cdot}4 \text{ m/s}^2$$

$\therefore$ net piston force,

$$F = pa - Rf_p = 840 \times 10^3 \times \frac{\pi}{4} \times 0{\cdot}2^2 - 135 \times 111{\cdot}4 = 11\,350 \text{ N}$$

The force on the crankshaft bearing at O is equal to the force Q in the connecting rod, which is applied to the crank at C. From the triangle of forces,

$$Q = F \sec\phi$$

$$\sin\phi = \frac{150 \sin 30^\circ}{625} = 0{\cdot}12$$

$$\therefore \phi = 6{\cdot}9^\circ$$

$$\therefore Q = 11\,350 \sec 6{\cdot}9^\circ = 11\,440 \text{ N} \quad \text{or} \quad \underline{11{\cdot}44 \text{ kN}}$$

(c) Torque reaction on frame,

$$T = F \times \mathrm{OM} \qquad \text{from equation 3.18}$$
$$= 11\,350 \times 0{\cdot}0907 = 1029\,\mathrm{Nm} \quad \text{or} \quad \underline{1{\cdot}029\,\mathrm{kN\,m}}$$

Example 3.15 A petrol engine of cylinder diameter 100 mm and stroke 120 mm has a piston of mass 1·1 kg and a connecting rod of length 250 mm. When rotating at 2000 rev/min, and on the explosion stroke with the crank at 20° from the t.d.c. position, the gas pressure is 700 kN/m².

Find: (a) the resultant load on the gudgeon pin, and (b) the thrust on the cylinder wall.

Determine also the speed above which, the other conditions remaining constant, the gudgeon pin load would be reversed in direction. (U. Lond.)

Solution

Piston acceleration,

$$f_p = \omega^2 r\left(\cos\theta + \frac{\cos 2\theta}{n}\right)$$
$$= \left(\frac{2\pi}{60} \times 2000\right)^2 \times 0{\cdot}06\left(\cos 20^\circ + \frac{0{\cdot}06}{0{\cdot}25}\cos 40^\circ\right) = 2960\,\mathrm{m/s^2}$$

∴ net piston force,

$$F = pa - Rf_p = 700 \times 10^3 \times \frac{\pi}{4} \times 0{\cdot}1^2 - 1{\cdot}1 \times 2960 = 2244\,\mathrm{N}$$

In Fig. 3.82,

$$\sin\phi = \frac{0{\cdot}06 \sin 20^\circ}{0{\cdot}25} = 0{\cdot}0821$$
$$\therefore\ \phi = 4{\cdot}72^\circ$$

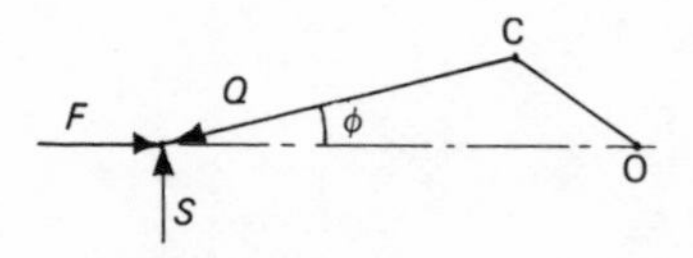

Fig. 3.82

(a) Resultant load on gudgeon pin,

$$Q = F\sec\phi = 2244 \sec 4{\cdot}72^\circ = \underline{2255\,\mathrm{N}}$$

(b) Thrust on cylinder wall,

$$S = F\tan\phi = 2244 \tan 4{\cdot}72^\circ = \underline{185\,\mathrm{N}}$$

For Q to be reversed, F must be reversed,

i.e.
$$R\omega^2 r\left(\cos\theta + \frac{r}{l}\cos 2\theta\right) > pa$$

i.e.
$$1{\cdot}1\left(\frac{2\pi}{60}N\right)^2 \times 0{\cdot}06\left(\cos 20^\circ + \frac{0{\cdot}06}{0{\cdot}25}\cos 40^\circ\right) > 700 \times 10^3 \times \frac{\pi}{4} \times 0{\cdot}1^2$$

from which
$$N > \underline{2600\,\mathrm{rev/min}}$$

Example 3.16 In a simple engine mechanism, AB is the crank and BC is the connecting rod. C is to the left of A and reciprocates horizontally along a line passing through A. AB = 60 mm, BC = 240 mm, AB turns at 600 rev/min and there is a reciprocating mass of 0·7 kg at C.
(a) Determine the crank angles, measured from i.d.c., from 0° to 360° for which the torque on the crankshaft due to the mass at C is zero.
(b) Determine the torques on the crankshaft due to the mass at C for crank angles of 20° and 210°, measured from i.d.c., stating clearly the sense in which they are acting. (U. Lond.)

Solution

(a) From equation 3.18,

$$T = F \times \text{OM}$$

Therefore the torque on the crankshaft is zero when either F or OM is zero.

OM is zero when $\theta = \underline{0°, 180° \text{ and } 360°}$

F is zero when $$R\omega^2 r\left(\cos\theta + \frac{\cos 2\theta}{n}\right) = 0$$

i.e. $$\cos\theta + \frac{\cos 2\theta}{4} = 0$$

i.e. $$2\cos^2\theta + 4\cos\theta - 1 = 0$$

from which $$\cos\theta = 0{\cdot}225$$

or $$\theta = \underline{77° \text{ and } 283°}$$

(b) $$T = R\omega^2 r\left(\cos\theta + \frac{\cos 2\theta}{n}\right) \times \text{OM}$$

$$= 0{\cdot}7\left(\frac{2\pi}{60} \times 600\right)^2 \times 0{\cdot}06\left(\cos\theta + \frac{\cos 2\theta}{4}\right) \times \text{OM}$$

$$= 166\left(\cos\theta + \frac{\cos 2\theta}{4}\right) \times \text{OM}$$

When $\theta = 20°$, $\text{OM} = 25{\cdot}7$ mm

$$\therefore\ T = 166\left(\cos 20° + \frac{\cos 40°}{4}\right) \times 0{\cdot}0257 = \underline{4{\cdot}82 \text{ N m}}$$

When $\theta = 210°$, $\text{OM} = 23{\cdot}3$ mm

$$\therefore\ T = 166\left(\cos 210° + \frac{\cos 420°}{4}\right) \times 0{\cdot}0233 = \underline{-2{\cdot}86 \text{ N m}}$$

The negative sign denotes an acceleration away from the crankshaft. In each case the torque is opposing the motion of the crank.

Problems: crank and connecting rod

Note Gravitational effects on piston mass are ignored unless the question specifies a vertical engine. Where necessary, atmospheric pressure has been taken as 100 kN/m^2.

37. A horizontal gas engine has a cylinder diameter of 250 mm and a stroke length of 500 mm. The length of the connecting rod is 1125 mm. When the engine is running at 250 rev/min and the crank is at 30° to the i.d.c. and moving outwards, the turning moment on the crankshaft is 5·25 kN m.

Determine for this position: (a) the effective pressure on the piston, (b) the sidethrust on the piston, and (c) the rubbing velocity at the surface of the crank pin which has a diameter of 100 mm. (I. Mech. E.) (*Answer:* 716·8 kN/m^2; 3·834 kN; 0·2535 m/s)

38. A vertical steam engine has a stroke of 600 mm and a cylinder diameter of 225 mm. The piston rod diameter is 50 mm and the length of the connecting rod is 1200 mm. The reciprocating parts have a mass of 112 kg. When the crank has rotated 60° from the t.d.c. position, the steam pressure on the top side of the piston is 560 kN/m^2 (gauge) and on the under side is 160 kN/m^2 (gauge).

For this position, calculate the thrust in the connecting rod, the thrust on the crosshead guide, the turning moment on the crankshaft and the radial force in the crank. The inertia forces may be neglected. (U. Lond.)

(*Answer:* 17·74 kN; 3·84 kN; 5·075 kN m; 5·335 kN)

39. The crank and connecting rod lengths of an air compressor are 150 mm and 600 mm respectively. The connecting rod has a mass of 183·5 kg and its centre of mass is 330 mm from the small end, the radius of gyration about the small end being 375 mm. The diameters of the gudgeon pin and crank pin are 67·5 mm and 105 mm respectively and the engine speed is 300 rev/min.

Determine, for a crank angle of 45° from the inner dead centre: (a) the K.E. of the connecting rod due to its angular velocity about its centre of mass, and (b) the surface rubbing speeds of the gudgeon and crank pins. (U. Lond.)

(*Answer:* 92·7 J; 0·193 m/s; 1·95 m/s)

40. In a vertical single-cylinder diesel engine, the reciprocating parts have a mass of 95 kg. At a point on the compression stroke when the angle between the crank pin and the t.d.c. is 120°, the air pressure in the cylinder is 1 bar gauge. The cylinder is 200 mm diameter, the connecting rod is 500 mm long, the stroke is 250 mm and the crankshaft is rotating at 240 rev/min.

Determine the crankshaft torque for this position. (*Answer:* 827 N m)

41. A horizontal reciprocating engine mechanism has a crank OA, 0·225 m long and a connecting rod AB, 1 m long. Viewing the mechanism with B to the left of O the crank rotates clockwise, and when it is at 30° to the i.d.c. and moving outwards, the angular velocity of the connecting rod is 5 rad/s.

For these conditions, determine: (a) the angular velocity of the crank, (b) the linear velocity of the centre point of the rod relative to that of the piston, and (c) the radial thrust and turning moment on the crankshaft per 100 N of effective force at B. (I. Mech. E.) (*Answer:* 25·5 rad/s; 2·5 m/s; 82 N; 13·5 N m)

42. A single-cylinder gas engine has a crank radius of 100 mm and the connecting rod length is 400 mm. The mass of the piston is 40 kg and the crankshaft rotates at 240 rev/min. The diameter of the cylinder is 120 mm and on the working stroke, the gas expands according to the law $pv^{1 \cdot 4} = 30$ where p is the gas pressure (absolute) in N/m^2 and v is the volume occupied by the gas in m^3. The clearance volume is $0{\cdot}5 \times 10^{-3}$ m^3.

Determine the torque on the crankshaft on the working stroke when the crank makes an angle of 45° from the top dead centre position. (*Answer:* 275 N m)

43. An air compressor has a stroke of 200 mm and connecting rod of length 400 mm; the radius of gyration about the centre of mass is 125 mm. The connecting rod has a mass of 120 kg and the effective reciprocating mass is 90 kg. The crank rotates at 240 rev/min.

Determine: (a) the crank torque due to the inertia of the reciprocating mass, and (b) the kinetic energy of rotation of the connecting rod for a crank angle of 45°. (U. Lond.) (*Answer:* 334 N m; 19·1 J)

44. An engine mechanism has a 150 mm crank radius and 375 mm connecting rod, with a piston of mass of 10 kg. The crank rotates at 60 rev/min. Draw the velocity diagram and the acceleration diagram for the crank position of 45° from the o.d.c. State the acceleration of the piston and calculate the crank torque due to inertia. (U. Lond.) (*Answer:* 4·19 m/s^2; 3·14 N m)

45. A horizontal single-cylinder engine has a cylinder diameter of 100 mm, a stroke of 120 mm and a connecting rod length of 250 mm. The mass of the piston is 0·7 kg and that of the connecting rod is 2 kg. The centre of mass of the connecting rod is 100 mm from the big end bearing centre and the effect of the connecting rod mass may be allowed for by apportioning it between the two ends.

When the crank is 20° from the top dead centre position on the working stroke, the gas pressure is 700 kN/m^2.

(a) Determine, for this position, the crankshaft torque when the speed is 2000 rev/min.
(b) At what speed would the crankshaft torque be zero for this crank position, the gas pressure remaining unchanged? (*Answer:* 26·7 N m; 2227 rev/min)

46. The crankshaft of a vertical single-cylinder engine, stroke 250 mm, rotates at 300 rev/min. The reciprocating parts (including part of the connecting rod) have a mass of 100 kg. The connecting rod has a mass of 120 kg; it is 450 mm long, the centre of mass is 300 mm from the gudgeon pin axis and the radius of gyration about the same axis is 363 mm.

When the crank is 30° from the t.d.c. position and moving downwards, find: (a) the sidethrust on the cylinder walls due to the inertia of the reciprocating parts, and (b) the total kinetic energy of the connecting rod. (U. Lond.) (*Answer:* 1·74 kN; 739 J)

4

Cams

4.1 Cams and followers

A plane or disc cam consists of a plate rotating about an axis perpendicular to its plane, its profile being such as to impart a reciprocating or oscillating motion to a follower, which bears against the cam edge. The follower, which moves in the same plane as the cam, may be roller-ended, flat-ended or knife-edged. It is kept in contact with the cam profile by gravity if the accelerations required are sufficiently small (i.e. less than g) or otherwise by a spring.

The principal types of cam are (a) those whose profiles are designed to give a specified motion to the follower, e.g. uniform acceleration or S.H.M., and (b) those whose profiles consist of straight lines, circular arcs and other mathematical curves. In type (a), the cam profile is obtained by geometrical construction; in type (b), the follower displacement for any cam angle is obtained analytically or graphically and the velocity and acceleration are then determined by mathematical* or graphical differentiation.

4.2 Specified motion of follower

Uniform acceleration and deceleration Let the follower lift through a distance s in time t while the cam rotates through a total angle θ; let the acceleration be f_1 during an angle of rotation θ_1 and the deceleration be f_2 during the remaining angle of rotation θ_2.

Then, if the angular speed of the cam is ω,

$$t = \frac{\theta}{\omega}$$

Also
$$\text{mean velocity} = \frac{s}{t} = \frac{\omega s}{\theta}$$

$$\therefore \text{ maximum velocity, } v = \frac{2\omega s}{\theta}$$

$$f_1 = \frac{v}{t_1} = \frac{2\omega s/\theta}{\theta_1/\omega} = \frac{2\omega^2 s}{\theta\theta_1}$$

* See *Mechanics of Machines—Advanced Theory and Examples*, Second Edition, J. Hannah and R. C. Stephens, Edward Arnold (1972).

and $$f_2 = \frac{v}{t_2} = \frac{2\omega s/\theta}{\theta_2/\omega} = \frac{2\omega^2 s}{\theta\theta_2}$$

If $\theta_1 = \theta_2 = \theta/2$, $$f_1 = f_2 = \frac{4\omega^2 s}{\theta^2}$$

The displacement, velocity and acceleration diagrams are shown in Fig. 4.1.

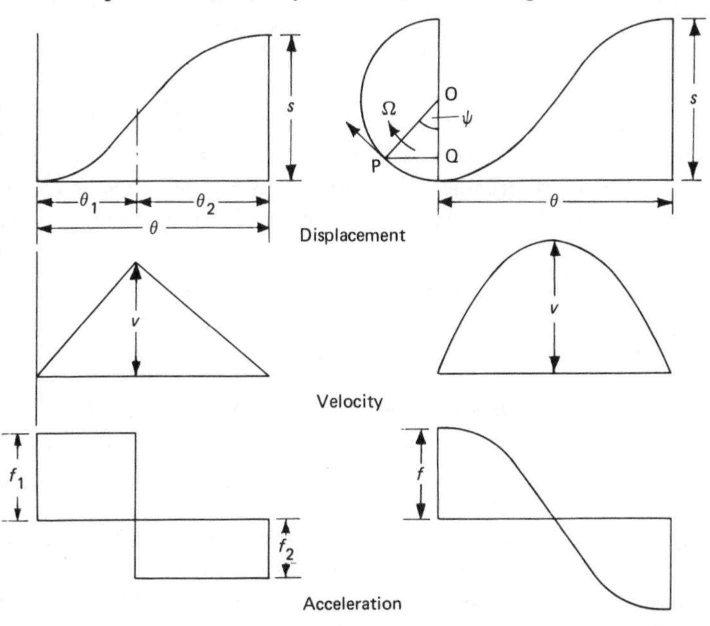

FIG. 4.1 FIG. 4.2

Simple harmonic motion The motion of the follower is identical with that of the projection Q, Fig. 4.2, on a diameter, of a point P, moving in a circular path with constant speed, the diameter being equal in length to the lift of the follower.

Let the follower lift through a distance s in time t while the cam rotates through an angle θ. Then, if Ω is the angular speed of OP,

$$\Omega = \frac{\pi}{t} = \frac{\omega\pi}{\theta}$$

$$\text{Tangential velocity of P} = \frac{\Omega s}{2} = \frac{\omega\pi s}{2\theta}$$

$$\therefore \text{ velocity of follower} = \text{velocity of Q} = \frac{\omega\pi s}{2\theta}\sin\psi$$

$$\text{Centripetal acceleration of P} = \Omega^2\frac{s}{2} = \frac{\omega^2\pi^2 s}{2\theta^2}$$

$$\therefore \text{ acceleration of follower} = \text{acceleration of Q} = \frac{\omega^2\pi^2 s}{2\theta^2}\cos\psi$$

$$\text{Maximum velocity, } v = \frac{\omega \pi s}{2\theta}$$

and

$$\text{maximum acceleration, } f = \frac{\omega^2 \pi^2 s}{2\theta^2}$$

The displacement, velocity and acceleration diagrams are shown in Fig. 4.2.

Examples

Example 4.1 A cam of minimum radius 50 mm, turning uniformly clockwise, is to give a lift of 30 mm to a follower which moves on a vertical straight line 20 mm to the right of the centre line of the camshaft, the follower being above the cam. The follower is fitted with a roller of 25 mm diameter and is to have the following motion

0°–180°	rising with S.H.M.
180°–270°	falling 15 mm with uniform acceleration
270°–360°	falling 15 mm with uniform deceleration

Draw the cam profile, full size, using intervals of 30°, and show the construction clearly. (U. Lond.)

Solution

The base circle, of radius 50 mm, the line of stroke of the follower, XX, and the follower roller in its lowest position are drawn, Fig. 4.3.

In constructing the cam profile, it is more convenient to regard the cam as fixed and the line of stroke of the follower to be rotating anticlockwise at constant speed round the cam. The subsequent positions of the follower stroke have therefore been set out at intervals of 30°.

The stroke of the follower, QQ, has also been divided geometrically into six divisions; the construction for S.H.M. being shown to the left and that for uniform acceleration and retardation to the right.* The various positions of the follower centres are then set off along the corresponding line of stroke; at 120°, for example, the distance $P_1 Q_1$ is made equal to PQ', and at 300° the distance $P_2 Q_2$ is made equal to PQ''. Using the centres so obtained, the follower circles are described and the cam profile drawn tangential to these circles.

Example 4.2 A cam with a roller follower is shown in Fig. 4.4., the roller being 40 mm diameter and carried on a lever pivoted at P. The base and nose circle radii of the cam are 40 mm and 20 mm respectively, the distance between their centres being 60 mm. The flanks of the cam are straight and tangential to both circles.

Plot a curve showing the angular displacement of the lever against that of the cam for the period of rise from the lowest position to the highest.

* See *Practical Geometry and Engineering Graphics*, W. Abbott, Blackie (1972), p. 62.

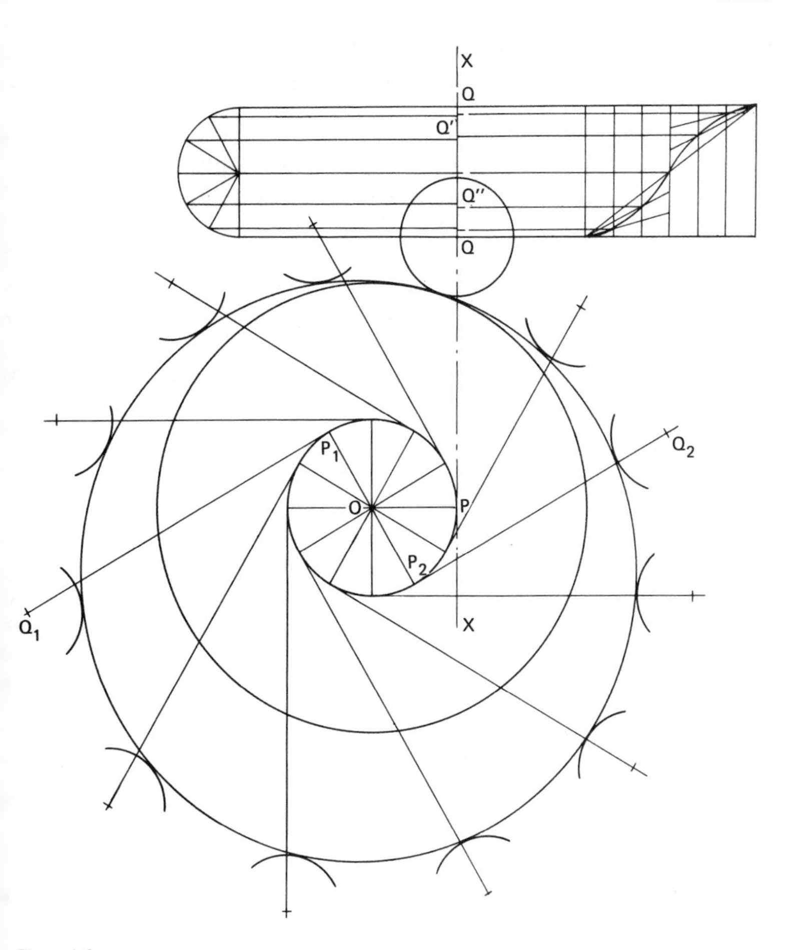

FIG. 4.3

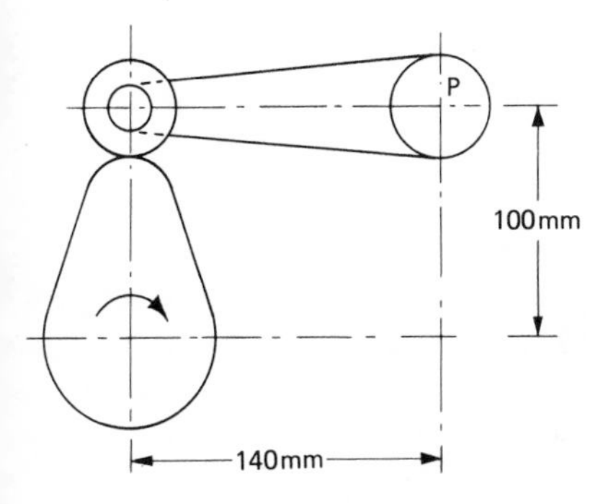

FIG. 4.4

If the velocity of the cam is uniform at 100 rev/min, what is the maximum angular velocity of the lever in rad/s during the period of lift? (U. Lond.)

Solution

The cam profile and follower are drawn for the given position, Fig. 4.5(a). Regarding the cam as fixed and the line OP to be rotating round the cam,

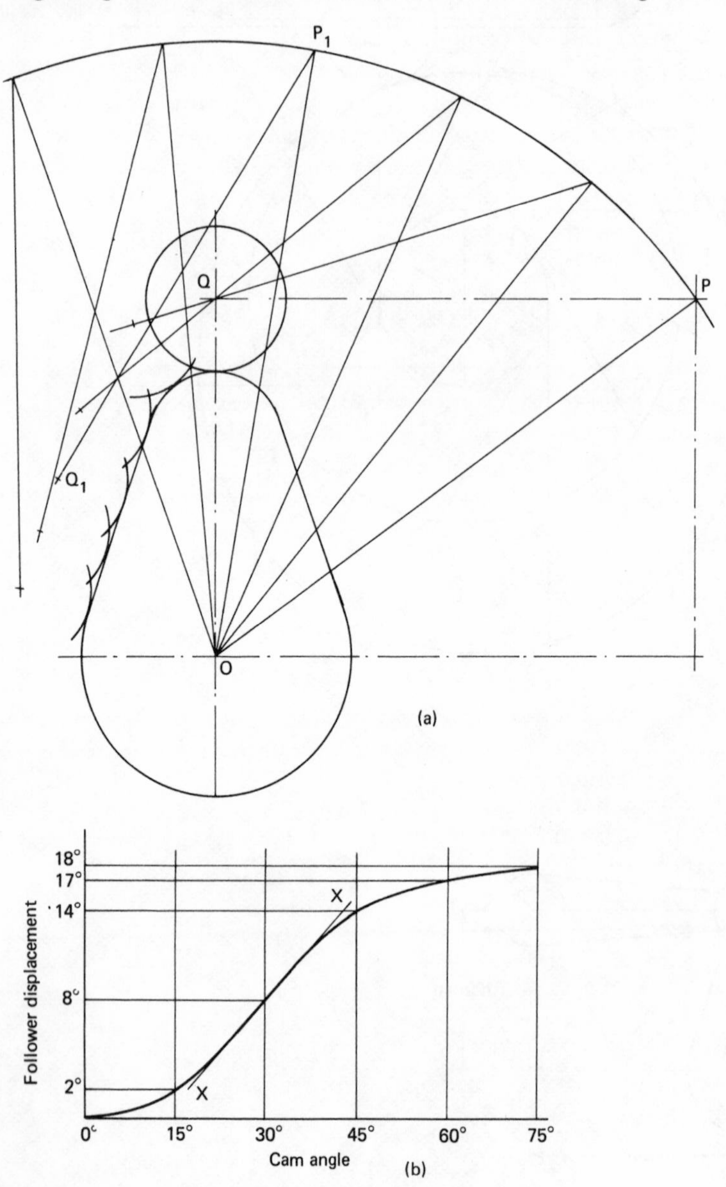

FIG. 4.5

subsequent positions of the follower pivot are obtained, angular displacements of 15° having been taken for the line OP.

From each position of the pivot, an arc of radius PQ is drawn and the centre of the follower is obtained by trial on this arc so that the follower circle is tangential to the cam profile. Thus, for a cam displacement of 45°, the position of the follower arm is represented by $P_1 Q_1$ and the angular displacement of the follower is then $\angle OPQ - \angle OP_1 Q_1$.

The angular displacements at various cam angles are shown graphically in Fig. 4.5 (b). The slope of this curve gives the ratio of the displacements of the lever and cam and hence the ratio of their angular velocities. Since the cam speed is uniform, the maximum angular velocity of the lever occurs at the point of maximum slope, i.e. at a cam angle of 30°.

$$\text{Slope of tangent, XX, to curve at this point} = 0{\cdot}485$$

$$\therefore \text{ maximum angular velocity of lever} = 0{\cdot}485 \times \left(\frac{2\pi}{60} \times 100\right) = \underline{5{\cdot}08 \text{ rad/s}}$$

Example 4.3 A cam rotates at 300 rev/min and operates a flat follower, the face of which is prependicular to the line of strcke. Draw to scale the velocity diagram for the follower during the lift and find the acceleration of the follower when the cam is at 35° below the top of the lift, which is at 80°. The shape of this portion of the cam profile is given in polar notation with origin at the centre of the base circle which is 65 mm diameter.

Angle	0°	10°	20°	30°	40°	50°	60°	70°	80°
Radius (mm)	32·5	33·0	33·5	35·0	37·5	40·8	44·5	48·3	48·8

(U. Lond.)

Solution

The cam profile, AA, is drawn from the given data, Fig. 4.6 (a). The positions of the follower end are then drawn tangential to the cam and perpendicular to the

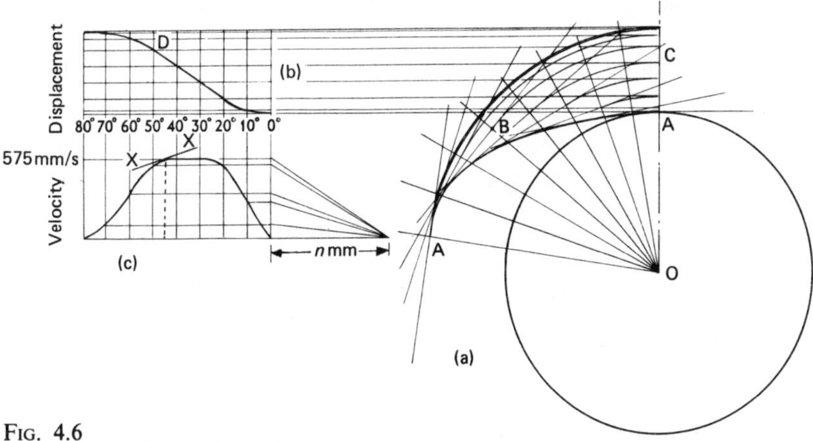

Fig. 4.6

radial construction lines which also serve as the successive positions of the follower axis.

The positions of the centre of the follower end are transferred to the vertical line of stroke and are then used to construct the displacement diagram, Fig. 4.6(b). Thus for a cam angle of 50°, OC is made equal to OB and a horizontal line through C then gives the corresponding point D in the displacement diagram. The velocity diagram, Fig. 4.6(c), is obtained by graphically differentiating the displacement diagram,* and the acceleration at 35° below the top of the lift is obtained from the slope of the tangent, XX, to the velocity diagram at this point.

Let the polar distance be n mm. Then, if the time scale is 1 mm $\equiv a$ s and the displacement scale is 1 mm $\equiv b$ mm, the scale of the velocity diagram is 1 mm $\equiv b/na$ mm/s.

Since the horizontal scale is in terms of crank angle, let 1 mm $\equiv \theta°$.

Then the time scale,

$$a = \left(\theta \times \frac{\pi}{180}\right) \bigg/ \left(\frac{2\pi}{60} \times 300\right) = \frac{\theta}{1800} \text{ s}$$

Thus the velocity scale is $\quad 1 \text{ mm} \equiv \dfrac{1800b}{n\theta} \text{ mm/s}$

By measurement,

$$\text{acceleration of follower at } 35° = \text{slope of tangent XX} \times \frac{\text{velocity scale}}{\text{time scale}}$$

$$\simeq \underline{8{\cdot}25 \text{ m/s}^2}$$

Example 4.4 The mechanism employed to reverse the direction of travel of the table of a planing machine is operated by a cam and lever as shown in Fig. 4.7. The cam is fixed to the table. At the end of the stroke the lever is raised through an angle of 30° while the cam moves horizontally a distance of 150 mm with uniform velocity. The lever is raised with uniform angular acceleration during the first 15° and with uniform angular retardation during the remaining 15°.

Draw the cam profile, full size. (U. Lond.)

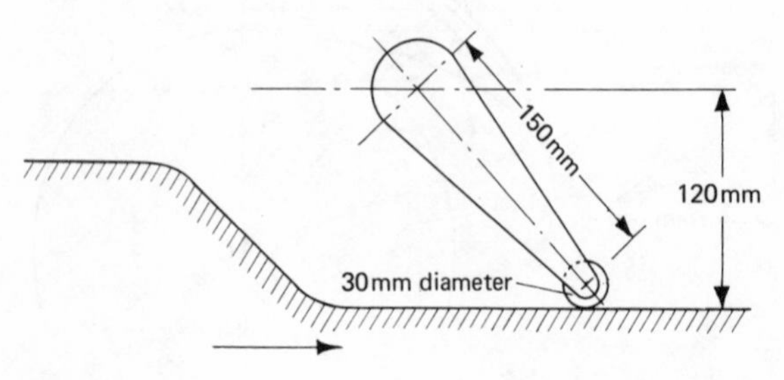

FIG. 4.7

* See *Practical Geometry and Engineering Graphics*, W. Abbott, Blackie (1972), p. 64.

Solution

Considering the cam as fixed and the fulcrum of the lever as moving to the left, the line of motion of the pivot, O, Fig. 4.8, is divided into six equal parts. The arc of movement of the roller, PQ, is then divided geometrically into six parts, corresponding to uniform acceleration and retardation of the follower; the construction has been applied to one half of the arc only, using the chord as a close approximation to the length of the arc.

Lines from the successive positions of O, parallel to the corresponding angular positions of the arm, are drawn to give the positions of the roller centres; thus $O_1 P_1$ is parallel to, and equal in length to, OP′. The roller circles are then constructed from these centres and the cam profile drawn tangential to the circles.

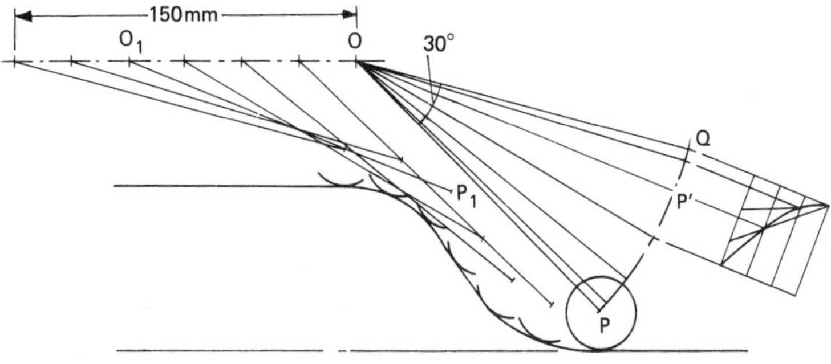

Fig. 4.8

Example 4.5 The flat-faced cam shown in Fig. 4.9 rotates about O at 60 rev/min. The roller follower has a mass of 0·8 kg. Calculate the vertical acceleration of the follower and the cam torque for both directions of rotation, when $\theta = 30°$. (U. Lond.)

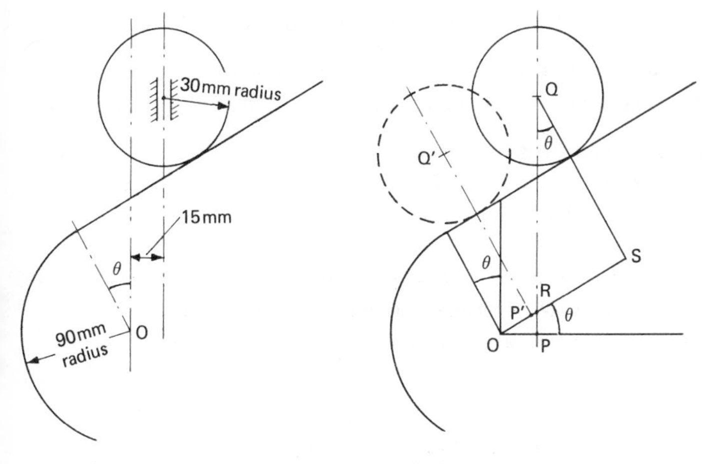

Fig. 4.9 Fig. 4.10

Solution

Referring to Fig. 4.10, the change of displacement, x, between positions Q′ and Q for a cam rotation θ is given by

$$x = \mathrm{PQ} - \mathrm{P'Q'} = (\mathrm{PR} + \mathrm{RQ}) - \mathrm{P'Q'}$$
$$= 15 \tan\theta + 120 \sec\theta - 120$$

$$\therefore\ v = \frac{\mathrm{d}x}{\mathrm{d}t} = (15 \sec^2\theta + 120 \sec\theta \tan\theta)\frac{\mathrm{d}\theta}{\mathrm{d}t}$$

$$= \omega(15 \sec^2\theta + 120 \sec\theta \tan\theta)$$

$$\therefore\ f = \frac{\mathrm{d}^2x}{\mathrm{d}t^2} = \omega(30 \sec^2\theta \tan\theta + 120 \sec^3\theta + 120 \tan^2\theta \sec\theta)\frac{\mathrm{d}\theta}{\mathrm{d}t}$$

$$= \omega^2 \sec\theta\,\{30 \sec\theta \tan\theta + 120(\sec^2\theta + \tan^2\theta)\}$$

$$\therefore\ \text{when } \theta = 30°,\quad f = (2\pi)^2 \frac{2}{\sqrt{3}}\left\{30\frac{2}{\sqrt{3}}\frac{1}{\sqrt{3}} + 120\left(\frac{4}{3} + \frac{1}{3}\right)\right\}$$

$$= \frac{8\pi^2}{\sqrt{3}} \times 220\ \mathrm{mm/s^2} = \underline{10\ \mathrm{m/s^2}}$$

$$\text{Inertia force on roller} = 0{\cdot}8 \times 10 = \underline{8\ \mathrm{N}}$$

When the cam rotates anticlockwise, the follower rises and accelerates upwards and thus the inertia force is downwards. When the cam rotates clockwise, the follower falls and decelerates downwards and thus the inertia force is again downwards. Therefore, for both directions of rotation,

$$\text{vertical force on roller} = 0{\cdot}8 \times 9{\cdot}81 + 8 = 15{\cdot}85\ \mathrm{N}$$

$$\therefore\ \text{force normal to flank (along SQ)} = 15{\cdot}85 \sec 30° = 18{\cdot}3\ \mathrm{N}$$

$$\therefore\ \text{cam torque} = 18{\cdot}3 \times \mathrm{OS} = 18{\cdot}3 \times (\mathrm{OR} + \mathrm{RS})$$
$$= 18{\cdot}3(0{\cdot}015 \sec 30° + 0{\cdot}12 \tan 30°) = \underline{1{\cdot}585\ \mathrm{N\,m}}$$

When the follower rises, the cam drives the follower. When the follower falls, the force exerted on the cam provides a torque assisting the rotation of the cam.

Example 4.6 A cam consists of a circular disc 320 mm diameter with centre B, and it rotates about a point A which is 80 mm from B. A roller follower of 80 mm diameter with centre C is constrained to move vertically (i.e. AC is a vertical).

Find, graphically or otherwise, the acceleration of C when AB makes 45° to the right of AC, the cam rotating anticlockwise at 60 rev/min.

What is the reaction on the cam if the roller has a mass of 0·9 kg?

(U. Lond.)

Solution

The acceleration of C, Fig. 4.11, may be obtained by constructing the displacement diagram for the roller and differentiating graphically twice.

The mechanism ABC, however, is kinematically identical with a simple

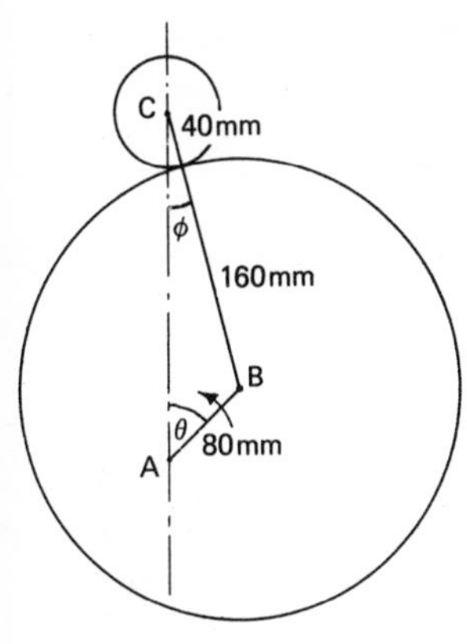

FIG. 4.11

reciprocating engine mechanism, AB being equivalent to the crank and BC to the connecting rod. The acceleration of C may therefore be obtained by the use of Klein's construction, Section 3.8, or by the analytical method of Section 3.9. In the latter case, however, the approximations based on the assumption that n was large are no longer justified since, in this case,

$$n = 200/80 = 2{\cdot}5$$

From Section 3.9,

$$x = (r+l)-(r\cos\theta + l\cos\phi)$$

and

$$\cos\phi = \sqrt{\left\{1-\left(\frac{\sin\theta}{n}\right)^2\right\}}$$

$$\therefore\ x = r(1-\cos\theta)+l\left[1-\sqrt{\left\{1-\left(\frac{\sin\theta}{n}\right)^2\right\}}\right]$$

$$= r\{1-\cos\theta+n-\sqrt{(n^2-\sin^2\theta)}\}$$

$$\therefore\ v = \frac{dx}{dt} = \omega r\left\{\sin\theta + \frac{\sin 2\theta}{2\sqrt{(n^2-\sin^2\theta)}}\right\}$$

$$\therefore\ f = \frac{d^2x}{dt^2} = \omega^2 r\left\{\cos\theta + \frac{\sin^4\theta + n^2\cos 2\theta}{(n^2-\sin^2\theta)^{3/2}}\right\}^*$$

$$= \left(\frac{2\pi}{60}\times 60\right)^2 \times 0{\cdot}08\left\{\cos 45^\circ + \frac{\sin^4 45^\circ + 2{\cdot}5^2\cos 90^\circ}{(2{\cdot}5^2-\sin^2 45^\circ)^{3/2}}\right\}$$

$$= \underline{2{\cdot}29\ \text{m/s}^2}$$

Inertia force on roller = $0{\cdot}9 \times 2{\cdot}29 = 2{\cdot}06$ N

* If the terms $\sin^4\theta$ and $\sin^2\theta$ are neglected, the approximate expression $f = \omega^2 r[\cos\theta + (\cos 2\theta)/n]$ is obtained, as in Section 3.9.

Therefore, in this case,

$$f = \left(\frac{2\pi}{60}\times 60\right)^2 \times 0{\cdot}08\left(\cos 45^\circ + \frac{\cos 90^\circ}{2{\cdot}5}\right) = \underline{2{\cdot}23\ \text{m/s}^2}$$

The roller is decelerating, so that the inertia force acts upwards against the dead load.

$$\therefore \text{ vertical force on roller} = 0{\cdot}9 \times 9{\cdot}81 - 2{\cdot}06 = 6{\cdot}77 \text{ N}$$

$$\therefore \text{ reaction on cam along CB} = \frac{6{\cdot}77}{\cos\phi} = 6{\cdot}77 \Big/ \sqrt{\left(1 - \frac{\sin^2 45^\circ}{2{\cdot}5^2}\right)} = \underline{7{\cdot}05 \text{ N}}$$

Problems

1. The exhaust valve of a diesel engine is operated by a cam designed to give a uniform acceleration of magnitude 1·5 times the uniform retardation during the opening period. During closure, the retardation is to be 1·5 times the acceleration. Opening and closing the valve each occupy 50° of cam rotation. If the minimum cam radius is 70 mm, valve lift 44 mm, and the follower, with a 38 mm diameter roller, operates in a line passing through the cam axis, draw the outline of the cam full size. (U. Lond.)

2. A push rod, operated by a cam, is to rise and fall with simple harmonic motion along an inclined straight path. The least radius of the cam is 50 mm and the push rod is fitted at its lower end with a roller 38 mm diameter. When in its lowest position, the roller centre is vertically above the cam axis. The maximum displacement of the roller is 50 mm in a direction 30° to the right of the vertical. The cam rotates at 100 rev/min in a clockwise direction. The time of lift is 0·15 s and the time of fall 0·10 s with a period of rest of 0·05 s at the upper position.

Draw the cam profile, full size. (U. Lond.)

3. The lift diagram for a follower operated by a cam is shown in Fig. 4.12; the follower is at rest for the remaining 210° of revolution. The follower moves in a straight line passing through the axis of rotation of the cam and carries a roller 45 mm diameter, the minimum radius of the cam being 70 mm.

Draw the cam profile, full size, leaving sufficient construction lines to indicate the method used. (U. Lond.)

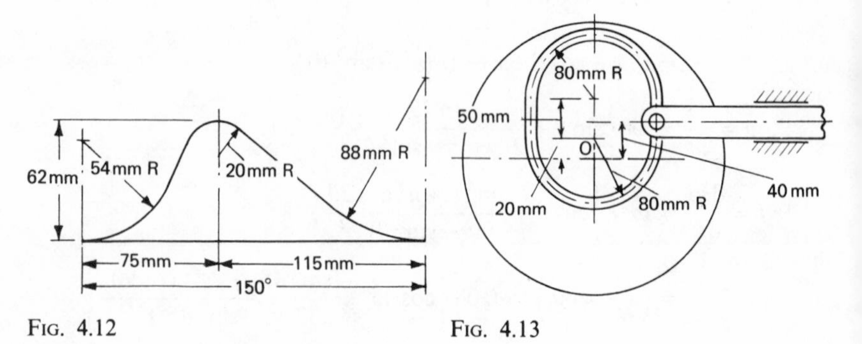

FIG. 4.12

FIG. 4.13

4. A box cam consists of a circular plate in which a slot is cut as shown in Fig. 4.13. The centre of rotation of the plate is at O and the slider is operated by a roller which fits into the slot without any slackness. Plot the displacement of the slider against cam angle, using a full-size scale for the displacement and a scale of 40 mm = 30° for the cam rotation. Take the given position as datum, and assume a clockwise direction of the cam. State the stroke of the slider.

If the speed of rotation is 1 rad/s, find the velocity of the slider when the cam has rotated 45° from the given position. (U. Lond.) (*Answer:* 25 mm; 0·0485 m/s)

5. A valve is required to lift 42 mm with equal uniform acceleration and retardation. The tappet which operates the valve is provided with a roller 48 mm diameter and its line of stroke intersects the axis of the cam at right angles. The cam turns through 60° during each stroke of the tappet and the shortest distance between the axis of the roller and the cam is 72 mm. Draw one-half of the outline of the cam.

If the cam turns at a uniform speed of 210 rev/min, find the acceleration and the maximum velocity of the valve. (I. Mech. E.) (*Answer:* 7·4 m/s^2; 1·76 m/s)

6. A circular disc of 180 mm diameter is mounted eccentrically on a shaft and used as a cam. The line of stroke of the follower passes through the shaft axis, the follower has a roller end, 45 mm diameter, and a stroke of 75 mm. The camshaft rotates uniformly at 400 rev/min.

Find, graphically or otherwise: (a) the acceleration of the follower when it is at its maximum displacement from the shaft centre line, and (b) the velocity and acceleration of the follower when the cam has rotated 60° from the position specified in (a). (U. Lond.) (*Answer:* 87·1 m/s^2; 1·59 m/s; 22 m/s^2)

7. A circular cam, 144 mm diameter, rotates clockwise at a uniform speed of 90 rev/min about a horizontal axis which is 18 mm from the centre of the cam. It actuates a follower fitted with a roller 36 mm diameter, the line of stroke of the follower being vertical, and displaced 24 mm to the right of the camshaft axis.
(a) Construct a lift diagram on a base of cam angle.
(b) Determine the maximum velocity of lift. (U. Lond.) (*Answer:* 0·193 m/s)

8. A cam operates a flat or palm follower. The follower moves in a straight line which is normal to its working face. The profile of the cam is formed by a base circle of 72 mm diameter joined by two tangents to a nose of 18 mm radius. The lift of the follower is 24 mm, and the speed of the cam is 30 rev/min about an axis which contains the centre of the base circle. Draw the displacement–time and velocity–time curves for the period of lifting only. (U. Lond.)

9. A cam rotates at 300 rev/min and operates a roller follower, which moves in a vertical line through the centre of the base circle of the cam (72 mm diameter). The shape of the cam profile for the first 90° is given in polar notation with origin at the centre of the base circle.

Angle	0°	15°	30°	45°	60°	75°	90°
Radius (mm)	36	36	39	45	54	57	72

Draw, full size, the profile of this part of the cam and construct the displacement curve for the roller follower, which is 24 mm diameter. State the maximum velocity of the follower. (U. Lond.) (*Answer:* 1·35 m/s)

10. A valve of an engine is operated by a cam and tappet. The line of motion of the valve is vertical; it does not pass through the camshaft axis but is offset 6 mm. The lower end of the tappet is fitted with a roller 18 mm diameter. The lift of the valve is 7·5 mm and occurs in 50° of camshaft rotation with a period of rest of 15°, followed by lowering during the next 50°. The valve is at rest for the remainder of the revolution.

If the acceleration of the valve is to be numerically constant, draw, four times full size, the cam profile if the least radius of the cam is 21 mm.

Find also the acceleration and maximum velocity of the valve when the camshaft speed is 500 rev/min. (U. Lond.) (*Answer:* 108 m/s^2; 0·9 m/s)

11. A cam rotating on a horizontal axis operates a vertical rod which is fitted with a roller 38 mm diameter at its lower end. The axis of the rod does not pass through the cam axis, but is offset 18 mm. The rod rises during 60° rotation of the cam, falls during the following 60°, and remains at rest for the following 240°. The rod moves with simple harmonic motion during both strokes, the lift being 50 mm.

Taking the least radius of the cam to be 62 mm, draw, full size, the profile of the cam. (U. Lond.)

12. A valve of a petrol engine is moved vertically by a cam which is in contact with the lower end of the valve stem. The valve is raised 5 mm while the cam rotates 45° and is lowered the same distance in the following 45°, the motion being simple harmonic. During the remaining 270° of cam rotation, the valve remains in its lowest position. The valve, which has a mass of 0·2 kg, is pressed down upon the cam by a spring which exerts a force of 900 N when the valve is in its highest position and 200 N when the valve is in its lowest position. The camshaft speed is 2000 rev/min.

Find the greatest speed of the valve and the force exerted by the cam on the valve at the highest and lowest positions. (U. Lond.) (*Answer:* 2·095 m/s; 549 N; 551 N)

13. The cam shown in Fig. 4.14 rotates uniformly in an anticlockwise direction about the fixed centre O, and actuates a follower through the roller of centre Q. The path of the roller centre lies on the straight line OX, contact between roller and cam being maintained by a spring.

Plot, on a cam angle base, the displacement of the centre Q for the first 240° from the position shown; mark on the curve the points reached by Q when the roller and cam are in contact at the 'change-over' points, A, B and C, and state the cam angle in each case. (U. Lond.) (*Answer:* 52°; 138°; 212°)

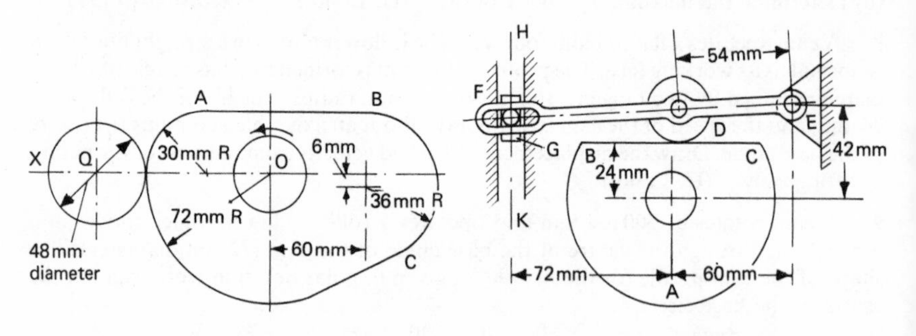

Fig. 4.14

Fig. 4.15

14. A cam A consists of a circular disc 96 mm diameter with a flat edge BC cut on it as shown in Fig. 4.15. The follower consists of a roller D, 24 mm diameter, mounted on an arm EF which is hinged at E and which moves a block G sliding in vertical guides HK. There is no interference between the arm EF and the cam. In the position shown, BC is horizontal.

(a) Find the angles through which the cam must rotate clockwise from the position shown in order to place G: (i) at its lowest position, and (ii) in its highest position.
(b) Find the total travel of G.
(c) Find the height of G above its lowest position when the cam has rotated 45° clockwise from the position shown.

A graphical solution is suggested. (U. Lond.)

(*Answer:* 10·5°; 69°; 61 mm; 15·8 mm)

15. Figure 4.16 shows an arm AB, 78 mm long, hinged to the underside of a slider which moves horizontally at 48 mm/s. The arm carries a roller 24 mm diameter, which rolls along the stationary profile CDE. This profile consists of two circular arcs of equal radii.

Draw a graph of angular displacement of AB against time. Find the average and the maximum angular velocities of AB while the roller rises from E to C. (U. Lond.)

(*Answer:* 0·31 rad/s; 0·86 rad/s)

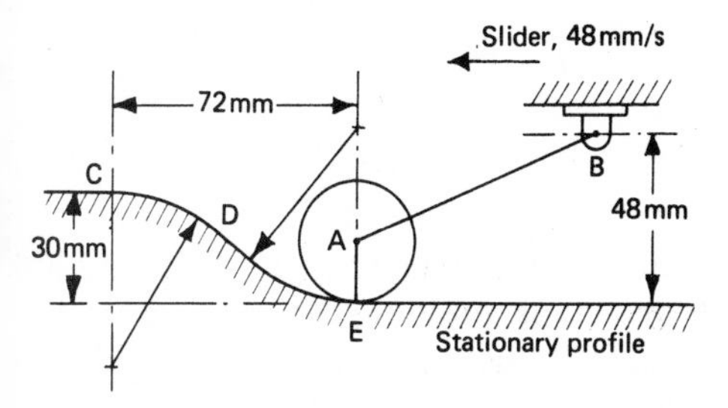

FIG. 4.16

16. A cam profile consists of two circular arcs, of radii 24 mm and 12 mm, joined by straight lines, giving the follower a lift of 12 mm. The follower is a roller of 24 mm radius and its line of action is a straight line passing through the camshaft axis. When the camshaft has a uniform speed of 500 rev/min, find the maximum value of the velocity and acceleration of the follower while in contact with the straight side of the cam. (U. Lond.) (*Answer:* 1·184 m/s; 198 m/s^2)

5

Crank effort diagrams

5.1 Introduction

The torque on an engine crankshaft varies considerably throughout the working cycle, due to variations in the crank position, the pressure in the cylinder and the inertia of the moving parts. If the crankshaft torque is plotted against the crank angle, a turning moment or crank effort diagram is obtained.

Figure 5.1 shows a crank effort diagram for a single-cylinder four-stroke engine; the net area of the diagram, shown shaded, represents the work done during the cycle and the average height represents the mean torque exerted, shown by the line AE. If the resisting torque is uniform, this is equal to the mean engine torque if the mean speed is to remain constant.

If the engine torque exceeds the resisting torque, such as between points A and B, the engine speeds up, and if it is less than the resisting torque, such as between points B and C, the engine slows down. At the points of intersection, A, B, C, etc., the engine and load torque are equal and hence there is no acceleration or deceleration of the crankshaft; thus the speed is a maximum or minimum at these points.

For multi-cylinder engines, the total torque for any crankshaft position is the algebraic sum of the torques exerted by the various cranks. These are not equal,

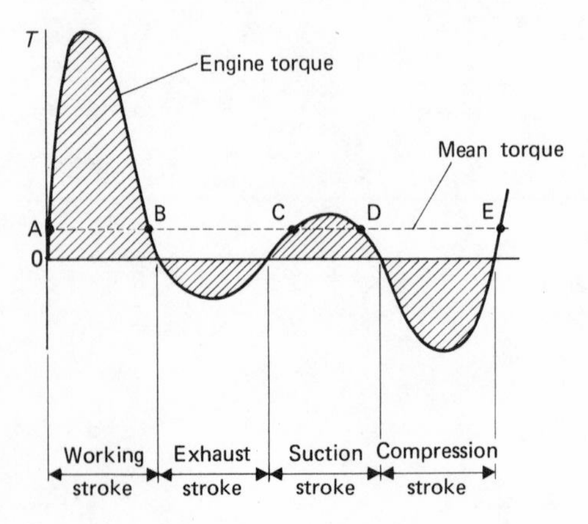

FIG. 5.1

since the cranks will have different angular positions relative to the crankshaft.

For presses and punches, etc., driven by electric motors, the input torque is constant and the resisting torque is fluctuating, so that the significance of the curved and straight lines in Fig. 5.1 becomes interchanged.

5.2 Fluctuation of speed and energy

Over a complete cycle, the sum of the areas of the loops above and below the mean engine torque line are equal. To keep the fluctuation of speed due to the variation of net torque within acceptable limits, a flywheel is fitted to the engine or machine shaft, its function being to act as a reservoir, absorbing energy as the speed increases and releasing it as the speed falls.

If the energy of the flywheel at point A, Fig. 5.2(a), is E, then

$$\text{energy of flywheel at point B} = E+a$$
$$\text{energy of flywheel at point C} = E+a-b$$
$$\text{energy of flywheel at point D} = E+a-b+c$$

and

$$\text{energy of flywheel at point E} = E+a-b+c-d = E$$

The energy at point E must be the same as the energy at point A since the cycle is then repeated.

By inspection of the areas of the loops above and below the mean torque line, the points of maximum and minimum energy may be obtained and these correspond with points of maximum and minimum speeds respectively. Thus for the engine cycle considered, the maximum and minimum speeds will occur where the engine torque curves cut the mean torque line on the working stroke, as shown in Fig. 5.2(b).

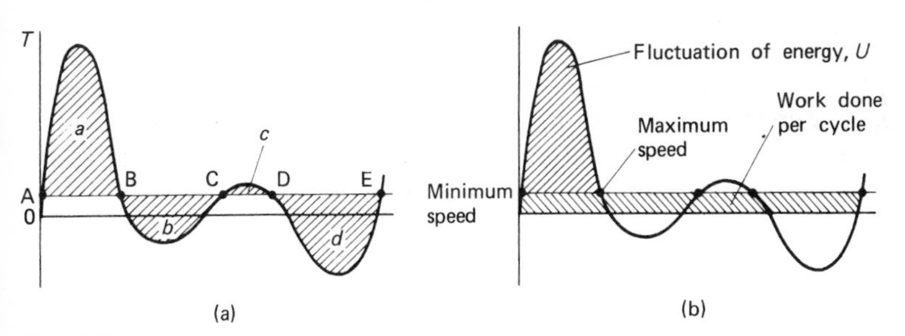

FIG. 5.2

The excess energy available between the points of minimum and maximum speeds is called the *fluctuation of energy* and this represents the difference between the kinetic energies of the system at these points.

Thus, if the fluctuation of energy is U and ω_1 and ω_2 are the maximum and minimum speeds respectively during the cycle, then

$$U = \tfrac{1}{2}I\omega_1^2 - \tfrac{1}{2}I\omega_2^2 \tag{5.1}$$

where I is the moment of inertia of the rotating parts.

The work done per cycle is represented by the area of the rectangle below the mean torque line and the ratio of fluctuation of energy to work done per cycle is called the *coefficient of fluctuation of energy.*

Equation 5.1 may be written

$$U = \tfrac{1}{2}I(\omega_1^2 - \omega_2^2)$$
$$= \tfrac{1}{2}I(\omega_1 + \omega_2)(\omega_1 - \omega_2)$$

The fluctuation of speed, $\omega_1 - \omega_2$, is small in comparison with the mean speed ω and, assuming that the variations above and below the mean speed are equal,

$$\omega_1 + \omega_2 = 2\omega$$
$$\therefore\ U = \tfrac{1}{2}I \times 2\omega(\omega_1 - \omega_2)$$
$$= I\omega(\omega_1 - \omega_2)$$

The ratio $(\omega_1 - \omega_2)/\omega$ is called the *coefficient of fluctuation of speed.* Hence

$$\frac{\omega_1 - \omega_2}{\omega} = \frac{U}{I\omega^2} \qquad (5.2)$$

Examples

Example 5.1 The turning-moment diagram for a petrol engine is drawn to the following scales: turning moment, 1 mm ≡ 5 N m; crank angle, 1 mm ≡ 1°. The turning-moment diagram repeats itself at every half-revolution of the engine and the areas above and below the mean turning-moment line, taken in order, are 295, 685, 40, 340, 960 and 270 mm². The rotating parts are equivalent to a mass of 36 kg at a radius of gyration of 150 mm.

Determine the coefficient of fluctuation of speed when the engine runs at 1800 rev/min. (U. Lond.)

Solution

The turning-moment diagram is shown in Fig. 5.3. By inspection, or by tabulating the energies at the various points A, B, C, etc., as in Section 5.2, the total energy is greatest at point B and least at point E.

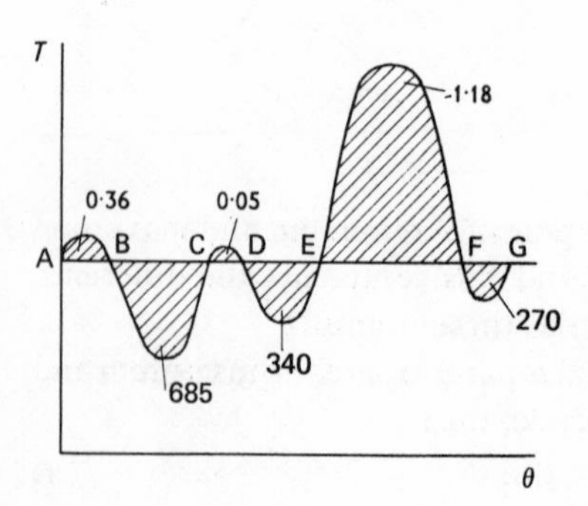

FIG. 5.3

The angle θ must be expressed in radians. Therefore

$$1\ \text{mm}^2 \text{ represents } 5 \times \pi/180 = \pi/36\ \text{J}$$

Therefore the fluctuation of energy between points B and E is given by

$$U = (-685 + 40 - 340) \times \pi/36 \quad \text{or} \quad (960 - 270 + 295) \times \pi/36$$
$$= 86\ \text{J}$$

$$\frac{\omega_1 - \omega_2}{\omega} = \frac{U}{I\omega^2} \qquad \text{from equation 5.2}$$

$$= 86 \Bigg/ \left[36 \times 0{\cdot}15^2 \times \left(\frac{2\pi}{60} \times 1800\right)^2\right]$$

$$= 0{\cdot}003 \quad \text{or} \quad \underline{0{\cdot}3 \text{ per cent}}$$

Example 5.2 The flywheel of a single-cylinder four-stroke diesel engine has a mass of 1 t with a radius of gyration of 0·8 m. The mechanical efficiency of the engine is 72 per cent when developing 30 kW at 400 rev/min. The area between the compression line and the atmospheric line of the indicator diagram is 0·52 of the net area of the diagram, and the suction and exhaust can be assumed to be at atmospheric pressure.

Determine the coefficient of fluctuation of speed of the engine at this load.

(U. Lond.)

Solution

The indicator diagram is shown in Fig. 5.4.

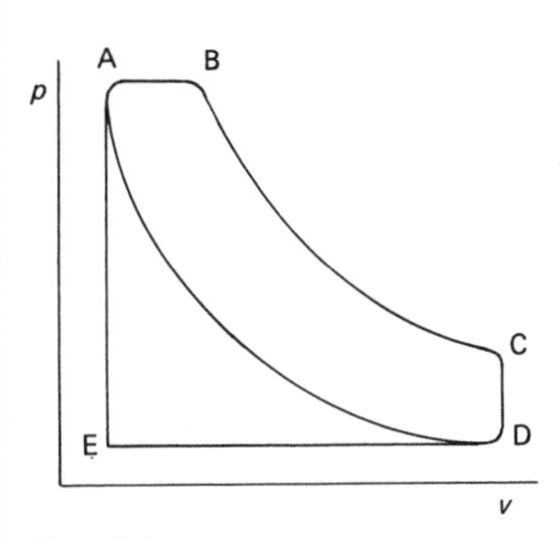

Fig. 5.4

$$\text{Output work per cycle} = \frac{30 \times 10^3 \times 60}{200} = 9000\ \text{J}$$

$$\therefore \text{ indicated work per cycle} = \frac{100}{72} \times 9000 = 12\,500\ \text{J}$$

$$\therefore \text{ average indicated work per stroke} = 12\,500/4 = 3125\ \text{J}$$

$$\text{Area EAD} = 0{\cdot}52 \times \text{area ABCD}$$

$\therefore$ work done during working stroke = area ABCDE

$$= 1{\cdot}52 \times \text{area ABCD}$$

$$= 1{\cdot}52 \times 12\,500 = 19\,000 \text{ J}$$

$\therefore$ excess energy absorbed by flywheel, $U = 19\,000 - 3125 = 15\,875$ J

$$\frac{\omega_1 - \omega_2}{\omega} = \frac{U}{I\omega^2} \qquad \text{from equation 5.2}$$

$$= 15\,875 \Bigg/ \left[1000 \times 0{\cdot}8^2 \times \left(\frac{2\pi}{60} \times 400\right)^2\right]$$

$$= 0{\cdot}0141 \quad \text{or} \quad \underline{1{\cdot}41 \text{ per cent}}$$

Example 5.3 The variation of crankshaft torque of a four-cylinder petrol engine may be approximately represented by taking the torque as zero for crank angles 0° and 180°, and as 260 N m for crank angles 20° and 45°, the intermediate portions of the torque graph being straight lines. The cycle is repeated in every half-revolution. The average speed is 600 rev/min. Supposing that the engine drives a machine requiring a constant torque, find the mass of the flywheel, of radius of gyration 250 mm, which must be fitted in order that the total variation of speed shall be 1 per cent. (U. Lond.)

Solution

The crankshaft torque diagram is shown in Fig. 5.5.

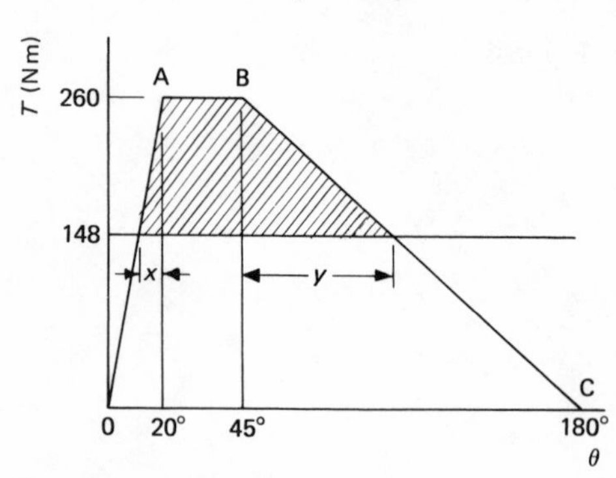

FIG. 5.5

Work done in $\frac{1}{2}$ revolution = area OABC

$$= 260 \times \left(25 \times \frac{\pi}{180}\right) + \frac{260}{2} \times \left(155 \times \frac{\pi}{180}\right)$$

$$= 465 \text{ N m}$$

$$\therefore \text{ mean torque} = 465/\pi = 148 \text{ N m}$$

$$x + y = \frac{112}{260} \times 155 = 66{\cdot}75°$$

Therefore the fluctuation of energy, represented by the shaded area in Fig. 5.5, is given by

$$U = 112 \times \left(25 \times \frac{\pi}{180}\right) + \frac{112}{2} \times \left(66{\cdot}75 \times \frac{\pi}{180}\right) = 114\ \text{J}$$

$$\frac{\omega_1 - \omega_2}{\omega} = \frac{U}{I\omega^2} \qquad \text{from equation 5.2}$$

$$\therefore\ I = 114 \Big/ \left[0{\cdot}01 \times \left(\frac{2\pi}{60} \times 600\right)^2\right] = 2{\cdot}888\ \text{kg m}^2$$

$$\therefore\ m = I/k^2 = 2{\cdot}888/0{\cdot}25^2 = \underline{46{\cdot}2\ \text{kg}}$$

Example 5.4 An engine working on the two-stroke cycle has three cylinders with cranks at 120°. The turning moment for any cylinder is assumed to increase uniformly from zero to a maximum while the crank turns 90°, to fall uniformly to zero over the next 90° and to remain zero over the remainder of the revolution.

If the engine develops 15 kW per cylinder when running at a mean speed of 400 rev/min, draw the turning-moment diagram for one cylinder and from it construct the combined diagram. Determine the variation in the kinetic energy of the flywheel and its required mass for a radius of gyration of 0·3 m to limit the total speed variation to 2 rev/min. (I. Mech. E.)

Solution

OABC is the diagram for one cylinder, Fig. 5.6.

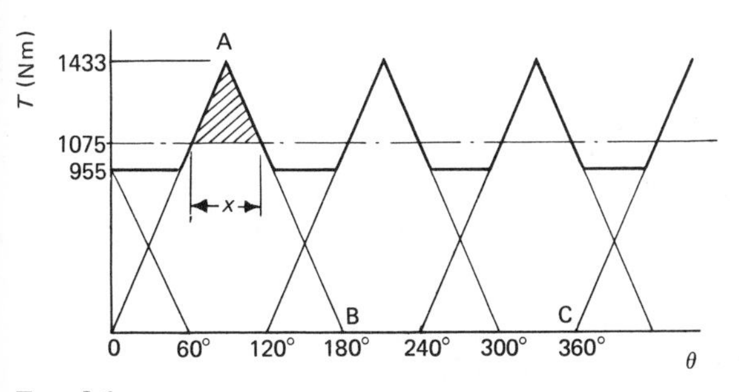

Fig. 5.6

$$\therefore\ \text{work done per cycle per cylinder} = \frac{15 \times 10^3 \times 60}{400} = 2250\ \text{J}$$

$$\therefore\ 2250 = \tfrac{1}{2} T_{max} \times \pi$$

$$\therefore\ T_{max} = 1433\ \text{N m}$$

$$\text{Mean torque for three cylinders} = \frac{3 \times 2250}{2\pi} = 1075\ \text{N m}$$

$$x = \frac{1433 - 1075}{1433} \times \pi = \frac{\pi}{4} \text{ rad}$$

Therefore the fluctuation of energy, represented by the shaded area in Fig. 5.6, is given by

$$U = \frac{(1433 - 1075) \times \pi/4}{2} = 140{\cdot}6 \text{ J}$$

$$U = I\omega(\omega_1 - \omega_2) \qquad \text{from equation 5.2}$$

i.e. $$140{\cdot}6 = m \times 0{\cdot}3^2 \times 400 \times 2 \times (2\pi/60)^2$$

$$\therefore\ m = \underline{178 \text{ kg}}$$

Example 5.5 A single-cylinder internal-combustion engine working on the four-stroke cycle develops 75 kW at 360 rev/min. The fluctuation of energy can be assumed to be 0·9 times the energy developed per cycle.

If the coefficient of fluctuation of speed is not to exceed 1 per cent and the maximum centrifugal stress in the flywheel is to be 5·5 MN/m^2, estimate the mean diameter and the cross-sectional area of the rim. Cast iron has a density of 7·2 Mg/m^3. (U. Lond.)

Solution

$$\omega = \frac{2\pi}{60} \times 360 = 12\pi \text{ rad/s}$$

$$\text{Work done per cycle} = \frac{75 \times 10^3 \times 60}{180} = 25\,000 \text{ J}$$

Therefore the fluctuation of energy is given by

$$U = 0{\cdot}9 \times 25\,000 = 22\,500 \text{ J}$$

$$\frac{\omega_1 - \omega_2}{\omega} = \frac{U}{I\omega^2} \qquad \text{from equation 5.2}$$

$$\therefore\ I = \frac{22\,500}{0{\cdot}01 \times (12\pi)^2} = 1584 \text{ kg m}^2$$

$$\text{Centrifugal stress, } \sigma = \rho v^2 = \rho\omega^2 r^2 *$$

where r is the mean rim radius,

i.e. $$5{\cdot}5 \times 10^6 = 7{\cdot}2 \times 10^3 \times 144\pi^2 \times r^2$$

$$\therefore\ r = 0{\cdot}732 \text{ m}$$

i.e. $$\text{mean diameter} = \underline{1{\cdot}464 \text{ m}}$$

$$I = mk^2 = \rho a \times 2\pi r \times r^2$$

where a is the cross-sectional area of the rim,

i.e. $$1584 = 7{\cdot}2 \times 10^3 \times a \times 2\pi \times 0{\cdot}732^3$$

$$\therefore\ a = \underline{0{\cdot}0892 \text{ m}^2}$$

* See *Strength of Materials*, Section 1.8, R. C. Stephens, Edward Arnold (1970).

Example 5.6 A single-cylinder four-stroke engine of 26 kW indicated power runs at 400 rev/min and drives a machine through gearing at 1200 rev/min. The moments of inertia of the rotating parts of the engine and the machine are 370 kg m^2 and 12 kg m^2 respectively. The excess energy developed by the engine is 80 per cent of the indicated work per cycle.

Calculate the coefficient of fluctuation of speed of the engine and then estimate how much additional flywheel rim mass, at 200 mm radius, is required on the machine to reduce this coefficient to 0·005. (U. Glas.)

Solution

$$\text{Equivalent moment of inertia of engine} = 370 + (1200/400)^2 \times 12^*$$
$$= 478\ \text{kg m}^2$$

$$\text{Work done per cycle} = \frac{26 \times 10^3 \times 60}{200} = 7800\ \text{J}$$

Therefore the fluctuation of energy is given by

$$U = 0{\cdot}8 \times 7800 = 6240\ \text{J}$$

$$\frac{\omega_1 - \omega_2}{\omega} = \frac{U}{I\omega^2} \qquad \text{from equation 5.2}$$

$$= 6240 \Big/ \left[478 \times \left(\frac{2\pi}{60} \times 400 \right)^2 \right]$$

$$= 0{\cdot}007\,44 \quad \text{or} \quad \underline{0{\cdot}744\ \text{per cent}}$$

If the coefficient of fluctuation of speed is reduced to 0·005,

$$\text{equivalent moment of inertia} = 478 \times \frac{0{\cdot}007\,44}{0{\cdot}005}$$
$$= 711{\cdot}5\ \text{kg m}^2$$

$$\therefore\ 711{\cdot}5 = 370 + (1200/400)^2 \{12 + m \times 0{\cdot}2^2\}$$

from which $\underline{m = 648\ \text{kg}}$

Example 5.7 The torque required by a machine rises parabolically from zero to a maximum of 150 N m and then falls again to zero in half a revolution. The torque remains at zero for the remainder of the revolution, after which the cycle is repeated, as shown in Fig. 5.7.

If the torque supplied by the driving motor is constant, find: (a) the power required at a mean speed of 240 rev/min, and (b) the moment of inertia of a flywheel attached to the machine shaft to keep the machine speed within $\pm$ 2 rev/min from the mean speed.

* See Section 1.24.

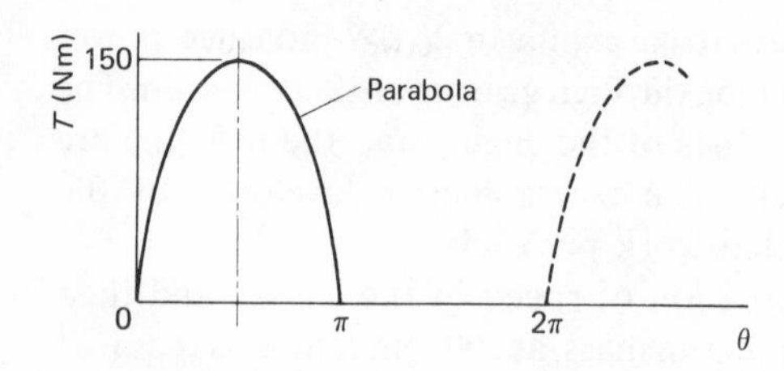

Fig. 5.7

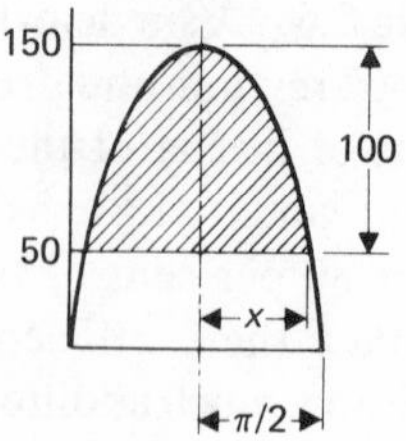

Fig. 5.8

Solution

(a) The area of a parabola is $\frac{2}{3}$ the area of the containing rectangle, so that

$$\text{mean torque} = \frac{\frac{2}{3} \times 150 \times \pi}{2\pi} = 50\,\text{N m}$$

$$\therefore \text{ power required} = 50\left(\frac{2\pi}{60} \times 240\right) = \underline{1258\,\text{W}}$$

(b) From the geometry of the parabola, Fig. 5.8,

$$\frac{100}{150} = \left(\frac{x}{\pi/2}\right)^2$$

from which $$x = 1{\cdot}283\,\text{rad}$$

The fluctuation of energy is then given by

$$U = \tfrac{2}{3}(2 \times 1{\cdot}283) \times 100 = 171\,\text{J}$$

$$U = I\omega(\omega_1 - \omega_2) \qquad \text{from equation 5.2}$$

$$\therefore\ 171 = I \times 240 \times 4 \times \left(\frac{2\pi}{60}\right)^2$$

from which $$I = \underline{16{\cdot}24\,\text{kg}\,\text{m}^2}$$

Problems

1. Distinguish between the functions of the governor and the flywheel of an engine.

An engine runs at 100 rev/min and a curve of the turning moment plotted on a crank angle base showed the following areas alternately above and below the mean turning-moment line: 780, 400, 520, 620, 260, 460, 340 and 420 mm^2. The scales used were 1 mm = 400 N m and 1 mm = 1° crank angle.

If the total fluctuation in speed is limited to 1·5 per cent of the mean speed, determine the mass of the flywheel necessary if the radius of gyration is 1·05 m. (U. Lond.)

(*Answer:* 3464 kg)

2. The turning-moment diagram for an engine is drawn on a base of crank angle and the mean resisting torque line added. The areas above and below the mean line are +4400, −1150, +1300 and −4550 mm^2, the scales being 1 mm = 100 N m torque and 1 mm = 1° crank angle.

Find the mass of flywheel required to keep the speed between 297 rev/min and 303 rev/min, if its radius of gyration is 0·525 m. (U. Lond.) (*Answer:* 1460 kg)

3. The turning-moment diagram of an engine, which has been drawn to scales of 1 mm to 50 N m and 1 mm to 1° of rotation of the crankshaft, shows that the greatest amount of energy which has to be stored by the flywheel is represented by an area of 2250 mm^2. The flywheel is to run at a mean speed of 240 rev/min with a total speed variation of 2 per cent. If the mass of the flywheel is to be 450 kg, determine suitable dimensions for the rim, the internal diameter being 0·9 of the external diameter.

Neglect the inertia of the arms and hub of the wheel. Cast iron has a density of 7·2 Mg/m^3. (U. Lond.) (*Answer:* 1·235 m external diameter; 275 mm thick)

4. A vertical diesel engine running at 350 rev/min develops 600 kW and has 4 impulses per revolution. If the fluctuation of energy is 25 per cent of the work done during each impulse, estimate the cross-sectional area of the rim of the flywheel required to keep the speed within 2 rev/min of the mean speed when the mean peripheral speed of the rim is 1350 m/min. Cast iron has a density of 7·2 Mg/m^3. (U. Lond.)

(*Answer:* 40 000 mm^2)

5. Establish an expression for the radius of gyration of an annular ring of given external and internal radii.

A disc-type cast-iron flywheel for a small oil engine is limited by engine framing and clutch requirements to the following sizes: rim, 400 mm external and 350 mm internal diameters and 50 mm long; hub, 100 mm outside diameter, 50 mm bore and 38 mm long; web, 12 mm thick. The turning-moment diagram for the engine shows that the maximum fluctuation of energy is 480 J. Determine the coefficient of speed fluctuation for a mean speed of 1500 rev/min.

Take the density of cast iron as 7·2 Mg/m^3. (U. Lond.)

(*Answer:* $\sqrt{[\frac{1}{2}(R^2+r^2)]}$; 3·86 per cent)

6. A motor driving a punching machine exerts a constant torque of 675 N m on the flywheel, which rotates at an average speed of 120 rev/min. The punch operates 60 times per minute, the duration of the punching operation being 0·20 s. It may be assumed that the punching effort is equivalent to a constant torque on the flywheel, and that friction and other losses may be neglected.

Deduce the value of the resisting torque and sketch graphs of input and output torques on a base representing the angle of rotation of the flywheel. Hence deduce the moment of inertia of the flywheel required if the speed variation from maximum to minimum is not to exceed 10 rev/min. (U. Lond.)

(*Answer:* 3·375 kN m; 516 kg m^2)

7. A machine press is worked by an electric motor, delivering 2·25 kW continuously. At the commencement of an operation, a flywheel of moment of inertia 50 kg m^2 on the machine is rotating at 250 rev/min. The pressing operation requires 4·75 kJ of energy and occupies 0·75 s. Find the maximum number of pressings that can be made in 1 h and the reduction in speed of the flywheel after each pressing. Neglect friction losses. (U. Lond.) (*Answer:* 1705; 23·5 rev/min)

8. A single-cylinder gas engine, working on the four-stroke cycle, develops 11 kW at 300 rev/min. The work done on the gases during the compression stroke is 0·7 times the work done by the gases during the power stroke. The turning-moment diagrams for the compression and power strokes may be taken as triangles and the turning moment during the suction and exhaust strokes is negligible.

(a) Determine the maximum fluctuation of energy.

(b) If the speed is not to fluctuate more than 2·5 per cent above or below the mean speed, find the mass of flywheel required, assuming its radius of gyration to be 450 mm. (U. Lond.) (*Answer:* 13·58 kJ; 1358 kg)

9. The motion of a machine shaft is resisted by a load torque which increases uniformly from 100 N m to 300 N m during a half-revolution, remains constant for the next revolution, decreases uniformly to 100 N m during the next half-revolution, and again remains constant for the next two revolutions, the cycle being repeated. The shaft is driven at a mean speed of 500 rev/min by an electric motor which provides a constant torque. The moment of inertia of the motor is 2 kg m^2 and fitted to the shaft is a flywheel having mass 80 kg and radius of gyration 0·8 m^2.

Determine: (a) the coefficient of fluctuation of speed, and (b) the mean motor power. (C.E.I.) (*Answer:* 0·0707 per cent; 9·163 kW)

10. (a) An automatic stamping machine stamps out components at the rate of one every half-second. The torque required at the motor is 10·0 N m for 0·2 s followed by 1·0 N m for 0·3 s, and the variation in torque with time is repeated every half-second. Given that the mean operating speed of the motor is 3000 rev/min, determine the mean power of the motor.

(b) A flywheel is fixed to the motor shaft in order to limit the speed fluctuations to $\pm$ 1·0 per cent. Determine the moment of inertia of the flywheel, given that the moment of inertia of the remainder of the machine is negligible. (C.E.I.)

(*Answer:* 1·445 kW; 0·172 kg m^2)

11. An electric motor drives a wrapping machine through a simple drive shaft at a mean speed of 40 rad/s. The total inertia referred to the drive shaft is 0·20 kg m^2. During one $\frac{1}{4}$ revolution of the shaft, the torque required by the machine increases uniformly with displacement from zero to 50 N m. During the next $\frac{1}{4}$ revolution, the machine requires no torque. A uniform torque of 25 N m is required during the next $\frac{1}{4}$ revolution, and no torque during the next $\frac{1}{4}$ revolution. The cycle is repeated continuously. The motor may be supposed to generate a uniform torque.

At what angular positions do the maximum and minimum speeds of the shaft occur? Give reasons for your answers. Determine these maximum and minimum speeds. Also find the speeds at the beginning and end of the third $\frac{1}{4}$ revolution. (C.E.I.)

(*Answer:* $\frac{1}{16}$ rev; $\frac{1}{4}$ rev; 395·2 rev/min; 368·8 rev/min; 392·3 rev/min; 368·8 rev/min)

12. A single-cylinder four-stroke internal combustion engine develops 30 kW at 300 rev/min. The turning-moment diagram for the expansion and compression strokes may be taken as two isosceles triangles, on bases 0 to π and 3π to 4π radians respectively, and the net work done during the exhaust and inlet strokes is zero. The work done during compression is negative and is one-quarter of that during expansion.

Sketch the turning-moment diagram for one cycle and find the maximum value of the turning moment during expansion.

If the load remains constant, mark on the diagram the points of maximum and minimum speeds. Also find the moment of inertia, in kg m^2, of a flywheel to keep the speed fluctuation within $\pm$ 1·5 per cent of the mean speed. (U. Lond.)

(*Answer:* 10·18 kN m; 444 kg m^2)

13. Figure 5.9 shows the variation with time of the torque required on the driving shaft of a machine during one cycle of operations. The shaft is direct-coupled to an electric motor which exerts a constant torque and runs at a mean speed of 1500 rev/min. The rotating parts are equivalent to a flywheel of mass 18 kg with a radius of gyration of 250 mm.

Determine: (a) the power of the motor, neglecting friction, and (b) the percentage fluctuation of speed. (U. Lond.) (*Answer:* 2·075 kW; 6·275 per cent)

14. A machine is engaged on repetitive work, each cycle of operations occupying one revolution of its shaft. The driving torque required on the machine shaft is 30 N m during 100° of its rotation, 12 N m during the next 120° and 3 N m for the remainder of

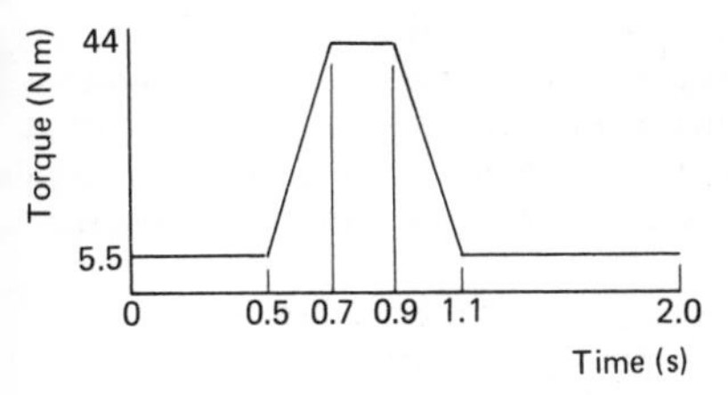

FIG. 5.9

the revolution. The rotating parts on the machine shaft have moment of inertia 3 kg m^2. The machine is driven through gearing by an electric motor whose speed is 4 times that of the machine shaft. The rotating parts on the motor shaft have moment of inertia 0·1 kg m^2.
(a) Assuming that the motor develops a constant torque, find its magnitude.
(b) If the maximum speed of the motor during the cycle of operations is 500 rev/min, find its minimum speed. (U. Lond.) (*Answer:* 3·375 N m; 481·2 rev/min)

15. A gas engine develops 22·5 kW at 270 rev/min. It has hit-and-miss governing and there are 125 explosions per minute. The flywheel has a mass of 900 kg and a radius of gyration of 0·675 m. If it is assumed that the work done is identical for each working cycle, that the work done by the gases on the explosion stroke is 2·4 times the work done on the gases during compression stroke, and that the work done on the other two strokes is negligible, find the maximum fluctuation of speed of the flywheel as a percentage of the mean speed. (I. Mech. E.) (*Answer:* 4·82 per cent)

16. Calculate the percentage fluctuation of speed in a four-stroke cycle engine from the following data: cylinder diameter, 200 mm; stroke, 400 mm; speed, 250 rev/min; mass of rotating parts, 1200 kg; radius of gyration of rotating parts, 0·6 m; mean pressure in the cylinder during the expansion stroke, 750 kN/m^2, and during the compression stroke, 170 kN/m^2 above atmosphere. The pressure during suction and exhaust may be neglected and the resistance at the crankshaft assumed constant. (U. Lond.)
(*Answer:* 2·57 per cent, assuming turning-moment diagrams for expansion and compression to be rectangular)

17. A rolling mill is to be driven by an electric motor giving a steady output of 150 kW, independent of speed. A heavy flywheel is directly coupled to the motor to take peak loads and is in the form of a solid cast-iron disc, 250 mm thick.

During a particular period of 9·5 s, the power absorbed by the mill is 375 kW for 2 s, 60 kW for 4 s, and 300 kW for 3·5 s.

Find the minimum diameter of flywheel required if the speed is allowed to fall, during the 9·5 s, from 500 to 450 rev/min. Ignore the inertia of the motor armature and mill and take the density of cast iron as 7·2 Mg/m^3. (U. Lond.) (*Answer:* 1·912 m)

18. An engine has three single-acting cylinders, the cranks being spaced 120° apart. For each cylinder, the crank effort diagram consists of a triangle:

Angle	0°	60°	180°	180°–360°
Torque (N m)	0	200 (max)	0	0

Find: (a) the mean torque, and (b) the moment of inertia of the flywheel in kg m^2 necessary to keep the speed within 180 ± 3 rev/min. (U. Lond.)
(*Answer:* 150 N m; 2·21 kg m^2)

19. A shaft fitted with a flywheel rotates at 250 rev/min and drives a machine, the resisting torque of which varies in a cyclic manner over a period of three revolutions. The torque rises from 675 N m to 2700 N m in a uniform manner during $\frac{1}{2}$ revolution and remains constant for the following 1 revolution. It then falls uniformly to 675 N m during the next $\frac{1}{2}$ revolution and remains constant for 1 revolution, the cycle being then repeated.

If the driving torque applied to the shaft is constant and the flywheel has a mass of 450 kg and a radius of gyration of 0·6 m, find the power necessary to drive the machine and the percentage fluctuation of speed. (U. Lond.)

(*Answer:* 44·2 kW; $\pm$ 3·58 per cent)

)rs

6.1 Function of a governor

The function of a governor is to control the mean speed of an engine, as distinct from that of a flywheel, which controls only cyclic fluctuations in speed. In the centrifugal types of governor, the effect of centrifugal force on the rotating balls causes a sleeve to rise until equilibrium is obtained. The sleeve controls the fuel supply by means of a linkage and any change in engine speed produces a change in the sleeve position, which adjusts the fuel supply accordingly.

6.2 Dead weight governors

Watt governor, Fig. 6.1(a) The Watt governor is basically a conical pendulum with the lower links attached to a sleeve of negligible mass. If the mass of the ball is m and the centrifugal force upon it is F, then, taking moments about the instantaneous centre for the lower link, I, to eliminate the effect of the tension in the upper arm,

$$F \times XZ = mg \times IZ$$

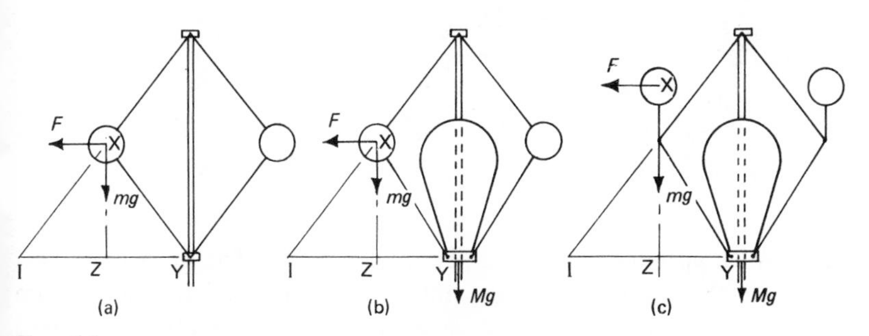

FIG. 6.1

Porter governor, Fig. 6.1 (b) The Watt governor is only suitable for a speed range of about 60–80 rev/min, and for higher speeds it is necessary to add a mass to the sleeve to increase the speed range for a given range of ball radius; this is the basis of the Porter governor. If the mass on the sleeve is M, then, taking moments about I to eliminate the effect of the tension in the upper arm and the reaction at the sleeve,

$$F \times \text{XZ} = mg \times \text{IZ} + \frac{Mg}{2} \times \text{IY}$$

Proell governor, Fig. 6.1 (c) The Proell governor is similar to the Porter governor, except that the balls are attached to extensions to the lower arms. This has the effect of reducing the change of speed necessary for a given sleeve movement. Taking moments about I,

$$F \times \text{XZ} = mg \times \text{IZ} + \frac{Mg}{2} \times \text{IY}$$

6.3 Spring-loaded governors

Hartnell governor, Fig. 6.2 (a) The governor is assembled with the central spring initially compressed and this compression can be adjusted to give any required equilibrium speed for a given ball radius. If F is the centrifugal force acting on the ball, P is the force exerted on the sleeve by the spring and M is the mass of the sleeve, then, taking moments about the fulcrum, O, of the bell-crank lever,

$$F \times a = \frac{P + Mg}{2} \times b$$

When the ball arm is not vertical, the moment of the ball weight about the fulcrum and the changes in length of the moment-arms are usually neglected.

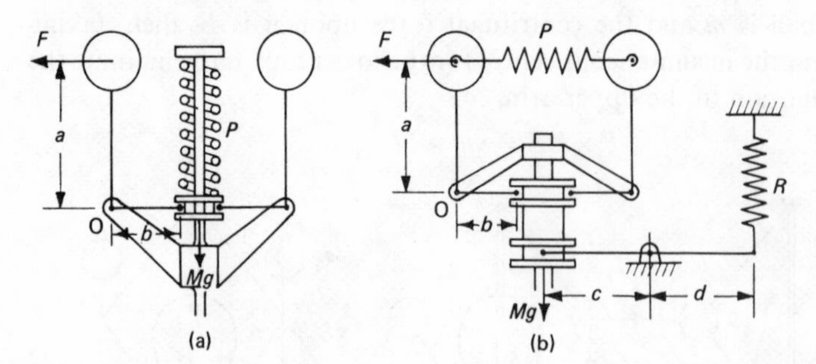

FIG. 6.2

Governor with spring-connected balls, Fig. 6.2 (b) The governor is assembled with the connecting spring initially stretched and the equilibrium speed for a given ball radius is adjusted by the use of an auxiliary spring attached to the sleeve mechanism. If P is the tension in the main spring, R is the tension in the auxiliary spring and M is the mass of the sleeve, then, taking moments about O,

$$(F - P) \times a = \frac{Mg + (d/c)R}{2} \times b$$

6.4 Effort and power

The *effort* of a governor is the force exerted at the sleeve for a given fractional change of speed.

The *power* of a governor is the work done at the sleeve for a given fractional change of speed.

i.e. power = mean effort × sleeve movement

6.5 Sensitivity and friction

If the maximum and minimum speeds of a governor are ω_1 and ω_2 respectively and its mean speed is ω, the *sensitivity* of the governor is defined as $\omega/(\omega_1 - \omega_2)$.

If there is a friction force f between the sleeve and the spindle, the effective sleeve load becomes $(Mg + f)$ when the sleeve is rising and $(Mg - f)$ when the sleeve is falling. For any sleeve position, there is thus a range of speed over which the governor is insensitive. If, for any position, the maximum and minimum speeds before sleeve movement occurs are ω' and ω'' respectively and the speed in the absence of friction is ω, the *coefficient of insensitiveness* is defined as the ratio $(\omega' - \omega'')/\omega$.

6.6 Controlling force and stability

The radially inward, or centripetal, force acting on each rotating ball due to the sleeve weight, spring force, etc., is termed the *controlling force* and a graph of the variation of this force against radius is called a controlling force curve. Figure 6.3 (a) shows such a curve for a Porter-type governor and Fig. 6.3 (b) shows that for a Hartnell-type governor. At any equilibrium speed, ω, the controlling force is equal and opposite to the centrifugal force.

i.e. the controlling force, $F = m\omega^2 r$

$$\therefore \omega^2 = F/mr$$

A governor is *stable* if for each speed within the working range there is only one radius of rotation for equilibrium. Thus r must increase as ω increases, i.e. the ratio F/r must increase as ω increases. This condition is satisfied in the case

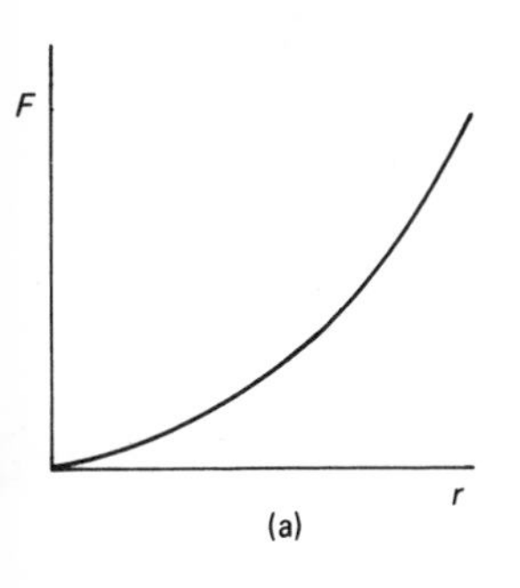

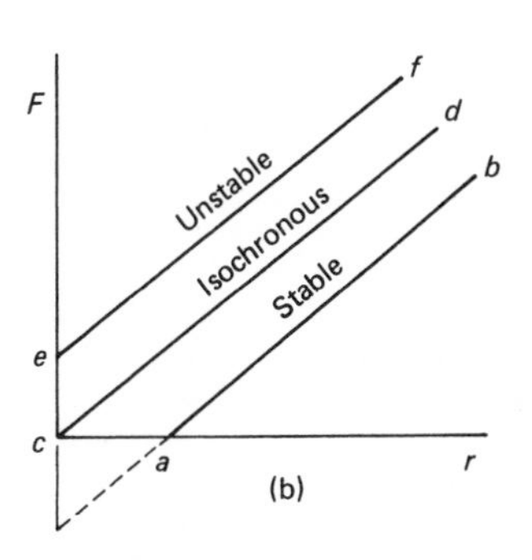

FIG. 6.3

of the Porter governor curve but, for the Hartnell governor, it is only satisfied if the straight line intercepts the vertical axis below the origin, such as the line *ab*.

A governor is *isochronous* if the equilibrium speed is the same for all radii of rotation. The ratio F/r is then a constant and corresponds to the controlling force line *cd*, passing through the origin.

For such a governor, the sleeve will move to one of its extreme positions immediately the speed deviates from the isochronous speed. The fuel inlet valve is therefore either fully open or fully closed and equilibrium is unlikely to be achieved.

A governor is *unstable* if the radius of rotation decreases as the speed increases, i.e. the ratio F/r decreases as r increases. This corresponds to a controlling force line which intercepts the vertical axis above the origin, such as the line *ef*.

For such a governor, the equilibrium speed at the lowest sleeve position is higher than at the highest position. The sleeve will therefore always move to one of the extreme positions and the engine speed will not settle between the two limiting speeds.

Examples

Example 6.1 Figure 6.4 (a) shows a Porter governor for which the speed range can be varied by means of the auxiliary spring S. The spring force is transmitted to the sleeve by the arm AB which is pivoted at A. The two balls each of mass 0·36 kg are supported by four links, C_1, C_2, C_3 and C_4, each 75 mm in length. The sleeve carries a mass of 0·9 kg.

The sleeve begins to rise when the balls revolve at 200 rev/min in a circle of 75 mm radius. The speed of the governor is not to exceed 220 rev/min when the sleeve has risen 10 mm from its original position.

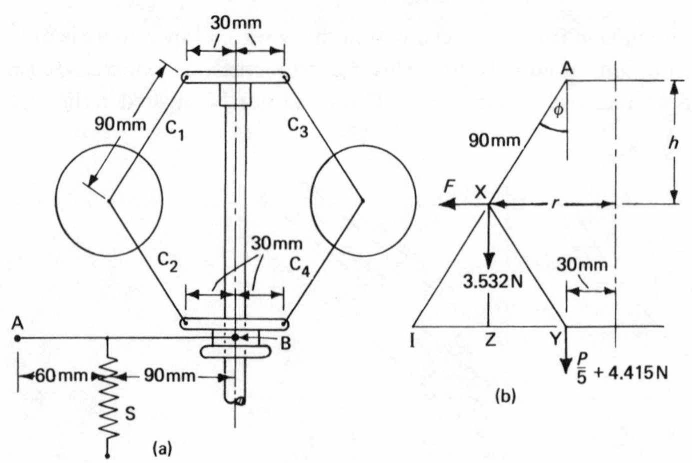

FIG. 6.4

Determine: (a) the necessary stiffness of the spring S, and (b) the tension in the link C_1 when the sleeve begins to rise. (U. Lond.)

Solution

If the force exerted by the spring S is P N,

then force on sleeve $= \frac{2}{5}P$ N

Therefore

$$\text{total load on sleeve} = \tfrac{2}{5}P + 0{\cdot}9 \times 9{\cdot}81 = \tfrac{2}{5}P + 8{\cdot}83\ \text{N}$$

The lower link C_2 is in equilibrium under the centrifugal force on the ball, F, the weight of the ball, 3·532 N, half the load on the sleeve, $\frac{1}{5}P + 4{\cdot}415$ N, the tension in the upper link C_1 and the horizontal reaction on the sleeve.

Taking moments about I, Fig. 6.4 (b),

$$F \times XZ = 3{\cdot}532 \times IZ + (\tfrac{1}{5}P + 4{\cdot}415) \times IY$$

i.e. $$0{\cdot}36\left(\frac{2\pi}{60}N\right)^2 \times r \times h = 3{\cdot}532(r - 0{\cdot}03) + (\tfrac{1}{5}P + 4{\cdot}415) \times 2(r - 0{\cdot}03)$$

i.e. $$0{\cdot}00395N^2 rh = (r - 0{\cdot}03)(12{\cdot}36 + 0{\cdot}4P) \qquad (1)$$

When $N = 200$ rev/min, $r = 0{\cdot}075$ m

and $h = \sqrt{(0{\cdot}09^2 - 0{\cdot}045^2)} = 0{\cdot}078$ m

Therefore, from equation (1), $P_1 = 20{\cdot}45$ N

When $N = 220$ rev/min, $h = 0{\cdot}078 - 0{\cdot}005 = 0{\cdot}073$ m

and $r = \sqrt{(0{\cdot}09^2 - 0{\cdot}073^2)} + 0{\cdot}03 = 0{\cdot}0826$ m

$\therefore$ from equation (1), $P_2 = 23{\cdot}9$ N

$$\therefore \text{spring stiffness} = \frac{P_1 - P_2}{\text{sleeve movement}} = \frac{23{\cdot}9 - 20{\cdot}45}{\frac{2}{5} \times 0{\cdot}01} = \underline{863\ \text{N/m}}$$

$$\text{Vertical reaction at A at 200 rev/min} = \frac{20{\cdot}45}{5} + 4{\cdot}415 + 3{\cdot}532 = 21{\cdot}04\ \text{N}$$

$$\therefore \text{tension in } C_1 = 21{\cdot}04 \sec\phi = 21{\cdot}04 \times 0{\cdot}09/0{\cdot}078 = \underline{13{\cdot}9\ \text{N}}$$

Example 6.2 A Porter governor has 300 mm arms and the rotating balls each have a mass of 1·8 kg. At the mean speed of 120 rev/min, the arms make 30° to the vertical. Determine the central dead load and the sensitivity of the governor if the sleeve movement is ± 25 mm. (U. Lond.)

Solution

Taking moments about I, Fig. 6.5,

$$F \times h = mg \times r + \frac{Mg}{2} \times 2r$$

i.e. $$1{\cdot}8\omega^2 rh = 1{\cdot}8 \times 9{\cdot}81r + M \times 9{\cdot}81r$$

i.e. $$0{\cdot}1835\omega^2 h = 1{\cdot}8 + M \qquad (1)$$

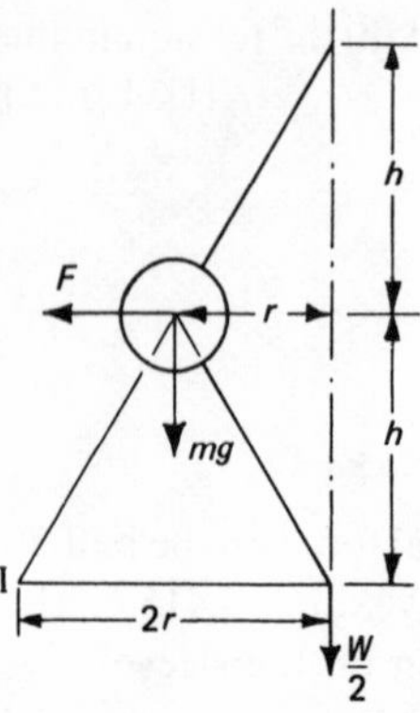

FIG. 6.5

At the mean speed, $\omega = \dfrac{2\pi}{60} \times 120 = 4\pi$ rad/s

and $h = 0{\cdot}3 \cos 30° = 0{\cdot}26$ m

Hence $M = \underline{5{\cdot}73 \text{ kg}}$

Equation (1) then gives $\omega = 6{\cdot}4/\sqrt{h}$

When sleeve rises 25 mm, $h = 0{\cdot}26 - 0{\cdot}0125 = 0{\cdot}2475$ m

$\therefore\ \omega_1 = 6{\cdot}4/\sqrt{0{\cdot}2475} = 12{\cdot}87$ rad/s

When sleeve falls 25 mm, $h = 0{\cdot}26 + 0{\cdot}0125 = 0{\cdot}2725$

$\therefore\ \omega_2 = 6{\cdot}4/\sqrt{0{\cdot}2725} = 12{\cdot}26$ rad/s

$$\therefore \text{ sensitivity} = \frac{\omega}{\omega_1 - \omega_2} = \frac{4\pi}{12{\cdot}87 - 12{\cdot}26} = \underline{20{\cdot}6}$$

Example 6.3 A governor of the Proell type is shown diagrammatically, with certain dimensions, in Fig. 6.6 (a). The central load acting on the sleeve has a

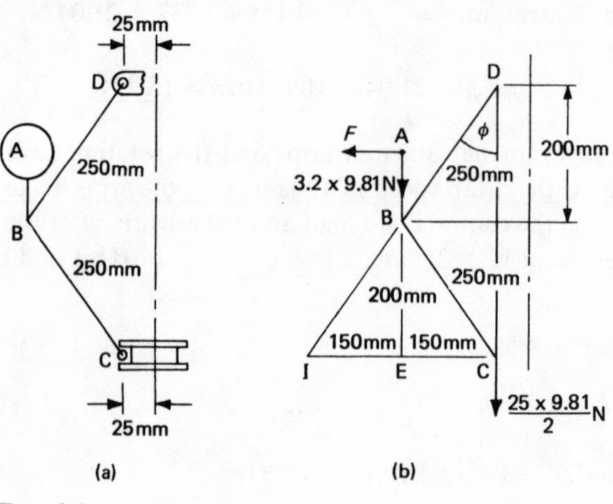

FIG. 6.6

mass of 25 kg and the two rotating masses each have a mass of 3·2 kg. When the governor sleeve is in mid-position, the arm AB of the cranked lever ABC is vertical and the radius of the path of rotation of the masses is 175 mm.

If the governor speed is to be 160 rev/min when in mid-position, find: (a) the length of the arm AB, and (b) the tension in the link BD.

Neglect friction. (U. Lond.)

Solution

Taking moments about I, Fig. 6·6(b),

$$F \times \text{AE} = 3{\cdot}2 \times 9{\cdot}81 \times 0{\cdot}3 + \frac{25 \times 9{\cdot}81}{2} \times 0{\cdot}6$$

i.e.
$$3{\cdot}2\left(\frac{2\pi}{60} \times 160\right)^2 \times 0{\cdot}175 \times (\text{AB} + 0{\cdot}2) = 82{\cdot}9$$

i.e.
$$\text{AB} = \underline{0{\cdot}327\ \text{m}}$$

$$\text{Vertical reaction at D} = 3{\cdot}2 \times 9{\cdot}81 + \frac{25 \times 9{\cdot}81}{2} = 154\ \text{N}$$

$$\therefore\ \text{tension in BD} = 154 \sec\phi = 154 \times 250/200 = \underline{192{\cdot}5\ \text{N}}$$

Example 6.4 A loose-ball governor is shown in Fig. 6.7(a). When in operation, the balls rise up the flange A and push up the sleeve against the spring, while the collar and lever on the top of the dome actuate the control gear. There are 12 steel balls, 50 mm diameter, and the dead load lifted (cover and control gear) is 25 kg. The ball centres are at 125 mm radius in the lowest position and the maximum rise of the sleeve is 20 mm. Calculate the initial compression and stiffness of the spring for a speed range of 350 rev/min to 364 rev/min. The mass of one ball is 0·535 kg. (U. Glas.)

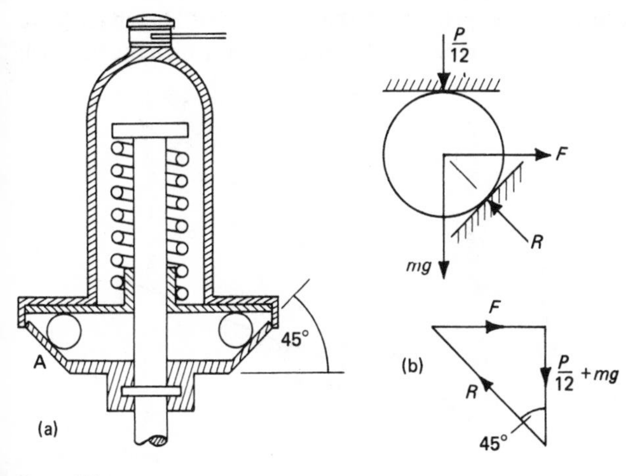

Fig. 6.7

Solution

Radius of rotation of balls at bottom speed = 125 mm

$\therefore$ radius of rotation of balls at top speed = 145 mm

If P is the total force exerted on the balls by the dead weight and the spring, one ball is in equilibrium under a vertical load, $P/12$, the centrifugal force on the ball, F, its weight, mg, and the reaction of the ball-race, R.

Then, from the triangle of forces, Fig. 6.7 (b),

$$\frac{P}{12} + mg = F = m\omega^2 r$$

i.e.
$$\frac{P}{12} + 0{\cdot}535 \times 9{\cdot}81 = 0{\cdot}535\left(\frac{2\pi}{60}N\right)^2 r$$

When $N = 350$ rev/min, $r = 125$ mm

$\therefore P = 1015$ N

$\therefore$ spring force $= 1015 - 25 \times 9{\cdot}81 = 770$ N

When $N = 364$ rev/min, $r = 145$ mm

$\therefore P = 1284$ N

$\therefore$ spring force $= 1284 - 25 \times 9{\cdot}81 = 1039$ N

$\therefore$ spring stiffness $= (1039 - 770)/0{\cdot}02 = \underline{13\,450 \text{ N/m}}$

$\therefore$ initial compression $= 770/13\,450 = 0{\cdot}0573$ m or $\underline{57{\cdot}3 \text{ mm}}$

Example 6.5 A form of emergency governor is shown in Fig. 6.8. The governor is contained in the main shaft A of the machine. The loose bolt BB has a mass of 0·15 kg and its centre of mass at G is 1·5 mm from the centre of rotation O. The movement of the bolt is opposed by a spring, giving a force of 16 kN/m. The travel of the bolt is 5 mm.

Calculate the initial force in the spring such that the bolt will begin to move outwards at 6600 rev/min. Find also the speed of rotation at which the bolt will

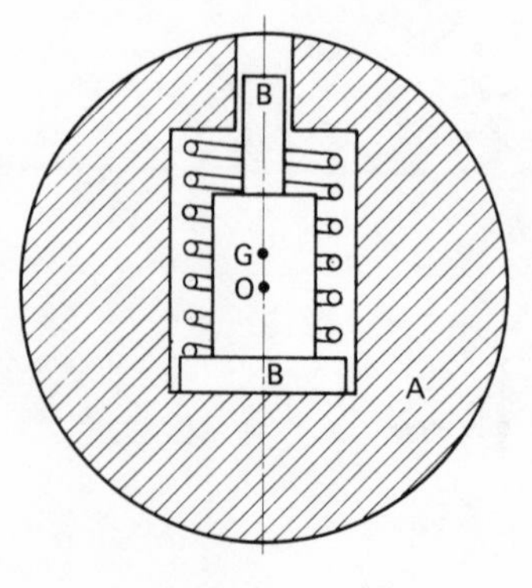

Fig. 6·8

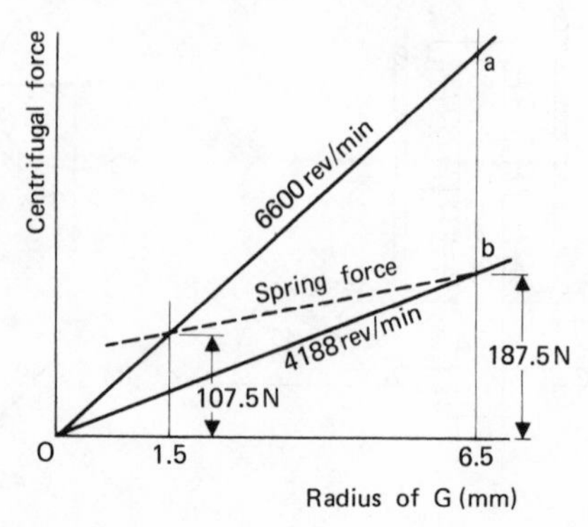

Fig. 6.9

eturn from the outer to the inner position and show that the governor is nstable in all positions. (U. Lond.)

olution

At 6600 rev/min,

$$\text{centrifugal force, } F = 0{\cdot}15\left(\frac{2\pi}{60} \times 6600\right)^2 \times 0{\cdot}0015$$

$$= \underline{107{\cdot}5 \text{ N}} = \text{initial spring force}$$

n the outer position,

$$\text{spring force} = 107{\cdot}5 + 0{\cdot}005 \times 16 \times 10^3 = 187{\cdot}5 \text{ N}$$

Therefore the bolt, whose centre of mass is now 6·5 mm from O, will move nwards when the centrifugal force falls below this spring force of 187·5 N,

e. when $$187{\cdot}5 > 0{\cdot}15\left(\frac{2\pi}{60} N\right)^2 \times 0{\cdot}0065$$

e. when $$N < \underline{4188 \text{ rev/min}}$$

In Fig. 6.9, graphs O*a* and O*b* are plotted, representing the centrifugal force n the bolt for different radii at speeds of 6600 rev/min and 4188 rev/min espectively. It will be seen that at 6600 rev/min the centrifugal force exceeds he spring force at all radii so that the bolt will immediately move to the outer top, and at 4188 rev/min the spring force exceeds the centrifugal force for all adii so that the bolt will immediately move to the inner stop. The governor is herefore unstable in all positions.

Example 6.6 A spring-controlled centrifugal governor has two rotating nasses, each of mass 2·7 kg, and the limits of their radii of rotation are 100 mm nd 125 mm. Each mass is directly controlled by a spring attached to it and to he casing of the governor, Fig. 6.10. The stiffness of each spring is 7 kN/m and he force in each spring, when the masses are in their mid-position, is 350 N. In ddition, there is an equivalent constant inward radial force $R = 70$ N acting n each mass, in order to allow for the dead weight of the mechanism.

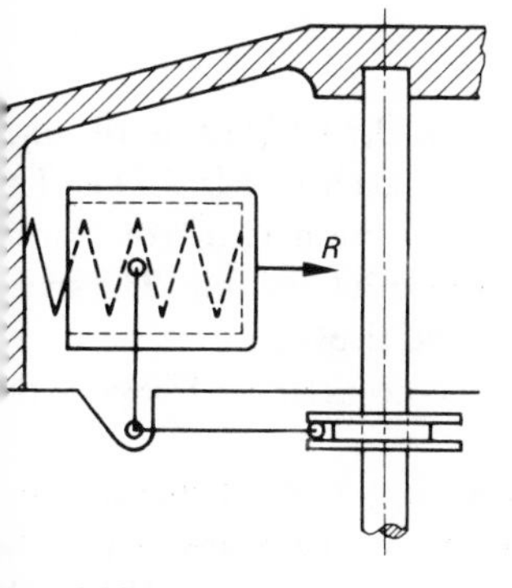

IG. 6.10

Neglecting friction, find the range of speed of the governor.

What would be the required force in each spring, when the masses are i their mid-position, for isochronism and what would then be the speed?

(U. Lond

Solution

If P is the *total* inward force on each rotating mass, then, for equilibrium,

$$\text{centrifugal force, } F = P$$

i.e.

$$2{\cdot}7\left(\frac{2\pi}{60}N\right)^2 \times r = P$$

or

$$N = 5{\cdot}81\sqrt{(P/r)} \quad (1$$

At maximum speed, $r = 0{\cdot}125$ m

and $P = 70 + 350 + 7000 \times 0{\cdot}0125 = 507{\cdot}5$ N

$$\therefore\ N_1 = 5{\cdot}81\sqrt{(507{\cdot}5/0{\cdot}125)} = \underline{370 \text{ rev/min}}$$

At minimum speed, $r = 0{\cdot}1$ m

and $P = 70 + 350 - 7000 \times 0{\cdot}0125 = 332{\cdot}5$ N

$$\therefore\ N_2 = 5{\cdot}81\sqrt{(332{\cdot}5/0{\cdot}1)} = \underline{335 \text{ rev/min}}$$

For isochronism, N is to remain constant as r changes and hence, fror equation (1), P/r is to remain constant.

If S is the spring force required at mid-position, then, equating the values c P/r at 125 mm and 100 mm radii,

$$\frac{70 + S + 7000 \times 0{\cdot}0125}{0{\cdot}125} = \frac{70 + S - 7000 \times 0{\cdot}0125}{0{\cdot}1}$$

from which $\underline{S = 717{\cdot}5 \text{ N}}$

Substituting in equation (1),

$$N = 5{\cdot}81\sqrt{\left(\frac{70 + 717{\cdot}5 + 7000 \times 0{\cdot}0125}{0{\cdot}125}\right)} = \underline{486 \text{ rev/min}}$$

Example 6.7 In a Hartnell governor, the length of the ball arm is 190 mn that of the sleeve arm is 140 mm, and the mass of each ball is 2·7 kg. Th distance of the pivot of each bell-crank lever from the axis of rotation i 170 mm, and the speed, when the ball arm is vertical, is 300 rev/min. The spee is to increase 0·6 per cent for a lift of 12 mm of the sleeve.

(a) Neglecting the dead load on the sleeve, find the necessary stiffness of th spring and the required initial compression.

(b) What spring stiffness and initial compression would be required if th speed is to remain the same for the changed position of the sleeve (i.e. th governor is to be isochronous)? (U. Lond

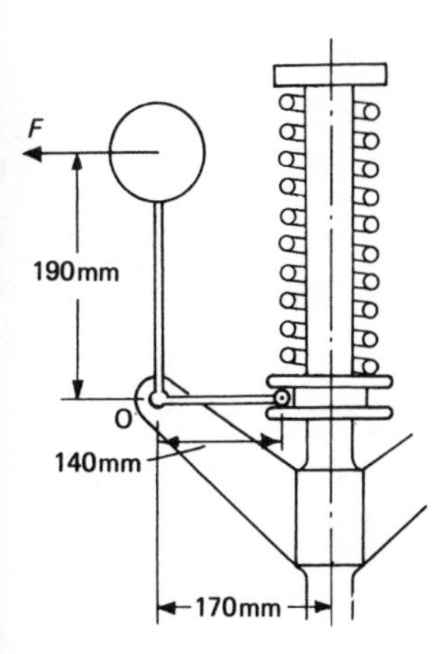

Fig. 6.11

Solution

Let P be the force exerted by the spring and F be the centrifugal force on one ball. Then, taking moments about O, Fig. 6.11,

$$F \times 0{\cdot}19 = \frac{P}{2} \times 0{\cdot}14$$

i.e.

$$2{\cdot}7\left(\frac{2\pi}{60}N\right)^2 \times r \times 0{\cdot}19 = \frac{P}{2} \times 0{\cdot}14$$

$$\therefore\ P = 0{\cdot}0804\,N^2 r$$

At 300 rev/min, $r = 0{\cdot}17\,\text{m}$

$$\therefore\ P_1 = 0{\cdot}0804 \times 300^2 \times 0{\cdot}17 = 1230\,\text{N}$$

For a 0·6 per cent increase in speed,

$$N = 301{\cdot}8\,\text{rev/min}$$

and $r = 0{\cdot}17 + 0{\cdot}012 \times 0{\cdot}19/0{\cdot}14 = 0{\cdot}1863\,\text{m}$

$\therefore$ at 301·8 rev/min, $P_2 = 0{\cdot}0804 \times 301{\cdot}8^2 \times 0{\cdot}1863 = 1364\,\text{N}$

$$\therefore\ \text{spring stiffness} = \frac{P_2 - P_1}{\text{compression of spring}} = \frac{1364 - 1230}{0{\cdot}012} = \underline{11\,160\,\text{N/m}}$$

$$\text{Initial compression} = \frac{\text{initial spring force}}{\text{stiffness}} = \frac{1230}{11\,160} = \underline{0{\cdot}11\,\text{m}}$$

For isochronism, the equilibrium speed remains constant for all radii of rotation. If P_3 and P_4 are the spring forces for isochronism corresponding to radii of rotation of 0·17 m and 0·1863 m respectively, then $P_3 = 0{\cdot}0804 \times 300^2 \times 0{\cdot}17 = 1231\ \text{N}$

and $P_4 = 0{\cdot}0804 \times 300^2 \times 0{\cdot}1863 = 1349\ \text{N}$

$$\therefore \text{ spring stiffness} = \frac{1349 - 1231}{0{\cdot}012} = \underline{9830\ \text{N/m}}$$

$$\text{Initial compression} = \frac{1231}{9830} = \underline{0{\cdot}1252\ \text{m}}$$

Problems

1. In a Porter governor, the upper and lower arms are each 200 mm long, and are each inclined at 30° to the vertical when the sleeve is in its lowest position. The points of suspension are each 36 mm from the axis of the spindle. The mass of each rotating ball is 3 kg, and that of the central load on the sleeve 20 kg. If the movement of the sleeve is 36 mm, find the range of speed of the governor. (U. Lond.)

(*Answer:* 170·6 rev/min to 185·4 rev/min)

2. For a Porter governor with equal arms, h is the vertical depth of the balls from the apex of the governor and ω is the angular velocity. Show from first principles that

$$\frac{dh}{h} = -2\frac{d\omega}{\omega}$$

If the mass of each ball is 2 kg and $h = 200$ mm, determine the friction effect at the sleeve for a speed range of 160 rev/min to 165 rev/min at this position. (U. Lond.)

(*Answer:* 3·56 N)

3. A Porter governor is arranged as shown in Fig. 6.12. The rotating masses are each 3 kg. The sleeve, of mass 1 kg, actuates a lever, which has a mass of 2 kg, its centre of mass being as marked, and which must exercise an operating pull, P, of 25 N. The frictional effect of the gear reduced to the sleeve may be taken as 18 N. Determine the central mass, M, with which the sleeve must be loaded so that the governor is on the point of moving out from the given position at 180 rev/min. Compare the speed at which it would begin to move in from this position. Any formula used must be established. (U. Lond.)

(*Answer:* 21·28 kg; 171 rev/min)

N.B. Moments must be taken about the top pivot of forces on the upper arm.

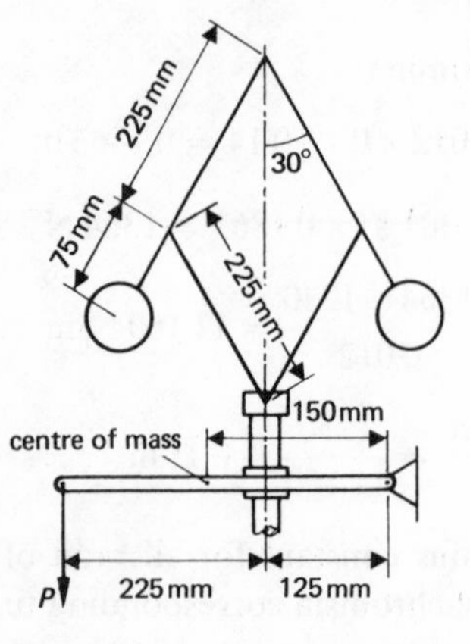

FIG. 6.12

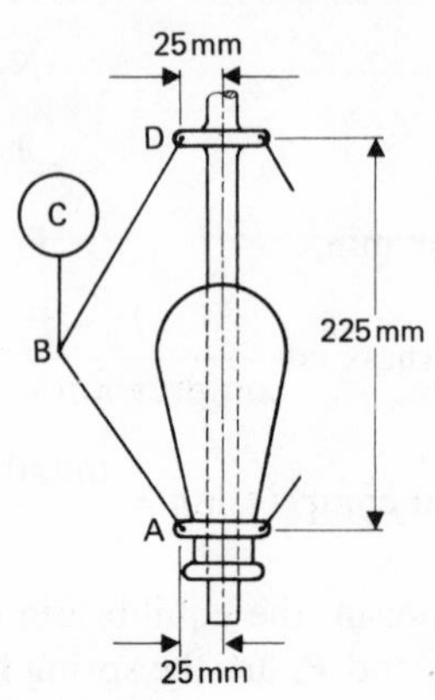

FIG. 6.13

4. Define sensitivity, stability and isochronism as applied to governors.

In the Proell governor shown in Fig. 6.13, the arm ABC is formed in one piece. Each ball has a mass of 1·8 kg and the effect of friction is equivalent to a force of 9 N at the sleeve. AB = 150 mm, BC = 50 mm and BD = 150 mm.

Find the magnitude of the central load on the sleeve, if the governor is to rise to its mean position as shown in the diagram when the speed is 120 rev/min, the portion BC of the lower arm then being vertical. The gravitational effect of the balls must be taken into account. (U. Lond.) (*Answer:* 3·17 kg)

5. A ball governor of the vertical compression-spring type runs at a mean speed of 180 rev/min. The bell-crank lever arms are 150 mm and 75 mm and the rotating balls have a mass of 2·75 kg each, the ball radius being 100 mm at mean speed.

If the speed range of the governor at mean speed is $\pm$ 1 rev/min, determine the friction effect at the sleeve. (Neglect the effect of gravity on the balls.)

What spring stiffness would be required to make the governor isochronous, if the sleeve lift is $\pm$ 12 mm from the mean position? (Assume the ball arm vertical at the mean position.) (U. Lond.) (*Answer:* 4·34 N; 7·82 kN/m)

6. In a spring-controlled governor of the Hartnell type, the arms of the bell-crank levers are horizontal and vertical when the sleeve is in its mid-position; the horizontal arm is 150 mm, the vertical arm is 190 mm long and the fulcrum is 175 mm from the vertical axis of the governor spindle. The vertical arms carry masses of 5·5 kg each. Find the strength of the central spring if the top and bottom speeds of the governor are 200 rev/min and 180 rev/min respectively and sleeve lift is 24 mm.

To increase the speed of the governor, an additional load is applied to the sleeve by means of a spring attached to the middle of a forked horizontal lever partially embracing a recess in the sleeve at the forked end and anchored at the other end. Find the strength of this spring when the top and bottom speeds are 260 rev/min and 230 rev/min respectively. (U. Lond.) (*Answer:* 15·5 kN/m; 50 kN/m)

7. A governor of the Hartnell type, with dimensions as shown in Fig. 6.14, runs at a mean speed of 300 rev/min, each ball has a mass of 2·3 kg, and a 3 per cent reduction in speed causes a sleeve movement of 6 mm. If the ball arm is vertical at the mean speed, and gravitational effects are ignored, find the spring stiffness in N/m. Neglect the mass

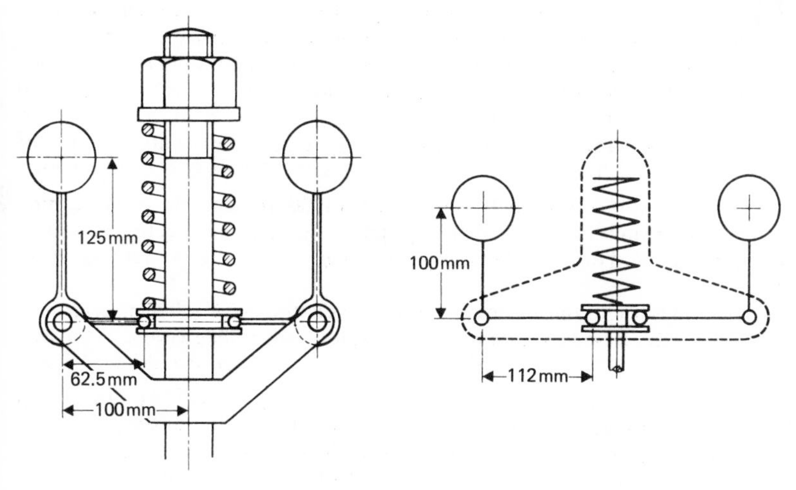

Fig. 6.14 Fig. 6.15

of the arms. By how much must the adjusting nut be screwed down to render the governor isochronous and what will be the resulting operational speed of the governor? (U. Lond.) (*Answer:* 26·1 kN/m; 15·17 mm; 359 rev/min)

8. Show that a governor will be stable if, when values of the controlling force, F, are plotted against the radius of rotation, R, the slope of the graph at any radius is greater than the corresponding value of F/R.

In a governor of the Hartnell type, the rotating masses are 1·8 kg each and the lengths of the arms of the bell-crank levers are shown in Fig. 6.15. If the speeds are required to be 295 rev/min and 305 rev/min, when the radii of rotation are 132 mm and 142 mm respectively, determine the stiffness of the spring and give the law connecting F and R. Neglect the moments due to the weights of the balls when the upright arms of the levers are not vertical. (U. Lond.)

(*Answer:* 5·37 kN/m; $F = 3430\,R - 230$ N, where R is in metres)

9. A Hartnell governor has two rotating balls, of mass 2·7 kg each. The ball radius is 125 mm in the mean position when the ball arms are vertical and the speed is 150 rev/min with the sleeve rising. The length of the ball arms is 140 mm and the length of the sleeve arms 90 mm. The stiffness of the spring is 7 kN/m and the total sleeve movement is $\pm$ 12 mm from the mean position.

Allowing for a constant friction force of 14 N acting at the sleeve, determine the speed range of the governor in the lowest and highest sleeve positions.

Neglect the obliquity of the ball arms. (U. Lond.)

(*Answer:* 154·3 rev/min to 161 rev/min; 122·5 rev/min to 133·6 rev/min)

10. A spring-controlled governor has two balls each of mass 9 kg, suspended on arms 300 mm long which are pivoted at the axis of the vertical governor spindle. The balls are connected by two springs, one on each side of the balls, which exert a radial pull on the balls. The springs have an unstretched length of 200 mm and the minimum speed of rotation of the governor is to be 150 rev/min, the radius of the governor circle being then 150 mm.

Determine the necessary stiffness of each spring, allowing for the effect of the weight of the balls, and the speed when the radius of the governor circle is 165 mm. (U. Lond.) (*Answer:* 1·41 kN/m; 161·5 rev/min)

11. Compare and contrast the functions of a governor and a flywheel on an engine.

In a centrifugal governor, the two operating masses, of mass 1·15 kg each, rotate in a circle of 175 mm diameter when in mid-position; the two controlling springs which connect directly the two operating masses in parallel have a stiffness of 875 N/m (each) and are extended 56 mm when the governor is in mid-position. Find the equilibrium speed for this position and for the case when the operating masses rotate in a circle of 225 mm diameter. (U. Lond.) (*Answer:* 297·5 rev/min; 361 rev/min)

12. When the governor shown in Fig. 6.16 rotates at 300 rev/min, the two operating masses B move in a circle of 225 mm diameter and the sleeve S is then at the right-hand stop T_1; when the speed drops to 250 rev/min, the sleeve moves 40 mm to the left-hand stop T_2.

If the two springs connecting the masses B each have a stiffness of 750 N/m, determine the operating masses B and the extension of the springs when the sleeve is at the left-hand stop. (U. Lond.) (*Answer:* 1·96 kg; 64·9 mm)

13. A spring-controlled governor has two balls, each of mass 2·3 kg. The mean speed is to be 500 rev/min and the variation $\pm$ 2 per cent. The extreme radii of the path of the balls are 110 mm and 85 mm. Find the controlling force at the balls in each case.

If the effect of friction be $\pm$ 45 N at each ball, find the highest and lowest speeds. (U. Lond.) (*Answer:* 721·5 N; 514·5 N; 525 rev/min; 468 rev/min)

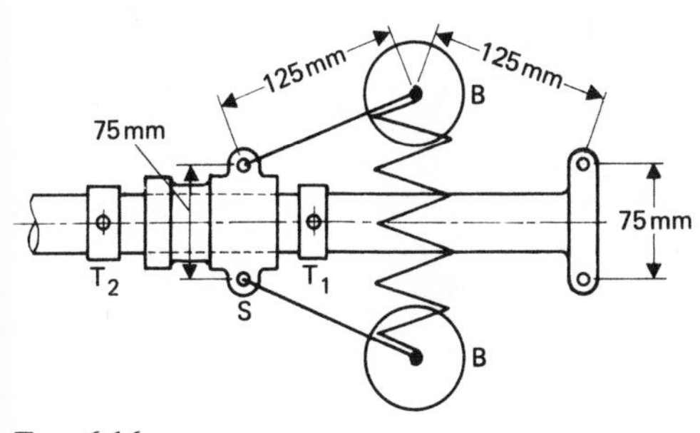

Fig. 6.16

14. The mass of each ball of a spring-loaded governor is 2·7 kg, the maximum and minimum radii of rotation are 100 mm and 60 mm and the corresponding values of the controlling force on each ball are 525 N and 250 N. Assuming the controlling force curve to be a straight line, find the equilibrium speeds for radii of rotation of 60, 85 and 110 mm.

If friction of the governor mechanism is equivalent to a force of 5 N at each ball, what are the extreme equilibrium speeds for a radius of rotation of 85 mm? (I. Mech. E.)
(*Answer:* 375 rev/min; 392 rev/min; 401 rev/min; 390 rev/min; 395 rev/min)

15. A spring-loaded governor is shown in Fig. 6.17. The two balls, each of mass 5·5 kg, are connected across by two springs A. A supplementary spring B provides an additional force at the sleeve through the medium of a lever which pivots about a fixed centre at its left-hand end. In the mean position, the radius of the governor balls is 150 mm and the speed is 600 rev/min. The tension in each spring A is then 1100 N. Find the tension in the spring B for this position.

If, when the sleeve moves up 20 mm, the speed is to be 630 rev/min, find the necessary stiffness of the spring B in N/m, if the stiffness on each spring A is 8 kN/m.

Neglect the moments produced by the weights of the balls. (U. Lond.)
(*Answer:* 3·165 kN; 39·4 kN/m)

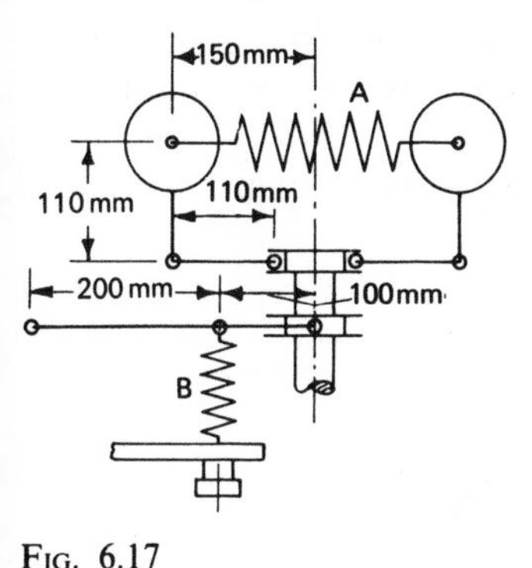

Fig. 6.17

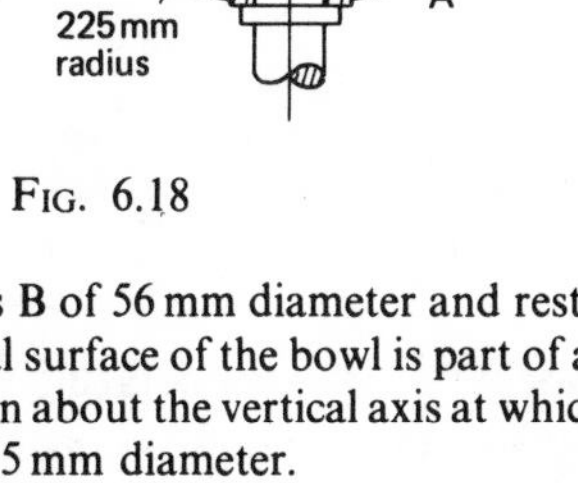

Fig. 6.18

16. In the bowl A, Fig. 6.18, are 8 steel balls B of 56 mm diameter and resting on the balls is the disc C, of mass 22 kg. If the internal surface of the bowl is part of a sphere of 225 mm radius, determine the speed of rotation about the vertical axis at which the balls will move with their centres in a circle of 265 mm diameter.

The density of steel is 7·8 Mg/m^3. (U. Lond.) (*Answer:* 172 rev/min)

7

Balancing

7.1 Static and dynamic balance

When a shaft carrying several eccentric masses is in *static* balance, the centre of mass of the system lies in the axis of the shaft so that the shaft and attached masses remain in any position in which it is placed. When the shaft rotates, however, centrifugal forces act upon the masses, and if these are not rotating in the same plane, couples also act upon the shaft. Therefore, for complete *dynamic* balance, (a) the resultant force acting upon the shaft must be zero and (b) the resultant couple acting upon the shaft must be zero.

7.2 Balancing of masses rotating in the same plane

If m_1, m_2, m_3, etc., Fig. 7.1, are the out-of-balance masses and r_1, r_2, r_3, etc., are the respective radii of rotation, then for dynamic balance, the vector sum of the centrifugal forces must be zero,

i.e. $\Sigma m\omega^2 r = 0$ where ω is the angular speed of the shaft

i.e. $\Sigma mr = 0$ since ω^2 is the same for each mass

If, therefore, a force polygon with sides representing the magnitudes and directions of the mass–arm products $m_1 r_1$, $m_2 r_2$, etc., is drawn, Fig. 7.2, the closing side represents the product of the balance mass, B, and its radius of rotation, b.

The condition $\Sigma mr = 0$ is also the condition for static balance.

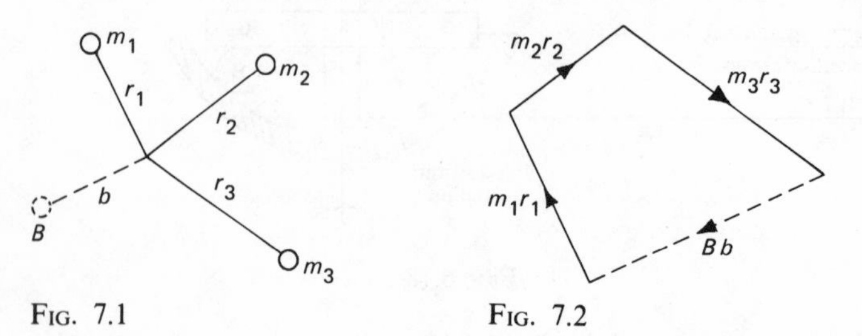

FIG. 7.1 FIG. 7.2

7.3 Balancing of masses rotating in different planes—Dalby's method

If the out-of-balance masses, m_1, m_2, m_3, etc., Fig. 7.3, are situated at distances l_1, l_2, l_3, etc., from a *reference plane*, the forces $m_1 r_1$, $m_2 r_2$, etc., may be

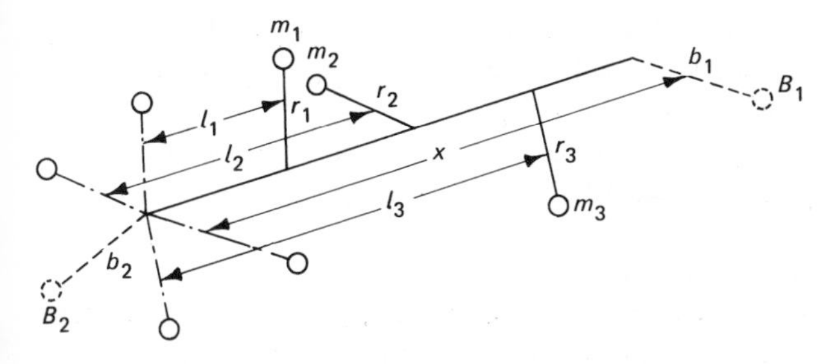

FIG. 7.3

transferred to the reference plane by the addition of couples of magnitude $m_1r_1l_1$, $m_2r_2l_2$, etc., acting in the planes containing the respective forces and the shaft axis. Then, for balance, the resultant force in the reference plane must be zero, i.e. $\Sigma mr = 0$, and the resultant couple in the reference plane must be zero, i.e. $\Sigma mrl = 0$.

These two conditions determine the necessary mass–arm products B_1b_1 and B_2b_2 for balance. Thus the closing side of the couple (mrl) polygon, Fig. 7.4, represents the magnitude and direction of the couple required for equilibrium, B_1b_1x, and the closing side of the force (mr) polygon, Fig. 7.5, represents the magnitude and direction of the force, B_2b_2, required in the reference plane for equilibrium.

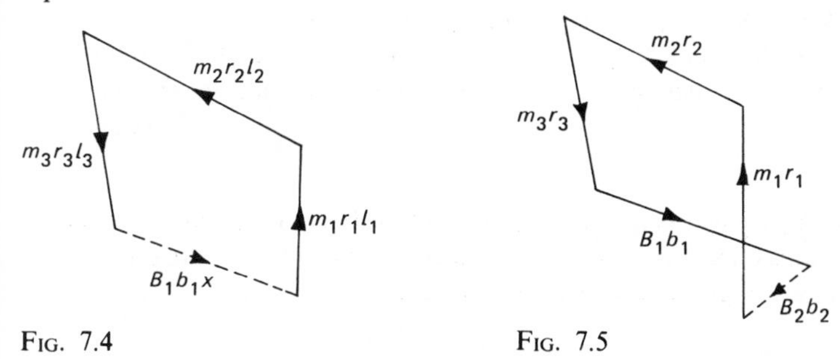

FIG. 7.4

FIG. 7.5

The reference plane is normally chosen to coincide with the plane of revolution of one of the unknown masses, thus eliminating the couple produced by this mass.

In constructing the couple polygon, it is usual to draw the couple vectors in the directions of the respective forces instead of 90° anticlockwise to them, as required by the normal convention for couple vectors. The shape of the polygon is unaffected and it then has sides parallel to those of the force polygon.

Where the reference plane divides the planes of revolution of the masses, the vectors for the couples to one side of the plane are drawn radially outwards and those to the other side, being regarded as negative, are drawn radially inwards.

When a shaft is not in static balance, the unbalanced couple varies with the position of the reference plane but when in static balance, i.e. when the force polygon closes, the rotating shaft is subjected to a pure rocking couple which is independent of the choice of reference plane.

A shaft in static balance is not necessarily in dynamic balance but, to be dynamically balanced, a shaft must be in static balance.

7.4 Dynamic forces at bearings

If a shaft is unbalanced, the dynamic forces at the bearings may be obtained by first determining the mass–arm products, B_1b_1 and B_2b_2, necessary in these planes to achieve balance. These products represent the forces, at a speed of 1 rad/s, which must be exerted *by* the bearings to hold the rotating shaft in position, and the forces *on* the bearings are in the opposite directions. The actual forces at a speed of ω rad/s are then given by $B_1b_1\omega^2$ and $B_2b_2\omega^2$.

For a shaft in static balance, the pure rocking couple generated gives bearing forces which are equal and opposite.

Examples

Example 7.1 A rotating shaft carries four masses, A, B, C and D, rigidly attached to it; the centres of mass are at 30 mm, 36 mm, 39 mm and 33 mm respectively from the axis of rotation; A, C and D are 7·5 kg, 5 kg and 4 kg; the axial distance between A and B is 400 mm and that between B and C is 500 mm; the eccentricities of A and C are at 90° to one another.

Find, for complete balance: (a) the angles between A, B and D, (b) the axial distance between the planes of revolution of C and D, and (c) the mass B.

(U. Lond.)

Solution

The plane of B, Fig. 7.6 (a), is chosen as the reference plane since it contains one of the unknown masses and distances measured to the right of it are regarded as positive. Using the given data, the following table is compiled.

Plane	*m* (*kg*)	*r* (*mm*)	*mr* (*kg mm*)	*l* (*m*)	*mrl* (*kg mm m*)
A	7·5	30	225	−0·4	−90
B	*m*	36	36*m*	0	0
C	5	39	195	0·5	97·5
D	4	33	132	0·5+*x*	66 + 132*x*

The couple polygon, Fig. 7.7, is constructed from the data in the *mrl* column, the direction of the couple due to A being downwards since it is negative.

By measurement, the closing side,

$$ad = 132{\cdot}6 = 66 + 132x \text{ kg mm m}$$

from which $$\underline{x = 0{\cdot}505 \text{ m}}$$

Also $$\underline{\theta = 47{\cdot}5°}$$

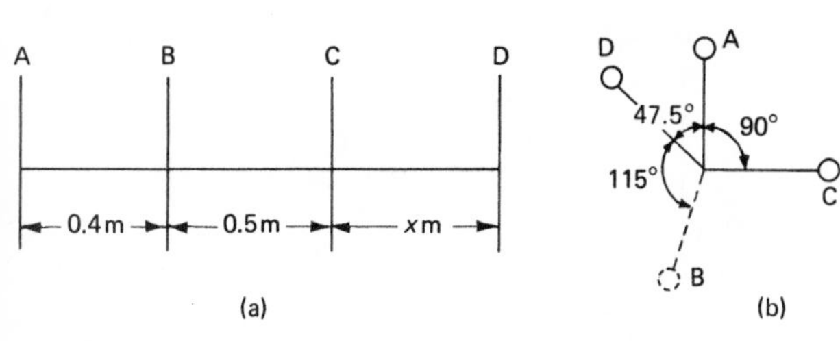

FIG. 7.6

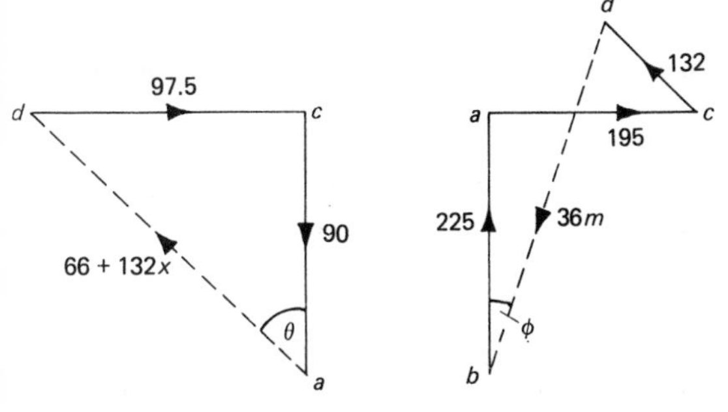

FIG. 7.7

FIG. 7.8

The force polygon, Fig. 7.8, is then constructed, using the data in the *mr* column and the known directions of the forces at A, C and D.

By measurement, the closing side,

$$db = 330 = 36m \text{ kg mm}$$

$$\therefore\ m = \underline{9{\cdot}16 \text{ kg}}$$

Also $$\phi = 17{\cdot}5^\circ$$

from which $$\text{angle between B and D} = \underline{115^\circ}$$

The relative positions of the masses are shown in Fig. 7.6 (b).

Example 7.2 Attached to a uniformly rotating shaft are four discs, A, B, C and D, spaced at equal intervals along the shaft, of mass 7·5 kg, 12·5 kg, 7 kg and 6 kg respectively; the centres of mass of the discs are at 4 mm, 3 mm, 5 mm and 8 mm respectively from the axis of rotation. An additional mass M may be attached to D at an effective radius of 60 mm from the axis of rotation.

Find the minimum value of the mass of M, and the relative angular positions of the centres of mass of all the masses to ensure complete dynamic balance for the rotating shaft. (U. Lond.)

Solution

The plane of D, Fig. 7.9 (a), is chosen as the reference plane. Using the given data, the following table is compiled.

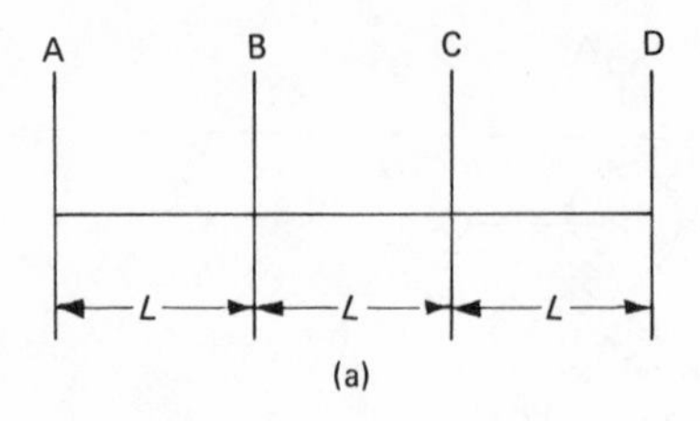

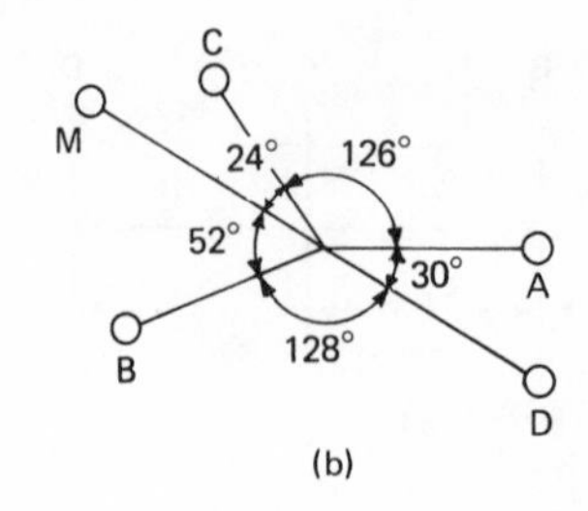

Fig. 7.9

Plane	m (kg)	r (mm)	mr (kg mm)	l (m)	mrl (kg mm m)
A	7·5	4	30	$3L$	$90L$
B	12·5	3	37·5	$2L$	$75L$
C	7	5	35	L	$35L$
D {	6	8	48	0	0
	M	60	$60M$	0	0

From the couple polygon, Fig. 7.10, the relative directions of A, B and C are obtained. The force polygon is then constructed, Fig. 7.11, from which the resultant mass–arm product in plane D is obtained.

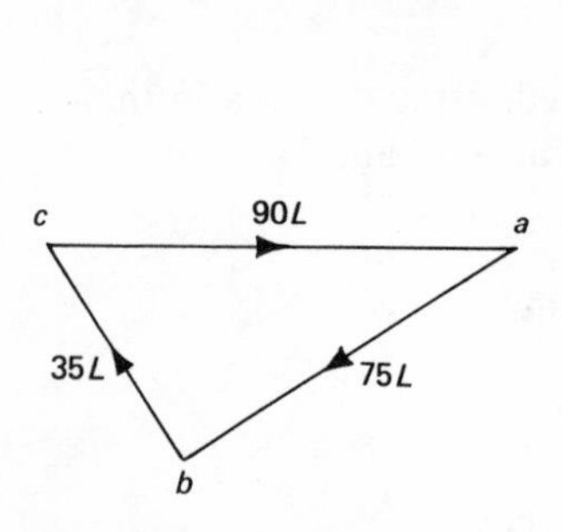

Fig. 7.10

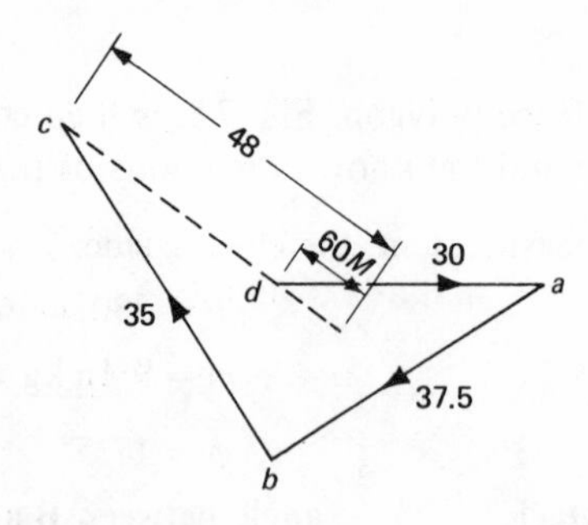

Fig. 7.11

For M to be a minimum, the force due to M must lie in the same straight line as that due to the 6 kg mass.

Thus, by measurement,

$$48 + 60M = cd = 29{\cdot}6 \text{ kg mm}$$

$$\therefore\ M = \underline{-0{\cdot}307 \text{ kg}}$$

i.e. its direction is opposite to that of the 6 kg mass. The relative angular positions of the masses are shown in Fig. 7.9 (b).

Example 7.3 A shaft rotating at 120 rev/min is supported in bearings A and B, 1·8 m apart, A being at the left-hand end. Two unbalanced rotating masses of 7·5 kg and 10 kg at radii of 75 mm and 50 mm respectively are situated between A and B at distances of 0·6 m and 0·9 m respectively from A. The angle between the radii is 60° when viewed along the shaft.

Find the magnitudes, directions and senses of the forces of the shaft on the bearings due to the combined action of the dynamical forces and gravity when the 7·5 kg mass is vertical and above the shaft and the 10 kg mass is on the right of the 7·5 kg mass when viewed from B to A. Show the results in an end view looking from B to A. (U. Lond.)

Solution

The plane of A, Fig. 7.12 (a), is chosen as the reference plane. Using the given data, the following table is compiled.

Plane	*m* (*kg*)	*r* (*mm*)	*mr* (*kg mm*)	*l* (*m*)	*mrl* (*kg mm m*)
A	—	—	R_1	0	0
B	—	—	R_2	1·8	$1{\cdot}8R_2$
C	7·5	75	562·5	0.6	337·5
D	10	50	500	0·9	450

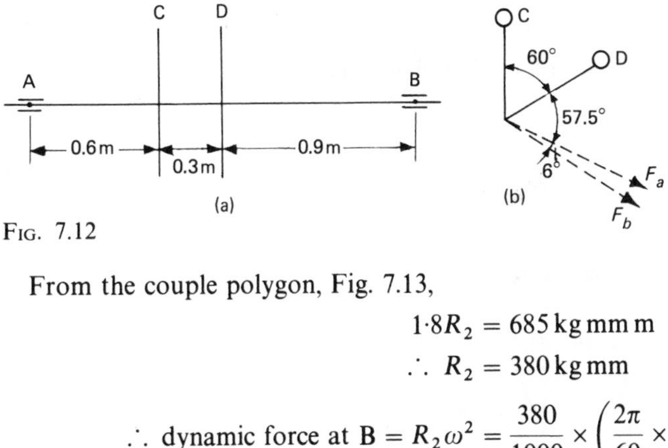

FIG. 7.12

From the couple polygon, Fig. 7.13,

$$1{\cdot}8R_2 = 685\,\text{kg mm m}$$

$$\therefore\ R_2 = 380\,\text{kg mm}$$

$$\therefore\ \text{dynamic force at B} = R_2\omega^2 = \frac{380}{1000} \times \left(\frac{2\pi}{60} \times 120\right)^2 = 60\,\text{N}$$

From the force polygon, Fig. 7.14,

$$R_1 = 545\,\text{kg mm}$$

$$\therefore\ \text{dynamic force at A} = R_1\omega^2 = \frac{545}{1000} \times \left(\frac{2\pi}{60} \times 120\right)^2 = 86\,\text{N}$$

The directions of R_1 and R_2 given by the couple and force polygons are those for balance. The unbalanced forces at the bearings are therefore opposite to these directions, i.e. in directions *ab*, Fig. 7.14, and *bd*, Fig. 7.13.

$$\text{Force on bearing B due to dead weight} = \frac{7{\cdot}5 \times 0{\cdot}6 + 10 \times 0{\cdot}9}{1{\cdot}8} \times 9{\cdot}81$$

$$= 71{\cdot}9\ \text{N}\quad \text{(downwards)}$$

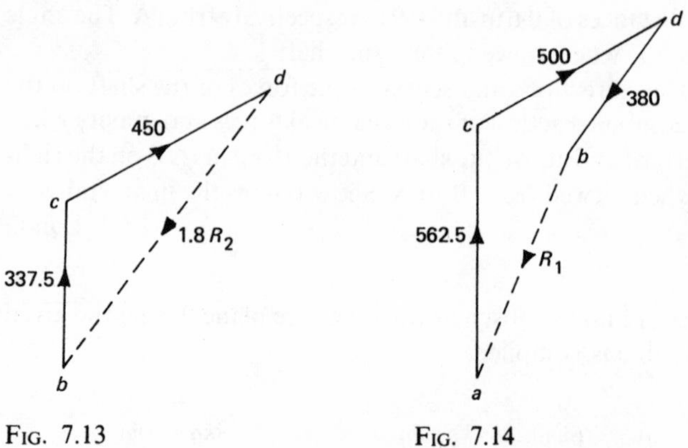

Fig. 7.13 Fig. 7.14

Force on bearing A due to dead weight $= 17{\cdot}5 \times 9{\cdot}81 - 71{\cdot}9$

$= 99{\cdot}7$ N (downwards)

Therefore resultant force at B $= F_b = \underline{36 \text{ N}}$ (Fig. 7.15)

and resultant force at A $= F_a = \underline{39{\cdot}2 \text{ N}}$ (Fig. 7.16)

The directions of the forces are shown in Fig. 7.12 (b).

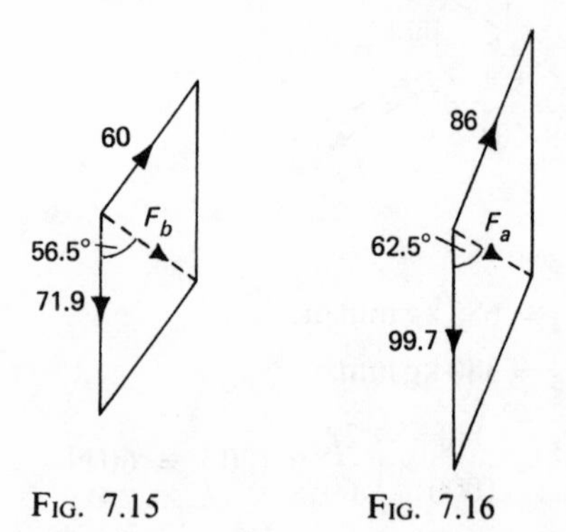

Fig. 7.15 Fig. 7.16

Example 7.4 A small three-throw crankshaft has cranks of radii 125 mm, set at 120° to each other, and equally spaced with a pitch of 250 mm. The revolving masses at crank radii are the same for each line and of amount 15 kg. The shaft is supported in two bearings symmetrically arranged with respect to the cranks and 850 mm apart. Determine the dynamical loads on the bearings for a speed of 500 rev/min.

The shaft is to be balanced by means of a mass at a radius of 187·5 mm in the plane of crank 1, and a mass at radius 250 mm attached to the flywheel situated 225 mm beyond the bearing adjacent to crank 3. Determine the magnitude of these balance masses and their angular positions relative to crank 1.

(U. Lond.)

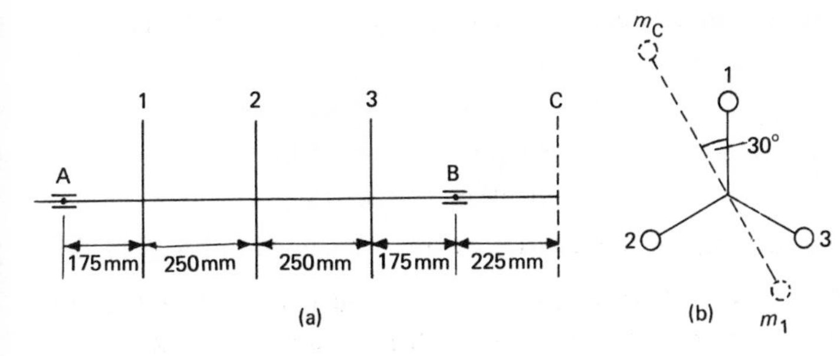

FIG. 7.17

Solution

Since the cranks are symmetrical, the forces are balanced.

The plane of A, Fig. 7.17 (a), is chosen as the reference plane. Using the given data, the following table is compiled.

Plane	*m* (*kg*)	*r* (*m*)	*mr* (*kg m*)	*l* (*m*)	*mrl* (*kg m*²)
A	—	—	*R*	0	0
1	15	0·125	1·875	0·175	0·328
2	15	0·125	1·875	0·425	0·797
3	15	0·125	1·875	0·675	1·266
B	—	—	*R*	0·850	0·85*R*

From the couple polygon, Fig. 7.18,

$$0{\cdot}85R = 0{\cdot}8125 \text{ kg m}^2$$

$$\therefore\ R = 0{\cdot}9566 \text{ kg m}$$

$$\therefore \text{ dynamic load at bearing B} = 0{\cdot}956 \times \left(\frac{2\pi}{60} \times 500\right)^2 = \underline{2620 \text{ N}}$$

The dynamic load on bearing A is equal and opposite to that on bearing B.

To balance the shaft, the balance masses in planes 1 and C must provide a couple equal to that provided by the reactions at the bearings.

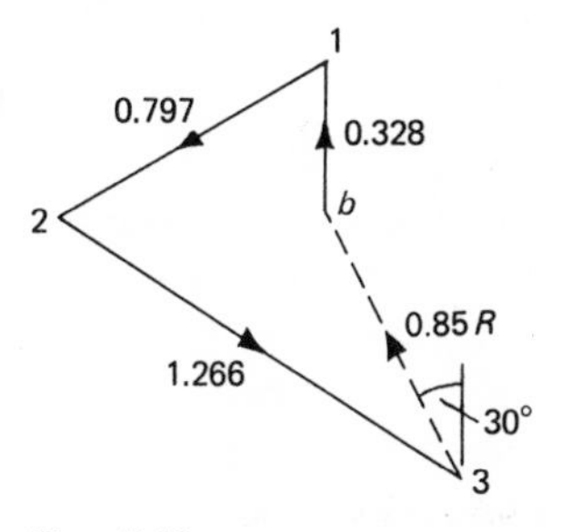

FIG. 7.18

If F is the mass–arm product in planes 1 and C, then, since the planes are 900 mm apart,

$$F = 0{\cdot}8125/0{\cdot}9 = 0{\cdot}903 \text{ kg m}$$

Therefore, in plane 1, $\quad m_1 = 0{\cdot}903/0{\cdot}1875 = \underline{4{\cdot}815 \text{ kg}}$

and, in plane C, $\quad m_C = 0{\cdot}903/0{\cdot}25 = \underline{3{\cdot}61 \text{ kg}}$

The angular positions of the balance masses are shown in Fig. 7.17 (b).

Example 7.5 A shaft 1·4 m long carries four eccentric loads A, B, C and D, spaced at 0, 0·45, 0·75 and 1·4 m from one end. The loads are respectively 7·5, 11, 15 and 6 kg, and the eccentricities are 36, 48, 54 and 96 mm. The directions of the eccentricities of B, C and D relative to A are 60, 200 and 270°. The shaft is carried in bearings E and F, which are 0·175 m and 1·0 m from A, E being between A and B.

(a) Determine the maximum and minimum vertical forces on bearing F when the shaft rotates at 90 rev/min.

(b) What is the maximum speed at which the shaft may run, to ensure that the vertical component of the load on F is always downwards? (U. Lond.)

Solution

The plane of E, Fig. 7.19 (a), is chosen as the reference plane. Using the given data, the following table is compiled.

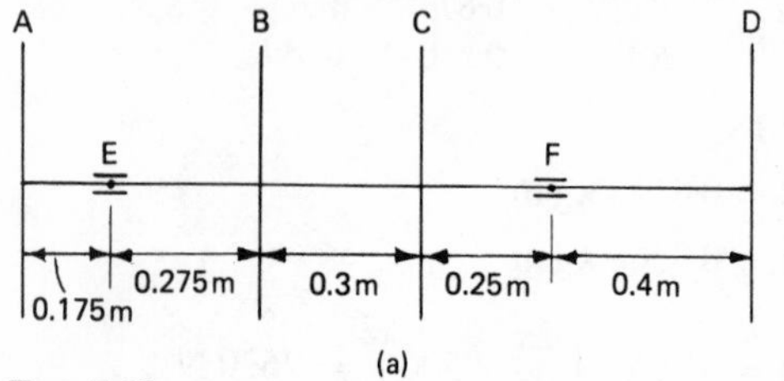

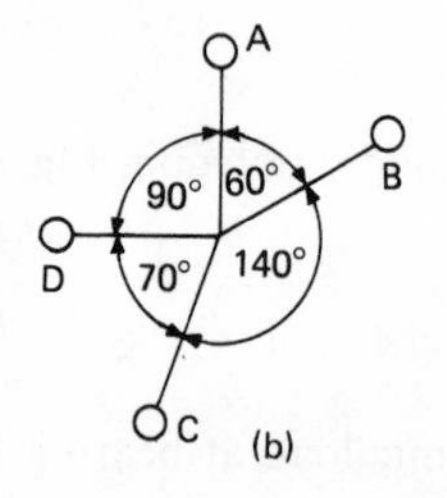

FIG. 7.19

Plane	m (kg)	r (mm)	mr (kg mm)	l (m)	mrl (kg mm m)
A	7·5	36	270	0·175	−47·3
E	—	—	R_e	0	0
B	11	48	528	0·275	145·2
C	15	54	810	0·575	465·8
F	—	—	R_f	0·825	$0{\cdot}825\,R_f$
D	6	96	576	1·225	705·5

From the couple polygon, Fig. 7.20,

$$0{\cdot}825 R_f = 848 \text{ kg mm m}$$

$$\therefore\ R_f = 1028 \text{ kg mm}$$

$$\therefore \text{ dynamic reaction at F} = \frac{1028}{1000} \times \left(\frac{2\pi}{60} \times 90\right)^2 = 91{\cdot}3 \text{ N}$$

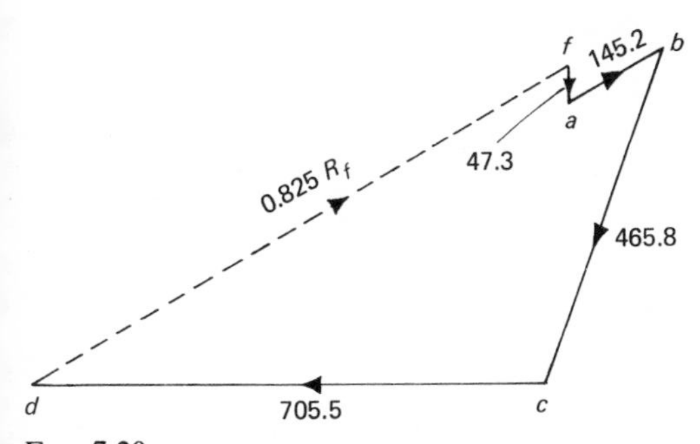

FIG. 7.20

If P_f is the static reaction at F, then, taking moments about plane E,

$$0{\cdot}825\,P_f + (7{\cdot}5 \times 9{\cdot}81 \times 0{\cdot}175) = (11 \times 0{\cdot}275 + 15 \times 0{\cdot}575 + 6 \times 1{\cdot}225) \times 9{\cdot}81$$

from which $\quad P_f = 210\text{ N}\quad$ downwards

$\therefore$ maximum vertical force at F $= 210 + 91{\cdot}3 = \underline{301{\cdot}3\text{ N}}$

and $\quad$ minimum vertical force at F $= 210 - 91{\cdot}3 = \underline{118{\cdot}7\text{ N}}$

For the vertical component of the force at F to be always downwards,

$$1{\cdot}028 \times \left(\frac{2\pi}{60}N\right)^2 \leqslant 210$$

i.e. $\quad N \leqslant \underline{136{\cdot}5\text{ rev/min}}$

Example 7.6 A shaft 1·3 m long carries three unbalanced wheels, A, B and C, spaced at 0, 0·7 and 1·3 m from one end respectively. The masses are 8, 10 and 6 kg, and the eccentricities of the centres of mass are 60, 50 and 40 mm respectively. The directions of the eccentricities of B and C relative to A are 60° and 270° respectively. The shaft is supported in bearings at X and Y which are 0·2 m and 1 m from A.

(a) Find the forces on the bearings, in magnitude and direction, when the shaft rotates at 90 rev/min.

(b) A mass is to be bolted to the wheel C at a radius of 100 mm to make the forces on the bearings equal and opposite. Find the mass required and its angular position.

Solution

(a) The plane of X, Fig. 7.21 (a), is chosen as the reference plane. Using the given data, the following table is compiled.

Plane	*m* (*kg*)	*r* (*mm*)	*mr* (*kg mm*)	*l* (*m*)	*mrl* (*kg mm m*)
A	8	60	480	−0·2	−96
B	10	50	500	0·5	250
C	6	40	240	1·1	264
X	—	—	R_x	0	0
Y	—	—	R_y	0·8	0·8 R_y

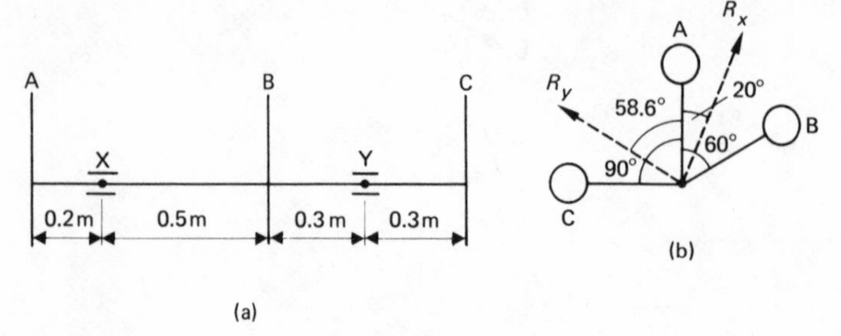

FIG. 7.21

From the couple polygon, Fig. 7.22,

$$0{\cdot}8\,R_y = 55{\cdot}6\,\text{kg mm m}$$

$$\therefore R_y = 69{\cdot}6\,\text{kg mm}$$

$$\therefore \text{ dynamic force at Y} = R_y\omega^2 = \frac{69{\cdot}6}{1000}\left(\frac{2\pi}{60}\times 90\right)^2 = \underline{6{\cdot}18\,\text{N}}$$

From the force polygon, Fig. 7.23,

$$R_x = 73{\cdot}8\,\text{kg mm}$$

$$\therefore \text{ dynamic force at X} = R_x\omega^2 = \frac{73{\cdot}8}{1000}\left(\frac{2\pi}{60}\times 90\right)^2 = \underline{6{\cdot}56\,\text{N}}$$

The directions of the forces *on* the bearings are as shown in Fig. 7.21 (b).

(b) If the bearing forces are to be equal and opposite, the forces in planes A, B and C must balance to give a pure rocking couple. To achieve this, the original point c must be moved to c', Fig. 7.24, necessitating the addition of the vector cc' to bc.

From the diagram,

$$cc' = 755\,\text{kg mm}$$

Thus the mass required at a radius of 100 mm is given by

$$m = 755/100 = \underline{7{\cdot}55\,\text{kg}}$$

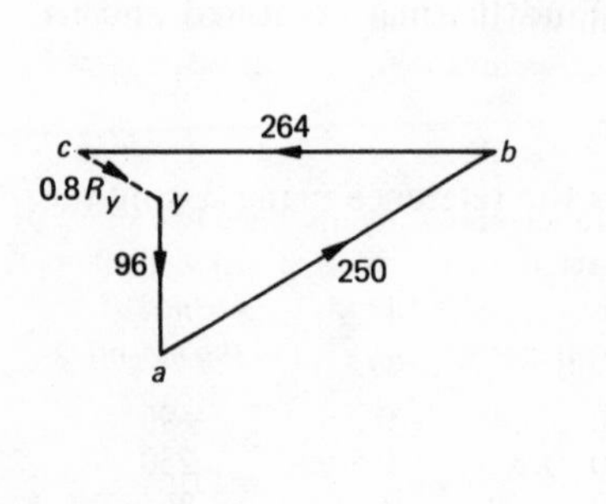

FIG. 7.22

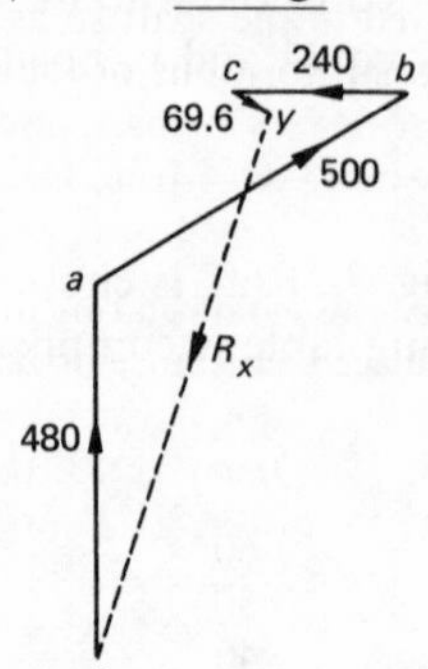

FIG. 7.23

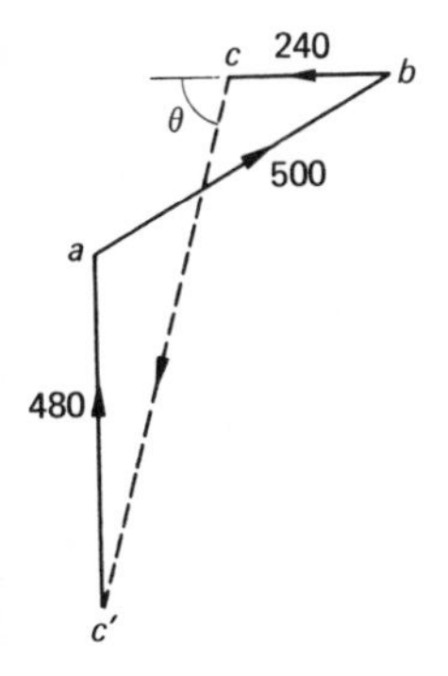

Fig. 7.24

The direction of this mass, θ, is 85° anticlockwise from the original eccentricity of the wheel C.

Problems

1. Two similar discs, A and B, are mounted on a shaft with their centre planes 200 mm apart. Masses C, D and E are attached to the disc A in the positions shown in the following table.

	Mass (kg)	*Radius (mm)*	*Angular position (degrees)*
C	2	100	0
D	1·5	125	60
E	3	87·5	135

Determine: (a) the unbalanced force on the shaft when its speed is 300 rev/min, and (b) the magnitude and angular position of a mass attached to disc B, at a radius of 150 mm, that will make the resultant radial force zero. What will be the rocking couple in case (b) when the speed is 120 rev/min? (I. Mech. E.)

(*Answer:* 360 N; 2·43 kg; 252·73°; 11·51 N m)

2. A shaft carries three pulleys, A, B and C, the pulleys B and C being 1·2 m and 2·1 m from A. The pulleys are out-of-balance to the extent of 0·03, 0·05 and 0·04 kg m respectively and are keyed to the shaft so as to give static balance.

Find the relative angular positions of the radii which define the unbalance. If the shaft is supported in bearings 1·8 m apart, find the dynamic load on each bearing when the shaft makes 300 rev/min. (I. Mech. E.)

(*Answer:* relative to A: B, 126·87°; C, 270°; 27·9 N)

3. Four masses, at equal radii and rotating in parallel planes, are attached to a shaft. In each of the end planes, spaced a distance of $2b$ apart, there is a mass of 12 kg. The inner planes, spaced a distance $2a$ apart, are symmetrical with the end planes and each contains a mass of 15 kg. If the masses on the inner planes are at right angles to each other, find the ratio of a to b and the relative angular positions of the other masses for complete balance. (U. Lond.)

(*Answer:* 0·53; 163° between each inner and outer mass)

4. A turbine rotor is found to be out of balance to the extent of 1·5 kg at 0·45 m radius in the plane AA and 2 kg at 0·6 m radius in the plane BB, the relative angular positions

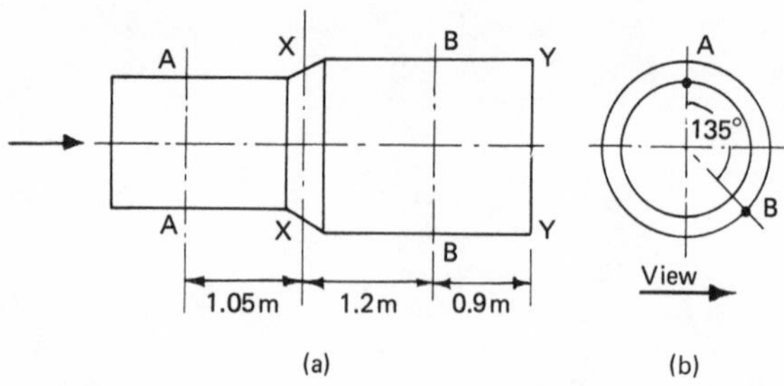

FIG. 7.25

being given in the end view, Fig. 7.25. It is desired to balance these masses by a mass in each of the planes XX and YY at radii of 0·525 m and 0·45 m respectively.

Determine the magnitude and positions of these masses and show their positions in an end view. (U. Glas.)

(*Answer:* X, 1·42 kg, 209·27° from A; Y, 2·12 kg, 329·1° from A)

5. A rough casting for a rotor has a mass of 225 kg and is mounted on centres 1·15 m apart, ready for machining; it is given static balance by two masses A and B in planes which are situated 0·5 m and 0·4 m respectively on either side of the plane containing the centre of mass. The masses A and B are 10 kg and 12 kg respectively and their centres of mass are at 90° to each other relative to the axis of the casting, and at 0·375 m and 0·45 m radius respectively.

Determine the eccentricity of the centre of mass of the casting, its position relative to that of mass A, and the forces on the centres when the rotor with attached masses A and B is run at 50 rev/min. (U. Lond.) (*Answer:* 29·38 mm; 124·78°; 68·2 N)

6. A horizontal shaft supported in bearings 1·2 m apart carries two masses, A and B, each of mass 6 kg, placed at distances of 0·3 m and 0·6 m respectively from one of the bearings. Static balance is obtained by the addition of a balance mass of 5 kg acting at a radius of 200 mm, the position of the balance mass being midway between A and B. If the radii of the centres of mass of A and B are 175 mm and 225 mm respectively, find the relative angular positions of the three masses, and also the magnitude of the unbalanced couple acting on the bearings when the shaft rotates at 100 rev/min. (U. Lond.)

(*Answer:* 130·7° between A and B; 129·57° between B and balance mass; 36·2 N m)

7. A shaft is supported in two bearings 2·4 m apart and projects 0·6 m beyond the bearings at each end. The shaft carries three pulleys, one at each end and one at the middle of its length. The end pulleys have masses of 90 kg and 50 kg and their centres of mass are at 3·75 mm and 5 mm respectively from the shaft axis. The centre pulley has a mass of 70 kg and its centre of mass is 6·25 mm from the shaft axis. If the pulleys are arranged so as to give static balance, determine the dynamic forces produced on the bearings when the shaft rotates at 300 rev/min. (U. Lond.) (*Answer:* 297·4 N)

8. Four pulleys, A, B, C and D, are mounted on a shaft. Due to carelessness in manufacture, the centres of mass of the pulleys do not lie on the shaft axis, but are displaced slightly from it, as indicated by the following table.

Pulley	*Mass (kg)*	*Displacement of centre of mass from the axis of rotation*
A	500	7·5 mm
B	750	6·0 mm at 30° to A
C	750	6·0 mm at 90° to A
D	500	4·5 mm at 150° to A

This shaft is carried in bearings at E and F, the pulleys and bearings being situated so that the axial distances along the shaft are AB = +0·6 m, AC = +0·9 m, AD = +1·5 m, AE = +0·3 m and AF = +1·2 m.

If the system rotates at 300 rev/min, determine the magnitude of the bearing reactions arising from the lack of balance, and the direction of these reactions relative to the out-of-balance force from pulley A. (U. Lond.)

(*Answer:* 8·54 kN at 198°; 5·76 kN at 296°)

9. The out-of-balance of a machine rotor is equivalent to 5 kg at 10 mm radius in one plane A, together with an equal mass at 15 mm radius in a second plane B. AB = 375 mm and the two radii are at 120°. Find the mass required in a third plane C at a radius of 125 mm and its angular position with respect to the given radii so that there is no resultant out-of-balance force. Find also the position of C along the axis for the residual couple to be a minimum and the value of this couple when the speed is 500 rev/min. (I. Mech. E.)

(*Answer:* 0·529 kg; 260° to A, 321·5 mm from A; 50·5 N m)

10. A three-throw crankshaft has double-webbed cranks of 150 mm radius set at 120° to each other and equally spaced with a pitch of 500 mm. The rotating masses at crank radius are: No. 1, 30 kg; No. 2, 40 kg; No. 3, 40 kg. Balance is to be effected by a balance mass attached to the outside web of crank No. 1, with a centre of mass 150 mm from the central plane of the crank and at a radius of 225 mm from the centre of the shaft; and also by removing material, at a radius of 750 mm, from a wheel fixed 750 mm beyond the central plane of crank No. 3.

Determine the masses to be fitted and removed and their angular positions relative to crank No. 1. (U. Lond.)

(*Answer:* 3·8 kg removed at 151·3° from No. 1; 7·49 kg added at 126° from No. 1)

11. Three rotating masses, A = 14 kg, B = 11 kg and C = 21 kg, are carried on a shaft, with centres of mass 275 mm, 400 mm and 150 mm respectively from the shaft axis. The angular positions of B and C are 60° and 135° respectively from A, measured in the same direction. The distance between the planes of rotation of A and B is 1·35 m, and between those of A and C is 3·6 m, B and C being on the same side of A.

Two balance masses are to be fitted, each with its centre of mass 225 mm from the shaft axis, in planes midway between those of A and B and of B and C. Determine the magnitude and angular position with respect to A of each balance mass. (U. Lond.)

(*Answer:* 36·7 kg between A and B, 187° to A; 29·3 kg between B and C, 310° to A)

12. A rotating shaft is to carry four discs, A, B, C and D, whose masses are 15 kg, 25 kg, 20 kg and 18 kg respectively; they are to be spaced at intervals of 300 mm, 375 mm and 225 mm along the shaft. The centres of mass of A, B and C are at 36 mm, 12 mm and 18 mm respectively from their axes of rotation.

(a) The three discs A, B and C are first attached to the shaft, at the spacing specified above and with their centres of mass in positions which give static balance. Find the relative angular positions of their centres of mass.

(b) The disc D is then attached to the shaft in such a position as to nullify the out-of-balance couple, when the reference plane passes through A. Find the eccentricity of the centre of mass of D and its angular position relative to that of A. (U. Lond.)

(*Answer:* (a) B, 141·1°, and C, 211·33°, relative to A; (b) 17·8 mm, 14°)

13. A shaft carries four wheels, A, B, C and D, equally pitched 250 mm apart. The unbalance (*mr*) values for A and C are, respectively, 0·05 kg m and 0·06 kg m, and the line of unbalance in C is at 90° to that of A, which may be taken as the reference direction. The out-of-balance amounts for B and D are initially unknown, but the complete rotor is dynamically balanced by adding a mass of 0·324 kg to wheel B at a radius of 0·6 m and at an angle of 215° to the reference, and by removing material of mass 0·08 kg from D at a radius of 0·45 m and at an angle of 120°.

Determine the initial and final unbalance values for B and D. (U. Lond.)

(*Answer:* B, initial 0·117 kg m, 44°; final 0·081 kg m, 202°;
D, initial 0·007 kg m, 9·5°; final 0·039 kg m, 310°)

14. A shaft turning at a uniform speed carries two uniform discs A and B of masses 10 kg and 8 kg respectively. The centres of mass of the discs are each 2·5 mm from the axis of rotation. The radii to the centres of mass are at right angles. The shaft is carried in bearings C and D between A and B such that AC = 0·3 m, AD = 0·9 m and AB = 1·2 m.

It is required to make the dynamic loading on the bearings equal and a minimum for any given shaft speed by adding a mass at a radius of 25 mm in a plane E.

Determine: (a) the magnitude of the mass in plane E and its angular position relative to the radius through the centre of mass in plane A, (b) the distance of plane E from plane A, and (c) the dynamic loading on each bearing when the mass in plane E has been attached and the shaft turns at 200 rev/min. (U. Lond.)

(*Answer:* 0·3 kg; 51·33°; 1·965 m; 6·97 N, bearing loads in same direction;
1·28 kg; 141·33°; 0·483 m; 13·94 N, bearing loads in opposite direction)

(The first set of answers gives a lower value for the minimum bearing loads.)

8
Friction

8.1 Friction between dry unlubricated surfaces

When one body slides on another, the tangential force resisting the motion is directly proportional to the normal force between the surfaces in contact. The ratio of friction force to normal force is termed the *coefficient of friction* and is denoted by μ. The coefficient of friction is a constant for any given pair of surfaces and is approximately independent of the area of contact, the sliding velocity and the intensity of pressure.

The force required to start a body moving is slightly more than that necessary to keep the body moving at uniform speed, which leads to the concept of the coefficients of static and kinetic friction. Values of these coefficients quoted in problems must be assumed to be those appropriate to the nature of the problem.

8.2 Motion on a horizontal plane

If a body slides on a horizontal surface at uniform speed under the action of a horizontal force P, Fig. 8.1, the friction force is equal to P and the normal force between the surfaces is the weight of the body, W.

Thus

$$P = \mu W \tag{8.1}$$

The reaction of the plane, R, is found from the triangle of forces and the angle ϕ between R and the normal to the plane is called the *friction angle*.

Thus

$$\tan \phi = \frac{P}{W} = \mu \tag{8.2}$$

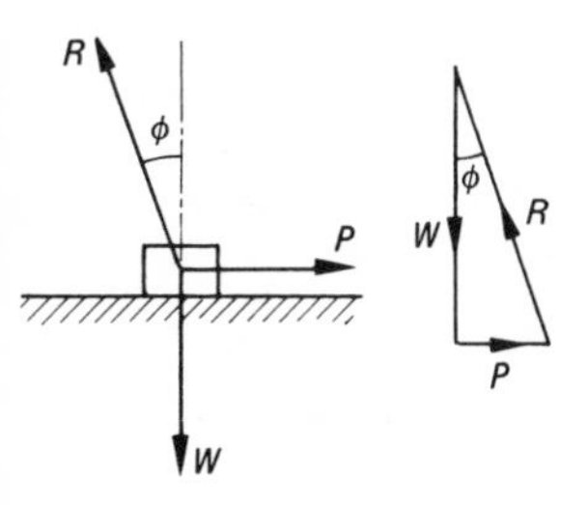

Fig. 8.1

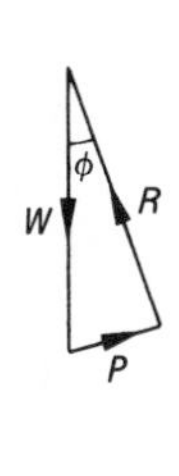

Fig. 8.2

If the force P is not parallel to the surfaces in contact as shown in Fig. 8.2, the reaction R is still inclined at the angle ϕ to the normal and P is then obtained from the triangle of forces.

The sense of the angle ϕ relative to the normal is such that the force R has a component which tends to oppose the motion of the body.

8.3 Motion on an inclined plane

Motion up plane If a body moves at uniform speed up a plane inclined at an angle α to the horizontal under the action of a force P, Fig. 8.3, the three forces W, P and R are represented by a triangle of forces in which the angle between R and W is $\alpha + \phi$.

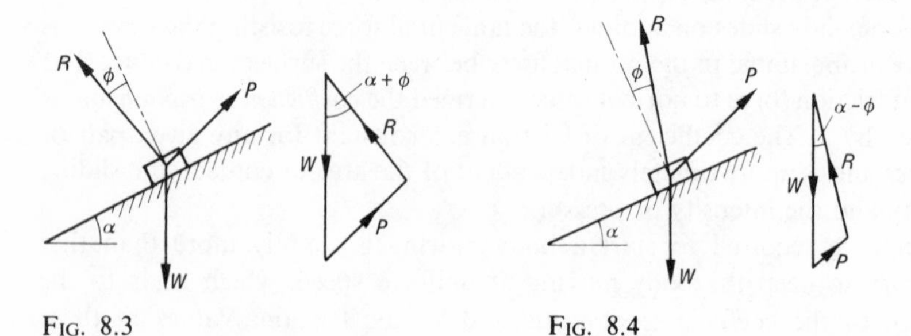

FIG. 8.3 FIG. 8.4

Motion down plane The nature of the triangle of forces depends on whether α is greater than, less than, or equal to the friction angle ϕ.

(a) $\alpha > \phi$, Fig. 8.4. The angle between R and W in the triangle of forces is $\alpha - \phi$ and the direction of P is such that it lowers the body down the plane at uniform speed.

(b) $\alpha < \phi$, Fig. 8.5. The angle between R and W in the triangle of forces is $\phi - \alpha$ and the direction of P is such that it pushes the body down the plane at uniform speed.

(c) $\alpha = \phi$, Fig. 8.6. The direction of R is vertical and so P is zero, i.e. no force is required to lower or push the body down the plane. Thus, when the angle of inclination is equal to the friction angle, the body will just slide down the incline at uniform speed.

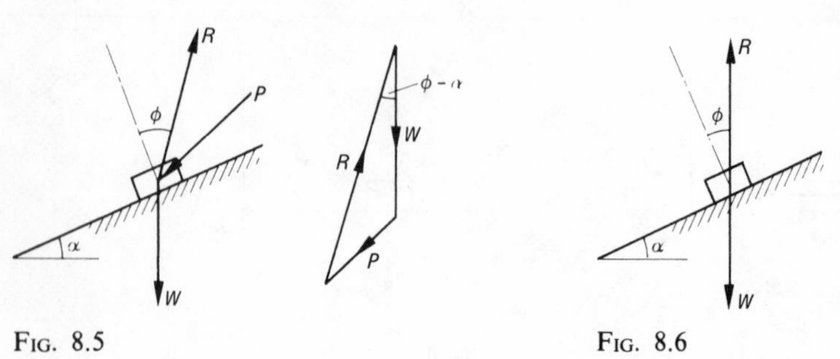

FIG. 8.5 FIG. 8.6

8.4 Special cases of motion on an inclined plane

Force *P* horizontal The triangle of forces becomes right-angled, Fig. 8.7, and hence

$$P = W \tan(\alpha + \phi) \qquad \text{for motion up the plane, Fig. 8.7(a)}$$

and

$$P = W \tan(\alpha - \phi) \qquad \text{for motion down the plane, Fig. 8.7(b)}$$

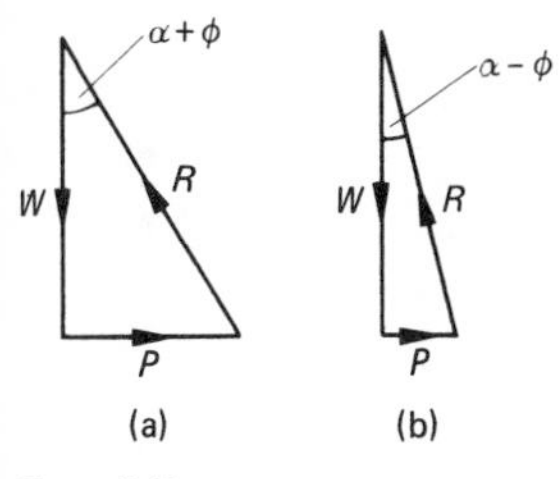

FIG. 8.7

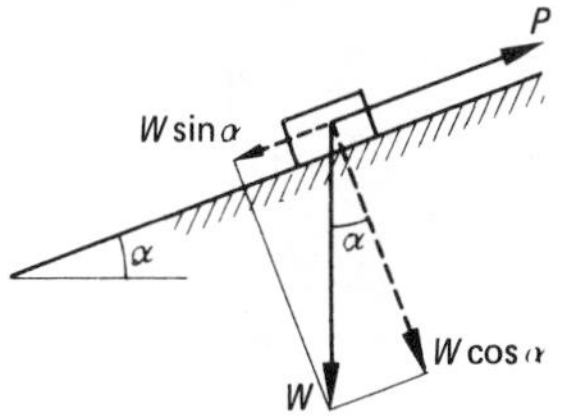

FIG. 8.8

Force *P* parallel to the plane P can be obtained, as before, from the triangle of forces but, alternatively, the force W can be resolved into components $W \sin\alpha$ and $W \cos\alpha$ which are parallel and perpendicular to the plane respectively, Fig. 8.8. Then

$$P = W \sin\alpha + \mu W \cos\alpha \qquad \text{for motion up the plane}$$

and

$$P = W \sin\alpha - \mu W \cos\alpha \qquad \text{for motion down the plane}$$

If $\mu W \cos\alpha > W \sin\alpha$ (i.e. $\phi > \alpha$), the direction of P in the latter case becomes reversed and must be applied to push the body down the plane with uniform speed.

If $\mu W \cos = W \sin\alpha$, P is zero and

$$\tan\alpha = \mu = \tan\phi$$

i.e. the plane is inclined at the friction angle to the horizontal.

8.5 Screw threads

Let W be the axial force against which the screw is turned, Fig. 8.9, and P be the tangential force at the mean thread radius necessary to turn the nut. The development of a thread is a plane inclined at the helix angle to the thread axis; if d is the mean thread diameter and p is the pitch, then

$$\tan\alpha = \frac{p}{\pi d} \tag{8.3}$$

Turning the nut on the screw is equivalent to moving the load W up or down the incline by means of a horizontal force P applied at the mean thread radius.

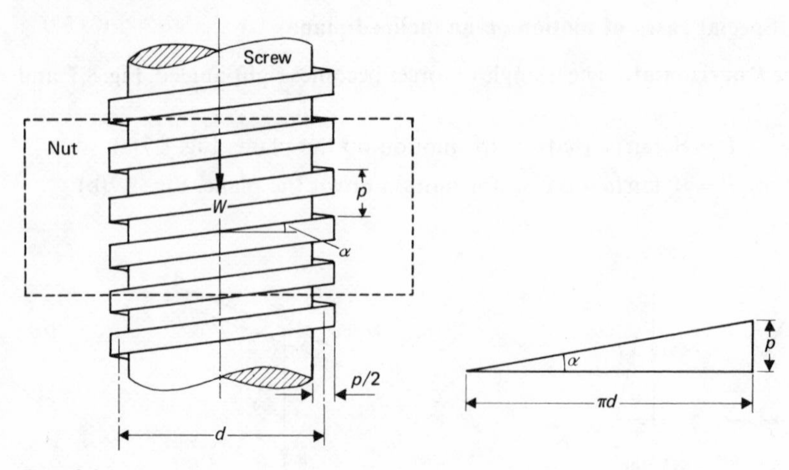

FIG. 8.9

For motion up the plane, i.e. raising the nut, the triangle of forces is shown in Fig. 8.10, from which

$$P = W \tan(\alpha + \phi) = W \frac{\tan \alpha + \tan \phi}{1 - \tan \alpha \tan \phi}$$

$$= W \frac{\tan \alpha + \mu}{1 - \mu \tan \alpha} \tag{8.4}$$

The torque is given by

$$T = P \times \frac{d}{2} = \frac{Wd}{2} \frac{\tan \alpha + \mu}{1 - \mu \tan \alpha} \tag{8.5}$$

For motion down the plane, i.e. lowering the nut, the triangle of forces is shown in Fig. 8.11, from which

$$P = W \tan(\alpha - \phi) = W \frac{\tan \alpha - \mu}{1 + \mu \tan \alpha} \tag{8.6}$$

and
$$T = \frac{Wd}{2} \frac{\tan \alpha - \mu}{1 + \mu \tan \alpha} \tag{8.7}$$

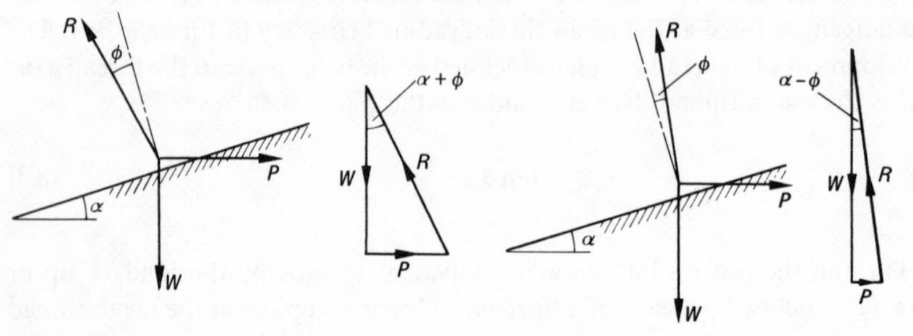

FIG. 8.10 FIG. 8.11

In most practical cases of screw threads, ϕ is greater than α, so that the direction of P becomes reversed.

In the case of multi-start threads, the pitch p must be replaced by the lead in calculating the helix angle α. The load W may be divided by the number of individual helices comprising the thread and the separate torques added together to give the total torque but this would be the same as assuming that the total load is caused by one thread.

8.6 Efficiency of a screw thread

For motion up the plane, i.e. against the load, the efficiency is the ratio of the work done by W to the work done by P,

i.e.
$$\eta = \frac{W \times p}{P \times \pi d} = \frac{\tan \alpha}{\tan(\alpha + \phi)} \tag{8.8}$$

For motion down the plane, i.e. with the load, the efficiency is the ratio of the work done by P to the work done by W, this being termed the reversed efficiency,

i.e.
$$\eta = \frac{P \times \pi d}{W \times p} = \frac{\tan(\alpha - \phi)}{\tan \alpha} \tag{8.9}$$

For the efficiency to be a maximum,

$$\frac{d}{d\alpha}\left(\frac{\tan \alpha}{\tan(\alpha + \phi)}\right) = 0$$

i.e.
$$\tan(\alpha + \phi)\sec^2 \alpha = \tan \alpha \sec^2(\alpha + \phi)$$

i.e.
$$\sin 2(\alpha + \phi) = \sin 2\alpha$$

$$\therefore\ 2(\alpha + \phi) = 2\alpha \quad \text{or} \quad \pi - 2\alpha$$

$$\therefore\ \alpha = \frac{\pi}{4} - \frac{\phi}{2}, \text{ ignoring the alternative solution}$$

$$\therefore\ \text{maximum efficiency} = \tan\left(\frac{\pi}{4} - \frac{\phi}{2}\right) \Big/ \tan\left(\frac{\pi}{4} + \frac{\phi}{2}\right) = \frac{1 - \sin \phi}{1 + \sin \phi} \tag{8.10}$$

For the reversed efficiency to be a maximum,

$$\frac{d}{d\alpha}\left(\frac{\tan(\alpha - \phi)}{\tan \alpha}\right) = 0$$

from which
$$\alpha = \frac{\pi}{4} + \frac{\phi}{2}$$

$$\therefore\ \text{maximum efficiency} = \tan\left(\frac{\pi}{4} - \frac{\phi}{2}\right) \Big/ \tan\left(\frac{\pi}{4} + \frac{\phi}{2}\right), \text{ as before}$$

8.7 Modification for V-threads

For a V-thread, the normal force between the nut and the screw is increased since the axial component of this force must equal W. Thus, if the semi-angle of the thread is β, Fig. 8.12, then

$$\text{normal force} = W \sec \beta$$

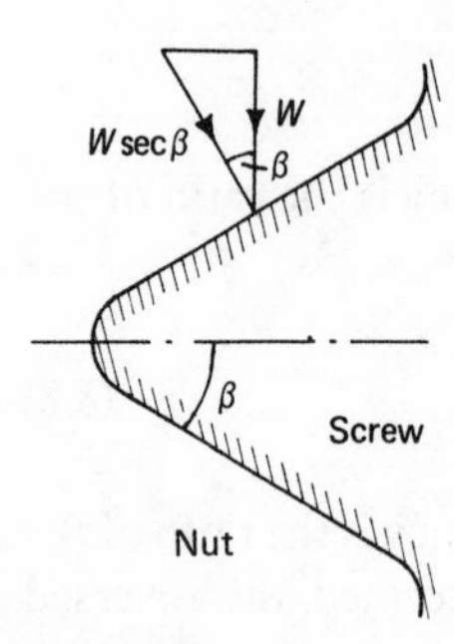

FIG. 8.12

The friction force is therefore increased in the ratio $\sec \beta : 1$, so that the V-thread is equivalent to a square thread having an equivalent, or *virtual*, coefficient of friction $\mu \sec \beta$.

8.8 Overhauling of a machine

Consider a machine in which a load W is just lifted a distance x by an effort P which moves a distance y in the same time. If the work done against friction is F, then

$$\text{efficiency} = \frac{Wx}{Py} \tag{8.11}$$

and

$$F = Py - Wx \tag{8.12}$$

If the machine just reverses, or *overhauls*, when the effort is removed, work done by the load in descending a distance $x = Wx$.

If the friction is assumed to be the same as when the load is raised,

$$Wx = F$$

Therefore, from equation 8.12,

$$Wx = Py - Wx$$

i.e.

$$Py = 2Wx$$

$$\therefore \eta = \frac{Wx}{2Wx} = \tfrac{1}{2} \quad \text{or} \quad 50 \text{ per cent}$$

Thus a machine will overhaul if the forward efficiency $\geqslant$ 50 per cent.

Examples

Example 8.1 Two wedges A and B, of mass 100 kg and 5 kg respectively, are in contact as shown in Fig. 8.13. If the coefficient of friction between all the contacting surfaces is 0·1, determine the least value of the force P to push the wedge A upwards. What would be its value if there were no friction?

(U. Lond.)

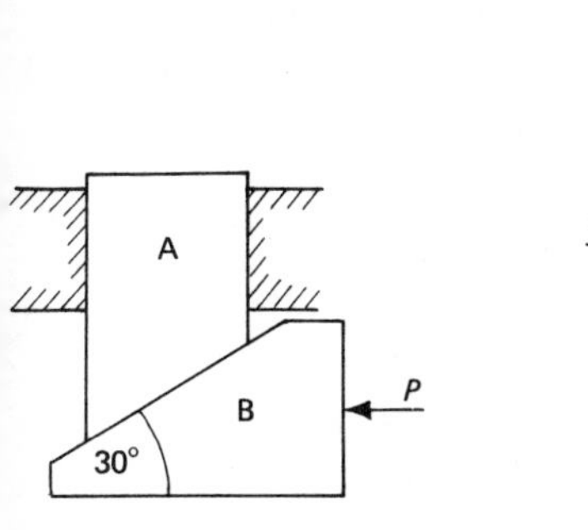

FIG. 8.13

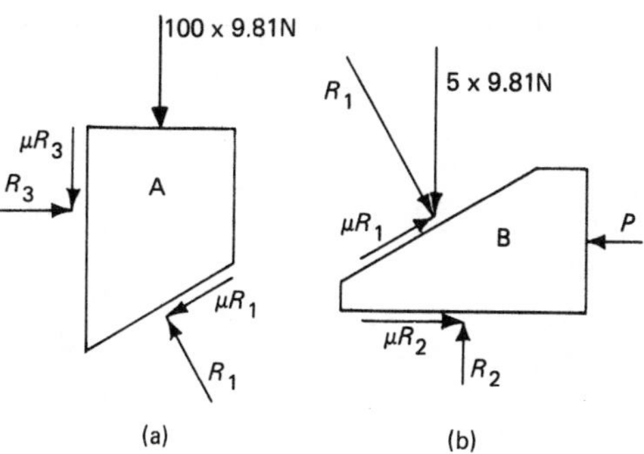

FIG. 8.14

Solution

The forces on the wedges when motion is about to take place are shown in Fig. 8.14, the effect of tilting being neglected.

Resolving forces vertically and horizontally for wedge A,

$$R_1 \cos 30^\circ = 100 \times 9{\cdot}81 + \mu R_3 + \mu R_1 \sin 30^\circ$$

i.e. $$0{\cdot}816 R_1 - 0{\cdot}1 R_3 = 981 \qquad (1)$$

and $$R_3 = \mu R_1 \cos 30^\circ + R_1 \sin 30^\circ$$

i.e. $$0{\cdot}5866 R_1 - R_3 = 0 \qquad (2)$$

and for wedge B,

$$5 \times 9{\cdot}81 + R_1 \cos 30^\circ = R_2 + \mu R_1 \sin 30^\circ$$

i.e. $$-0{\cdot}816 R_1 + R_2 = 490{\cdot}5 \qquad (3)$$

and $$P = R_1 \sin 30^\circ + \mu R_1 \cos 30^\circ + \mu R_2$$

i.e. $$0{\cdot}5866 R_1 + 0{\cdot}1 R_2 = P \qquad (4)$$

Therefore, from equations (1), (2), (3) and (4),

$$P = \underline{870 \text{ N}}$$

Without friction, equation (1) becomes

$$R_1 \cos 30^\circ = 981$$

and equation (4) becomes

$$P = R_1 \sin 30^\circ$$

Hence $$P = 981 \tan 30^\circ = \underline{566{\cdot}5 \text{ N}}$$

Example 8.2 A rotor rests with its journals in V-supports as shown in Fig. 8.15; the angle of the V is 2θ, the journal radius r and the coefficient of friction at the contacts μ. Show that provided contact at the point A is maintained, the value of the couple required to produce rotation is

$$T = \frac{\mu W r}{(1+\mu^2)\sin\theta}$$

and that contact ceases when the couple attains the value $Wr\cos\theta$.

(I. Mech. E.)

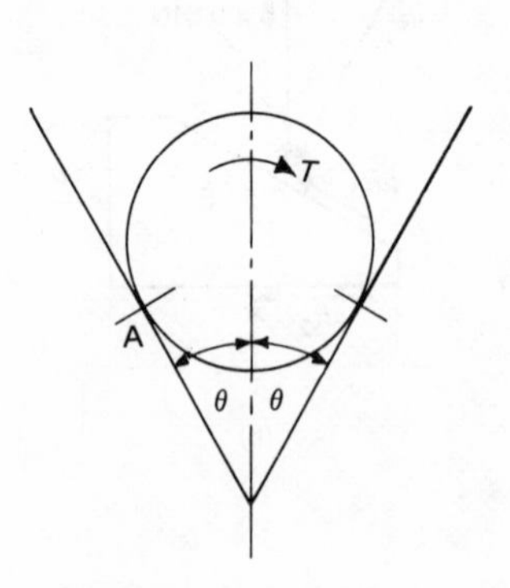

FIG. 8.15

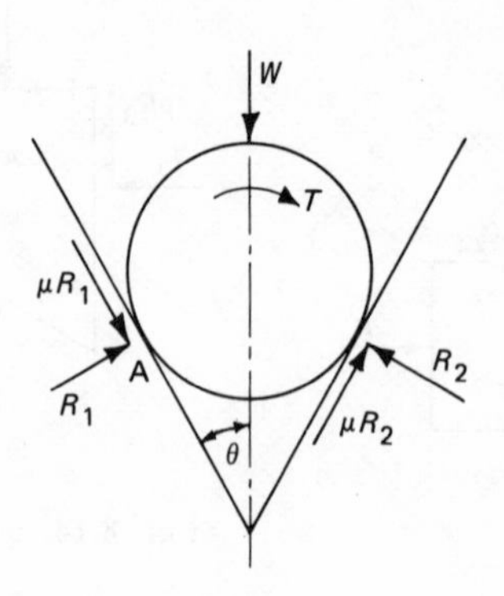

FIG. 8.16

Solution

The forces acting on the shaft when motion is about to take place are shown in Fig. 8.16.

Equating applied and resisting torques,

$$T = \mu(R_1 + R_2)r \tag{1}$$

Equating vertical forces, $W = (R_1 + R_2)\sin\theta + \mu(R_2 - R_1)\cos\theta$ (2)

Equating horizontal forces,

$$(R_2 - R_1)\cos\theta = \mu(R_1 + R_2)\sin\theta \tag{3}$$

Therefore, from equations (2) and (3),

$$W = (1+\mu^2)(R_1 + R_2)\sin\theta$$

Hence, from equation (1), $T = \dfrac{\mu W r}{(1+\mu^2)\sin\theta}$

When R_1 is zero, equation (1) becomes

$$T = \mu R_2 r$$

equation (2) becomes $W = R_2(\sin\theta + \mu\cos\theta)$

i.e. $R_2 = \dfrac{W}{\sin\theta + \mu\cos\theta}$

and equation (3) becomes $R_2\cos\theta = \mu R_2\sin\theta$

i.e.

$$\mu = \cot\theta$$

$$\therefore \; T = \cot\theta \times \frac{Wr}{\sin\theta + \cot\theta\cos\theta}$$

$$= \underline{Wr\cos\theta}$$

Note that the given conditions can only apply without acceleration of the shaft if μ has the value $\cot\theta$.

Example 8.3 In the friction gear shown in Fig. 8.17, the roller A is the driver and the disc B is the follower. The pressure distribution along the line of contact between roller and disc is uniform. When the power output on the follower is the greatest possible, show that the greatest possible efficiency is

$$\frac{2a+b}{2(a+b)}$$ (U. Lond.)

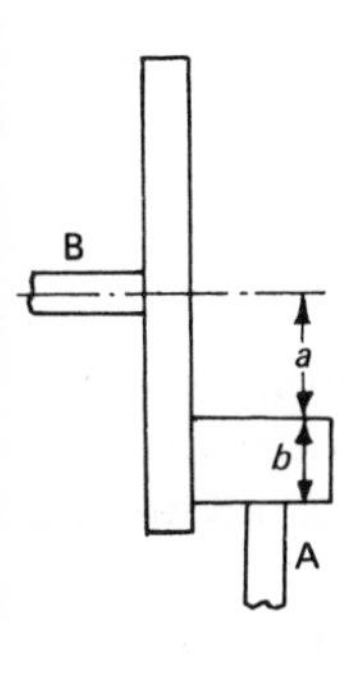

FIG. 8.17

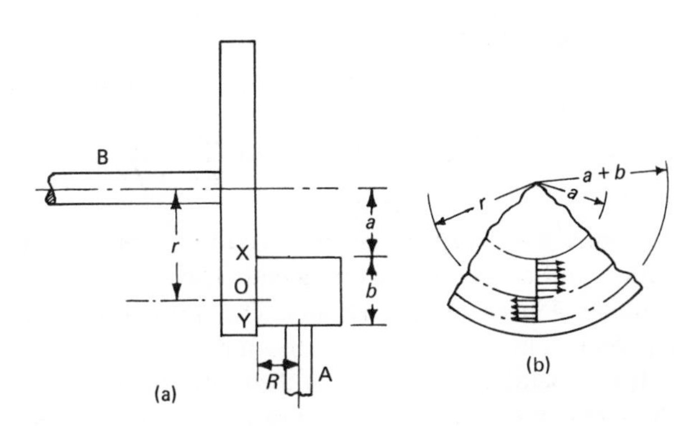

FIG. 8.18

Solution

If r is the radius at which there is no slip between the driver and the follower, Fig. 8.18 (a), then $\omega_A/\omega_B = r/R$. At points between O and X, the roller A is moving faster than are the corresponding points on B and between O and Y it is moving slower than are the corresponding points on B. Hence the friction forces acting on B are as shown in Fig. 8.18 (b).

If p is the force per unit length between the driver and follower,

$$\text{normal force on length OX} = p(r-a)$$

$$\therefore \; \text{friction force on length OX} = \mu p(r-a)$$

$$\therefore \; \text{friction torque on B due to length OX} = \mu p(r-a)\frac{(r+a)}{2}$$

$$= \frac{\mu p}{2}(r^2 - a^2)$$

Similarly,

$$\text{friction torque on B due to length OY} = \mu p(a+b-r)\frac{(a+b+r)}{2}$$

$$= \frac{\mu p}{2}([a+b]^2 - r^2)$$

$$\therefore \text{ resultant friction torque on B, } T_B = \frac{\mu p}{2}\{(r^2-a^2)-([a+b]^2-r^2)\}$$

$$\therefore \text{ power output} = T_B\omega_B = \frac{\mu p}{2}\omega_B\{2r^2-a^2-(a+b)^2\}$$

$$= \frac{\mu p}{2}\omega_A R\left\{2r - \frac{a^2}{r} - \frac{(a+b)^2}{r}\right\}$$

This has its greatest value when $r = a+b$.

$$\therefore \text{ greatest possible power output} = \frac{\mu p}{2}\omega_A Rb\left(\frac{2a+b}{a+b}\right)$$

For this condition, torque on A $= \mu pbR$

$$\therefore \text{ power input} = \mu pbR\omega_A$$

$$\therefore \text{ efficiency} = \frac{\mu p}{2}\omega_A Rb\left(\frac{2a+b}{a+b}\right)\Big/\mu pbR\omega_A = \underline{\frac{2a+b}{2(a+b)}}$$

Example 8.4 A spring is compressed by means of a screw which is co-axial with it and bears directly on its end. The screw has a 60° V-thread, of mean diameter 48 mm and pitch 8 mm. The spring has a stiffness of 35 kN/m, and has been compressed initially through 80 mm.

If the coefficient of friction at the screw thread is 0·1, find the work done on the screw in compressing the spring through a further 60 mm. (U. Lond.)

Solution

$$\text{Mean load on screw, } W = 35 \times 10^3 \times 0{\cdot}11 = 3850 \text{ N}$$

$$\text{Virtual coefficient of friction} = \mu \sec\beta = 0{\cdot}1 \times \sec 30^\circ = 0{\cdot}1155$$

$$\text{Tan }\alpha = \frac{p}{\pi d} = \frac{8}{\pi \times 48} = 0{\cdot}053$$

$$\therefore \text{ mean torque} = \frac{Wd}{2}\frac{\tan\alpha + \mu}{1-\mu\tan\alpha} \qquad \text{from equation 8.5}$$

$$= \frac{3850 \times 0{\cdot}048}{2}\frac{0{\cdot}053+0{\cdot}1155}{1-0{\cdot}1155\times 0{\cdot}053} = 15{\cdot}7 \text{ N m}$$

$$\text{Angle turned through} = 60/8 = 7{\cdot}5 \text{ rev} = 15\pi \text{ rad}$$

$$\therefore \text{ work done} = 15{\cdot}7 \times 15\pi = \underline{740 \text{ J}}$$

Example 8.5 A turnbuckle is used to tighten a steel wire rope, 46 m long and 80 mm² cross-sectional area, which is rigidly anchored at each end. The turnbuckle has right- and left-handed single-start square threads each of 8 mm pitch and mean diameter 28 mm. The coefficient of friction between the nut and the screw threads is 0·15.

If the turnbuckle has already been adjusted to put a load of 1·8 kN in the rope, determine the work done during further tightening of the rope by one revolution of the turnbuckle. How much of this work is wasted in overcoming friction? Take E for the wire rope as 140 GN/m^2.

Solution

For one turn, stretch of rope = 16 mm

$$\therefore \text{ increase in tension} = \frac{Eax}{l} = \frac{140 \times 10^9 \times 80 \times 10^{-6} \times 0{\cdot}016}{46} = 3900 \text{ N}$$

$$\therefore \text{ mean tension, } W = 1800 + \frac{3900}{2} = 3750 \text{ N}$$

$$\text{Tan}\,\alpha = \frac{p}{\pi d} = \frac{8}{\pi \times 28} = 0{\cdot}091$$

$$\therefore \text{ torque, } T = 2 \times \frac{Wd}{2}\,\frac{\tan\alpha + \mu}{1 - \mu\tan\alpha}$$

$$= 2 \times \frac{3750 \times 0{\cdot}028}{2}\,\frac{0{\cdot}091 + 0{\cdot}15}{1 - 0{\cdot}15 \times 0{\cdot}091} = 25{\cdot}65 \text{ N m}$$

$$\therefore \text{ work done on turnbuckle to tighten rope} = 25{\cdot}65 \times 2\pi = \underline{161{\cdot}3 \text{ J}}$$

$$\text{Work required to stretch rope} = 3750 \times 0{\cdot}016 = 60 \text{ J}$$

$$\therefore \text{ work done against friction} = 161{\cdot}3 - 60 = \underline{101{\cdot}3 \text{ J}}$$

Example 8.6 The cutter of a broaching machine is pulled by a square-threaded screw of 55 mm external diameter and 10 mm pitch; the operating nut takes the axial load of 400 N on a flat surface of 60 mm and 90 mm internal and external diameters respectively.

If the coefficient of friction is 0·15 for all contact surfaces on the nut, determine the power required to rotate the operating nut when the cutting speed is 6 m/min. (U. Lond.)

Solution

$$\text{Mean thread diameter, } d = 55 - 5 = 50 \text{ mm}$$

$$\text{Tan}\,\alpha = \frac{P}{\pi d} = \frac{10}{\pi \times 50} = 0{\cdot}0637$$

$$\text{Torque due to thread load and friction} = \frac{Wd}{2}\,\frac{\tan\alpha + \mu}{1 - \mu\tan\alpha}$$

$$= \frac{400 \times 0{\cdot}05}{2}\,\frac{0{\cdot}0637 + 0{\cdot}15}{1 - 0{\cdot}15 \times 0{\cdot}0637}$$

$$= 2{\cdot}16 \text{ N m}$$

$$\text{Torque due to collar friction} = \mu W r^*$$

$$= 0{\cdot}15 \times 400 \times \frac{0{\cdot}045 + 0{\cdot}03}{2} = 2{\cdot}25\ \text{N m}$$

$$\therefore \text{ total torque on nut} = 2{\cdot}16 + 2{\cdot}25 = 4{\cdot}41\ \text{N m}$$

$$\text{Speed of nut} = 6/0{\cdot}01 = 600\ \text{rev/min}$$

$$\therefore \text{ power} = \frac{2\pi TN}{60} = \frac{2\pi \times 4{\cdot}41 \times 600}{60} = \underline{280\ \text{W}}$$

Example 8.7 A worm wheel of 80 teeth is supported in bearings of 80 mm diameter for which the coefficient of friction is 0·015, and is driven by a double-threaded worm 70 mm diameter, 10 mm pitch (i.e. lead 20 mm). The thrust is carried on a collar of 100 mm mean diameter. The coefficients of friction for worm and collar are respectively 0·08 and 0·06. The worm may be assumed to have a square thread. The wheel-shaft has to deliver 15 kW at 20 rev/min.

Determine the input torque on the worm spindle and the overall efficiency.

(U. Lond.)

Solution

$$\text{Output torque on worm-wheel} = \frac{\text{power} \times 60}{2\pi N}$$

$$= \frac{15 \times 10^3 \times 60}{2\pi \times 20} = 7160\ \text{N m}$$

$$\text{Circular pitch of wheel} = 10\ \text{mm}$$

$$\therefore \text{ p.c.d. of wheel} = \frac{80 \times 0{\cdot}01}{\pi} = 0{\cdot}255\ \text{m}$$

$$\therefore \text{ tangential force on wheel} = 7160/0{\cdot}1275 = 56\,200\ \text{N}$$

This is also the axial force on the worm.

$$\text{Mean diameter of worm} = 0{\cdot}07 - 0{\cdot}005 = 0{\cdot}065\ \text{m}$$

$$\text{Tan}\,\alpha = \frac{\text{lead}}{\pi d} = \frac{0{\cdot}02}{\pi \times 0{\cdot}065} = 0{\cdot}098$$

$\therefore$ torque on worm due to thread load and friction

$$= \frac{Wd}{2} \frac{\tan\alpha + \mu}{1 - \mu \tan\alpha}$$

$$= \frac{56\,200 \times 0{\cdot}065}{2} \frac{0{\cdot}098 + 0{\cdot}08}{1 - 0{\cdot}08 \times 0{\cdot}098}$$

$$= 328\ \text{N m}$$

$$\text{Torque on worm due to collar friction} = \mu W r$$

$$= 0{\cdot}06 \times 56\,200 \times 0{\cdot}05 = 168{\cdot}6\ \text{N m}$$

* It is assumed that the friction force acts at the mean radius.

$$\text{Friction torque at wheel bearings} = 56\,200 \times 0{\cdot}015 \times 0{\cdot}04 = 33{\cdot}8\ \text{N m}$$

$$\therefore\ \text{torque on worm} = 33{\cdot}8 \times 1/40 = 0{\cdot}85\ \text{N m}$$

$$\therefore\ \text{total torque on worm} = 328 + 168{\cdot}6 + 0{\cdot}85 = \underline{497{\cdot}5\ \text{N m}}$$

$$\text{Speed of wheel} = 20\ \text{rev/min}$$

$$\therefore\ \text{speed of worm} = 20 \times 40 = 800\ \text{rev/min}$$

$$\therefore\ \text{input power} = \frac{497{\cdot}5 \times 800 \times 2\pi}{1000 \times 60} = 41{\cdot}7\ \text{kW}$$

$$\therefore\ \text{efficiency} = 15/41{\cdot}7 = 0{\cdot}359 \quad \text{or} \quad \underline{35{\cdot}9\ \text{per cent}}$$

Problems

1. A cylinder is held at rest on a slope of 20° by means of a cord in the vertical plane perpendicular to the axis of the cylinder and bisecting it. The end is attached to the curved surface of the cylinder, is tangential to it and parallel to the slope, i.e. inclined at 20° to the horizontal. If the cylinder has a mass 20 kg, find: (a) the tension in the cord, and (b) the minimum coefficient of friction between the sloping surface and the cylinder compatible with equilibrium. (I. Mech. E.) (*Answer:* 33·6 N; 0·182)

2. A body is pushed up a rough surface inclined at α to the horizontal. The inclination of the slope is such that the body would just slide down without acceleration if not supported (i.e. α is the angle of friction). Find the mechanical efficiency of the lift: (a) if the push is parallel to the incline, and (b) if it is horizontal (I. Mech. E.)
(*Answer:* 50 per cent; $\frac{1}{2}(1 - \tan^2 \alpha)$)

3. A body of mass m on a plane inclined at 20° to the horizontal and for which the coefficient of friction is μ, is acted upon by a force applied upwards and parallel to the plane. When this force has a value of 60 N, the body slides steadily downwards; when the value is 175 N, the body moves steadily upwards. Deduce from these results the values of m and μ.

A different body, of mass 50 kg and with a surface for which, on the same plane, the friction coefficient is 0·15, is to be moved by a force, P, directed at an angle of 15° to the plane, i.e. at 35° to the horizontal. Calculate the value of P which will cause steady upward movement, and also the value to which P must be reduced before downward movement becomes possible. Any formulae used should be established or explained by vector diagrams of forces. (I. Mech. E.) (*Answer:* 35 kg; 0·1785; 235·8 N; 106·4 N)

4. A cylindrical roller, 300 mm diameter, rests on a horizontal floor with its curved surface against the edge of a horizontal step 75 mm high and parallel to the axis of the roller. The coefficient of friction between the roller and step is 0·25. Find the magnitude of the torque about the axis of the roller which will suffice to rotate it in contact with the step if the roller has a mass of 50 kg. (I. Mech. E.) (*Answer:* 19·81 N m)

5. A rectangular sluice gate, 3 m high and 2·4 m wide, can slide up and down between vertical guides. Its vertical movement is controlled by a screw which, together with the weight of the gate, exerts a downward force of 4 kN in the centre line of the sluice. When nearly closed, the gate encounters an obstacle at a point 0·45 m from one end of the lower edge. If the coefficient of friction between the edges of the gate and the guides is 0·25, calculate the thrust tending to crush the obstacle. (U. Lond.)
(*Answer:* 3·555 kN)

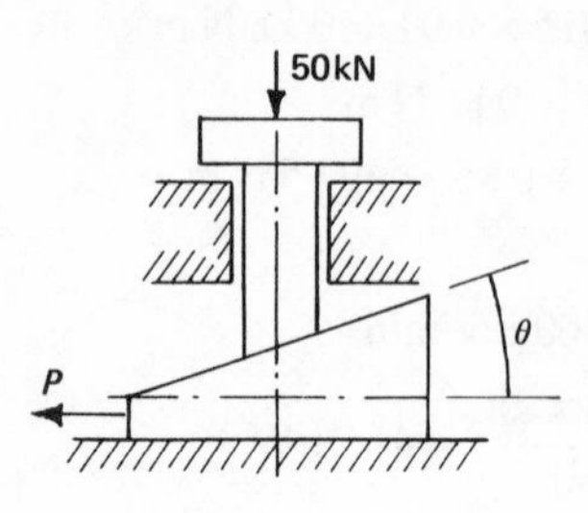

FIG. 8.19

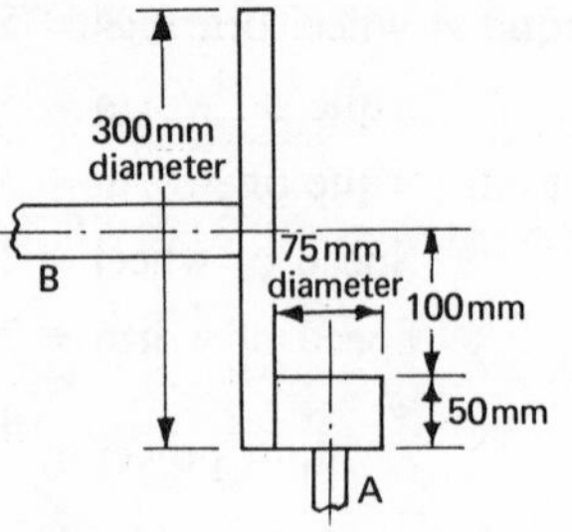

FIG. 8.20

6. The 50 kN load shown in Fig. 8.19 is raised by means of a wedge. Find the required force P, given that $\tan\theta = 0{\cdot}2$ and $\mu = 0{\cdot}2$ at all rubbing surfaces. (U. Lond.)

(*Answer:* 33·64 kN)

7. Shaft A drives B through the friction drive shown in Fig. 8.20. The normal pressure between wheel and disc is 1·75 kN/m along the line of contact, and the coefficient of friction is 0·2. When a torque of 0·55 N m is applied at shaft A, which rotates at 2000 rev/min, find: (a) the power transmitted to shaft B, and (b) the efficiency. (U. Lond.)

(*Answer:* 94·14 W; 81·7 per cent)

8. Figure 8.21 shows a friction drive in which roller D is the driver and disc F is the follower. It may be assumed that pressure is uniform along the line of contact and that the coefficient of friction is constant.

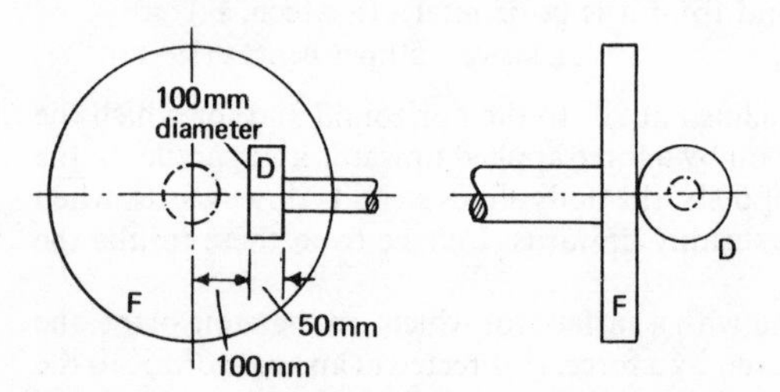

FIG. 8.21

(a) Show that the velocity ratio of D to F is 2·55 : 1, when no power is being transmitted.
(b) When power is being transmitted, show that the efficiency is a maximum for a velocity ratio of 3 : 1.
(c) Determine the efficiency under condition (b). (U. Lond.)

(*Answer:* 83·3 per cent)

9. A square-threaded screw of mean diameter 40 mm and having 160 threads per metre is used to raise a load of 7·5 kN. The nut, which rotates, has a bearing surface whose mean diameter is 56 mm. Find the effort required at the end of a lever of 300 mm effective length to raise the load when $\mu = 0{\cdot}08$. (U. Lond.) (*Answer:* 121 N)

10. A single-start thread has a mean diameter of 60 mm and a pitch of 12 mm. The section of the thread is of Acme form having a total angle of 29° between the faces. If $\mu = 0{\cdot}05$, find: (a) the torque necessary to overcome an axial load of 30 kN, and (b) the efficiency of the thread. (U. Lond.) (*Answer:* 104 N m; 55 per cent)

11. A screw jack is used horizontally in sliding a bedplate into position on its foundation. The plate has a mass of 4 t and μ between it and the foundation is 0·25. The screw of the jack has a mean diameter of 50 mm and a pitch of 12 mm; $\mu = 0{\cdot}1$. The axial thrust is carried on a collar of mean diameter 68 mm for which $\mu = 0{\cdot}15$. Determine the torque required on the jack and the efficiency of the operation. (U. Lond.)
(*Answer:* 93·6 N m; 20·05 per cent)

12. The saddle of a lathe is operated by means of a single-start square-threaded lead screw which has 160 threads per metre and an outside diameter of 32 mm. If the force required to move the saddle is 220 N, estimate the torque required to turn the lead screw if $\mu = 0{\cdot}12$. If a formula is used, the derivation must be given. (U. Lond.)
(*Answer:* 0·605 N m)

13. The table of a planing machine is traversed by means of a simple square-threaded screw of 12 mm pitch and 50 mm outside diameter. The thrust on the screw is taken by a collar bearing of 76 mm mean diameter.

If the pressure of the cut is 400 N, the total mass of the table and part being machined is 230 kg and the speed of cutting is 0·15 m/s, find the power employed.

Take the various coefficients of friction as follows: table in guides = 0·10, screw = 0·15, and collar bearing = 0.20. (U. Lond.) (*Answer:* 632 W)

14. A right- and left-hand screw has an outside diameter of 48 mm and square threads of 12 mm pitch. It is used to alter the distance apart of two nuts which press in opposite directions and do not turn against their loads. If the efficiency of the arrangement is found to be 33 per cent, find the coefficient of friction between the threads. (I. Mech. E.) (*Answer:* 0·18)

15. Two co-axial rods are connected by a turnbuckle, one rod having a right-handed and the other a left-handed V-thread. On both the rods, the mean diameter of the thread is 24 mm, the included angle of its profile is 55° and the pitch is 3·6 mm.

Taking μ to be 0·15, find the torque required to tighten the turnbuckle when the tensile force in the rods is 10 kN. (U. Lond.) (*Answer:* 52·47 N m)

16. A sluice gate, of mass 6 t, is subjected to a normal pressure of 2·5 MN. It is raised by means of a vertical screw which engages with a screwed bush fixed to the top of the gate. The screw is rotated by a 37 kW motor running at a maximum speed of 600 rev/min, a bevel pinion on the motor shaft gearing with a bevel wheel of 80 teeth keyed to the vertical screw. The screw is 125 mm mean diameter and 25 mm pitch. μ for the screw in the nut is 0·08 and between the gate and its guides is 0·10.

If friction losses, additional to those mentioned above, amount to 15 per cent of the total power available, determine the maximum number of teeth for the bevel pinion. (U. Lond.) (*Answer:* 14)

17. A worm running at 2000 rev/min is to drive a worm wheel at 100 rev/min and to transmit to it 15 kW. The worm is double-threaded, of 75 mm mean diameter and 12 mm pitch. The thrust of the worm is taken by a collar of mean diameter 75 mm. The coefficient of friction is 0·05 for both worm and collar. Find the input power. (U. Lond.) (*Answer:* 29·85 kW)

18. Working from first principles, derive the approximate rule that if the efficiency of a lifting machine is less than 50 per cent, the machine is self-locking, i.e. the load will not descend when the effort is removed.

A screw jack has a square thread of 64 mm mean diameter and 12 mm pitch. The load on the jack revolves with the screw. The coefficient of friction at the screw thread is 0·05.

(a) Find the tangential force required at 300 mm radius to lift a mass of 550 kg.

(b) State whether the jack is self-locking. If it is, find the torque necessary to lower the load. If it is not, find the torque which must be applied to keep the load from descending. (U. Lond.)

(*Answer:* 63·3 N; efficiency = 0·542; not self-locking; 1·66 N m)

19. A lifting jack with differential screw threads is shown diagrammatically, Fig. 8.22. The portion B screws into the fixed base C and carries a right-handed square thread of pitch 10 mm, the mean diameter of the thread being 58 mm. The part A is prevented from rotating and carries a right-handed thread of pitch 6 mm on a mean diameter of 32 mm, screwing into the part B. If μ for each thread is 0·15, find the torque necessary to be applied to the part B to raise a load of 450 kg. (U. Lond.) (*Answer:* 32·78 N m)

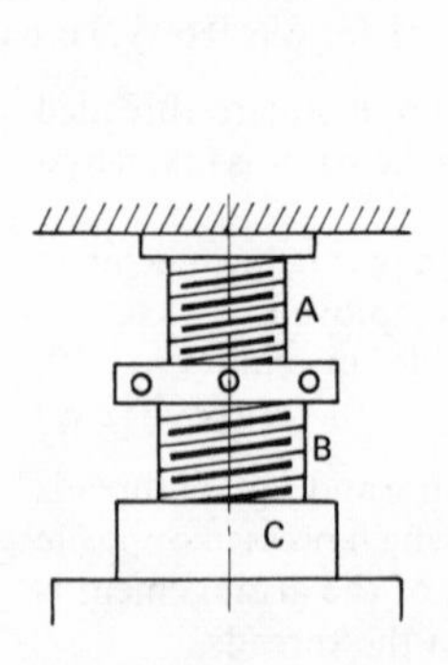

FIG. 8.22

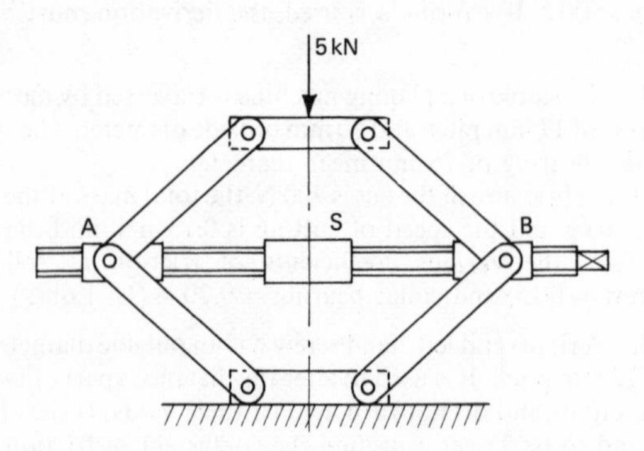

FIG. 8.23

20. The operating spindle S of the lifting jack shown in Fig. 8.23 is threaded at both ends to engage the nuts A and B. Nut A has a right-hand thread and B has a left-hand thread. Both threads are square in section, each having a pitch of 4 mm and a mean diameter of 20 mm. If μ for each nut is 0·15, and all the side links are inclined at 35° to the vertical, determine the torque required at the spindle: (a) to raise a load of 5 kN, and (b) to lower the same load.

Neglect pivot friction. (U. Lond.) (*Answer:* 15·1 N m; 6 N m)

21. A machine vice consists of a fixed jaw, a movable jaw and a spindle through which there passes a tommy bar of effective length 280 mm. The spindle screws into the fixed jaw, the screw thread being square, of 6 mm pitch and 32 mm outside diameter. A collar on the spindle, 44 mm mean diameter, bears on a suitably machined surface on the movable jaw. $\mu = 0·06$.

If the work in the vice jaws is gripped with a force of 10 kN, find the tangential force required at the end of the tommy bar: (a) to tighten the screw, and (b) to slacken it. (U. Lond.) (*Answer:* 112·5 N; 44 N)

(*Note* Except for collar friction, movable jaw would overhaul in case (b).)

9

Friction clutches

9.1 Introduction

A clutch enables two co-axial shafts to be engaged or disengaged while at rest or in relative motion. Friction clutches may be of plate, cone or centrifugal type; in each case slipping will occur until the two shafts are brought to the same speed. This feature permits gradual engagement of the driven shaft, such as in a car drive, and also limits the torque demanded from the driving shaft.

9.2 Plate clutches

In a plate clutch, the torque is transmitted by friction between one or more pairs of co-axial annular faces maintained in contact by an axial thrust. Both sides of each plate are normally effective, so that a single-plate clutch has two pairs of surfaces in contact.

Figure 9.1 shows a simplified form of multi-plate clutch. The inner plates are free to slide axially in grooves connected to the driving shaft and the outer plates are free to slide axially in grooves connected to the driven shaft. The axial force exerted on the plates by the toggle mechanism is transmitted through each plate and so, in a clutch having n pairs of surfaces in contact, the torque transmitted is n times that for a single pair.

A car clutch consists of a single plate, effective both sides.

Consider two flat annular surfaces, Fig. 9.2, maintained in contact by an axial thrust W.

Let T be the torque transmitted, p be the intensity of pressure between the surfaces, and r_1, r_2 and R be the outer, inner and mean radii of faces respectively.

$$\text{Normal force on elementary ring} = p \times 2\pi r\,\mathrm{d}r$$

$$\therefore \text{ total axial force, } W = 2\pi \int_{r_2}^{r_1} pr\,\mathrm{d}r \tag{9.1}$$

$$\text{Friction force on ring} = \mu p \times 2\pi r\,\mathrm{d}r$$

$$\therefore \text{ moment of friction force about axis} = \mu p \times 2\pi r^2\,\mathrm{d}r$$

$$\therefore \text{ torque transmitted, } T = 2\pi\mu \int_{r_2}^{r_1} pr^2\,\mathrm{d}r$$

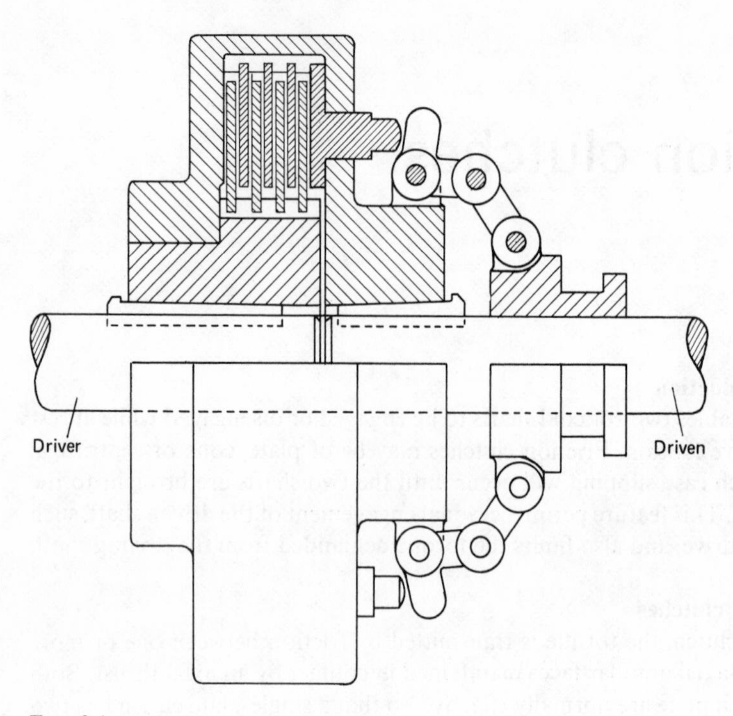

Fig. 9.1

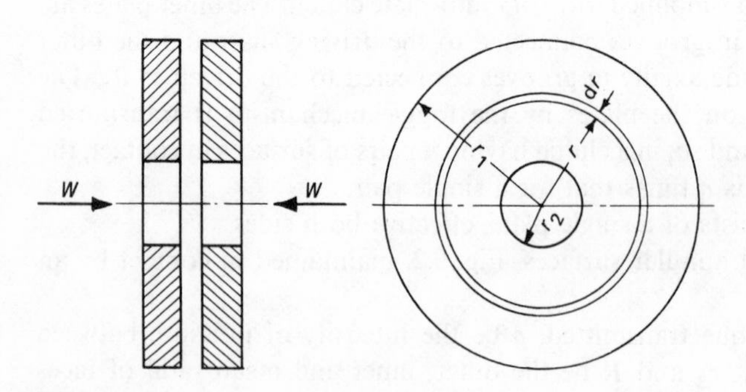

Fig. 9.2

If it is assumed that the pressure is uniform over the contact area, then p is a constant, so that

$$W = p \times \pi(r_1^2 - r_2^2)$$

and

$$T = \tfrac{2}{3}\pi\mu p(r_1^3 - r_2^2)$$

i.e.

$$T = \tfrac{2}{3}\mu W\left(\frac{r_1^3 - r_2^3}{r_1^2 - r_2^2}\right) \tag{9.2}$$

If it is assumed that the wear is uniform over the contact area, then since

$$\text{wear} \propto \text{pressure} \times \text{velocity}$$
$$\propto \text{pressure} \times \text{radius}$$

i.e.
$$pr = \text{constant } (= c)$$
$$\therefore W = 2\pi c(r_1 - r_2) \tag{9.3}$$

and
$$T = \pi\mu c(r_1^2 - r_2^2)$$

i.e.
$$T = \mu W\frac{(r_1 + r_2)}{2} = \mu WR \tag{9.4}$$

Equations 9.2 and 9.4 give the maximum torque which can be transmitted before the clutch slips.

The assumption applicable to a particular clutch depends upon its condition; for a new clutch the pressure will be approximately uniform but for a worn clutch the uniform wear theory is more appropriate.

The clutch theories apply equally to pivot, footstep and collar bearings. Since the uniform pressure theory gives a higher torque than the uniform wear theory, the latter should be used for clutches and the former for bearings in order that errors in assumption are on the safe side. However, in the case of annular surfaces, the difference between the two theories is very small (see Example 9.1) and the uniform wear theory is simpler to use.

9.3 Cone clutches

A cone clutch consists of one pair of friction faces only, the area of contact being a frustum of a cone, Fig. 9.3.

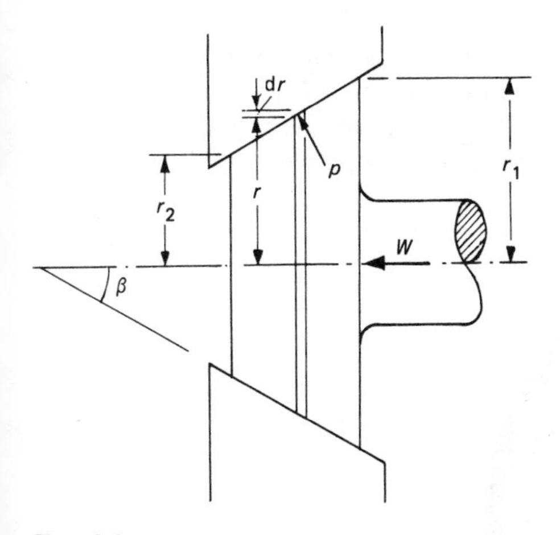

Fig. 9.3

If p is the normal pressure between the surfaces,

$$\text{normal force on elementary ring} = p \times 2\pi r\, \mathrm{d}r \operatorname{cosec} \beta$$

$$\text{Axial component of this force} = p \times 2\pi r\,\mathrm{d}r \operatorname{cosec}\beta \times \sin\beta$$
$$= 2\pi pr\,\mathrm{d}r$$

$$\therefore \text{ total axial force, } W = 2\pi \int_{r_2}^{r_1} pr\,\mathrm{d}r \qquad (9.5)$$

$$\text{Friction force on ring} = \mu p \times 2\pi r\,\mathrm{d}r \operatorname{cosec}\beta$$
$$\therefore \text{ moment of force about axis} = \mu p \times 2\pi r^2\,\mathrm{d}r \operatorname{cosec}\beta$$

$$\therefore \text{ torque transmitted, } T = 2\pi\mu \operatorname{cosec}\beta \int_{r_2}^{r_1} pr^2\,\mathrm{d}r$$

If p is assumed constant,

$$W = \mu p(r_1^2 - r_2^2) \qquad (9.6)$$

and

$$T = \tfrac{2}{3}\mu W\left(\frac{r_1^3 - r_2^3}{r_1^2 - r_2^2}\right)\operatorname{cosec}\beta \qquad (9.7)$$

If pr is assumed constant,

$$W = 2\pi c(r_1 - r_2) \qquad (9.8)$$

and

$$T = \pi\mu c(r_1^2 - r_2^2)\operatorname{cosec}\beta$$
$$= \frac{\mu W}{2}(r_1 + r_2)\operatorname{cosec}\beta$$

i.e.

$$T = \mu WR \operatorname{cosec}\beta \qquad (9.9)$$

9.4 Centrifugal clutches

A centrifugal clutch consists of a number of shoes which can move in radial guides and bear on the inside of an annular rim, Fig. 9.4. The outer surfaces of the shoes are covered with a friction material and as the speed rises the centrifugal force on the shoes causes them to transmit power by friction to the rim.

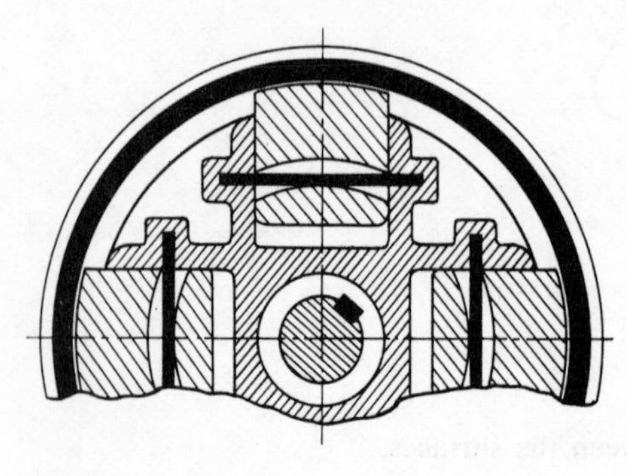

FIG. 9.4

Springs are fitted to keep the shoes clear of the rim at low speeds and thus allow the motor to gain speed before taking up the load.

Let n = number of shoes
F = centrifugal force on each shoe
P = inward force on each shoe exerted by spring
R = inside radius of rim
μ = coefficient of friction between shoe and rim

Then net radial force between each shoe and rim $= F - P$

$$\therefore \text{ friction force on rim} = \mu(F - P)$$

$$\therefore \text{ friction torque per shoe} = \mu R(F - P)$$

$$\therefore \text{ total friction torque} = n\mu R(F - P) \qquad (9.10)$$

Examples

Example 9.1 A plate clutch has for a friction surface one face of an annular disc of outer radius r_1 and inner radius r_2. Derive the expression for the friction torque in terms of the radii, the axial load and the coefficient of friction, assuming: (a) a uniform intensity of pressure, and (b) uniform wear. Calculate the ratio of the torques given by the above assumptions for values of r_2/r_1 equal to 0, $\frac{1}{4}$, $\frac{1}{2}$, $\frac{3}{4}$ and 1. Compare the assumptions in the light of these results. (U. Lond.)

Solution

For uniform pressure, $$T_p = \tfrac{2}{3}\mu W\left(\frac{r_1^3 - r_2^3}{r_1^2 - r_2^2}\right) \quad \text{from equation 9.2}$$

$$= \tfrac{2}{3}\mu W \frac{(r_1^2 + r_1 r_2 + r_2^2)^*}{(r_1 + r_2)}$$

For uniform wear, $$T_w = \frac{\mu W}{2}(r_1 + r_2) \quad \text{from equation 9.4}$$

$$\therefore \frac{T_p}{T_w} = \frac{4}{3}\frac{(r_1^2 + r_1 r_2 + r_2^2)}{(r_1 + r_2)^2} = \frac{4}{3}\left\{\frac{1 + (r_2/r_1) + (r_2/r_1)^2}{1 + 2(r_2/r_1) + (r_2/r_1)^2}\right\}$$

$\frac{r_2}{r_1}$	0	$\frac{1}{4}$	$\frac{1}{2}$	$\frac{3}{4}$	1
$\frac{T_p}{T_w}$	$\frac{4}{3}$	$\frac{28}{25}$	$\frac{28}{27}$	$\frac{148}{147}$	1

* It is necessary to factorise and cancel the term $(r_1 - r_2)$ since the expression is indeterminate when $r_1 = r_2$.

The torque given by the uniform pressure theory is always higher than that given by the uniform wear theory, although for values of $r_2/r_1 > \frac{1}{4}$, the difference is negligible.

Example 9.2 A plate clutch has three discs on the driving shaft and two discs on the driven shaft, providing four pairs of contact surfaces, each of 240 mm external diameter and 120 mm internal diameter. Assuming uniform pressure, find the total spring load pressing the plates together to transmit 25 kW at 1575 rev/min. Take $\mu = 0{\cdot}3$.

If there are six springs each of stiffness 13 kN/m and each of the contact surfaces has worn away by 1·25 mm, what is the maximum power that can be transmitted at the same rev/min, assuming uniform wear and the same coefficient of friction? (U. Lond.)

Solution

$$T = \frac{\text{power} \times 60}{2\pi N} = \frac{25 \times 10^3 \times 60}{2\pi \times 1575} = 151{\cdot}6 \text{ N m}$$

For uniform pressure, $T = 4 \times \frac{2}{3}\mu W\left(\frac{r_1^3 - r_2^3}{r_1^2 - r_2^2}\right)$ from equation 9.2

i.e. $$151{\cdot}6 = 4 \times \frac{2}{3} \times 0{\cdot}3W\left(\frac{0{\cdot}12^3 - 0{\cdot}06^3}{0{\cdot}12^2 - 0{\cdot}06^2}\right)$$

$$\therefore W = \underline{1355 \text{ N}}$$

$$\text{Total wear} = 8 \times 1{\cdot}25 = 10 \text{ mm}$$

$$\therefore \text{reduction in spring force} = 0{\cdot}01 \times 13 \times 10^3 \times 6 = 780 \text{ N}$$

$$\therefore \text{new axial load} = 1355 - 780 = 575 \text{ N}$$

For uniform wear, $T = 4\mu Wr$ from equation 9.4

$$= 4 \times 0{\cdot}3 \times 575 \times 0{\cdot}09 = 62 \text{ N m}$$

$$\therefore \text{maximum power} = \frac{62}{151{\cdot}6} \times 25 = \underline{10{\cdot}25 \text{ kW}}$$

Example 9.3 The mean diameter of the contact surfaces of a conical friction clutch is 300 mm and the width of the conical surface is 65 mm. The clutch is lined with material giving a coefficient of friction of 0·3 and the angle between the generator of the conical surface and its axis is 15°.

If the intensity of the normal pressure between the surfaces is limited to 70 kN/m², find the greatest power that can be transmitted at a speed of 1200 rev/min without slipping of the clutch and also the least axial force necessary to hold the clutch in engagement. (U. Lond.)

Solution

From Fig. 9.3, $r_1 - r_2 = 65 \sin 15° = 16{\cdot}83 \text{ mm}$

Also $$\frac{r_1 + r_2}{2} = 150\,\text{mm}$$

$$\therefore\ r_1 = 158{\cdot}4\,\text{mm} \quad \text{and} \quad r_2 = 141{\cdot}6\,\text{mm}$$

The greatest power that can be transmitted is given by the uniform pressure theory but this is only likely to be transmitted while the clutch is new. Hence, assuming uniform wear, the maximum pressure will occur at the minimum radius,

$$\therefore\ c = pr = 70 \times 10^3 \times 0{\cdot}1416 = 9912$$

$$\therefore\ W = 2\pi c(r_1 - r_2) \qquad \text{from equation 9.8}$$

$$= 2\pi \times 9912 \times 0{\cdot}01683 = 1048\,\text{N}$$

$$\therefore\ T = \mu W R \operatorname{cosec} 15^\circ \qquad \text{from equation 9.9}$$

$$= 0{\cdot}3 \times 1048 \times 0{\cdot}15 \times 3{\cdot}864 = 182{\cdot}3\,\text{N m}$$

$$\therefore\ \text{power} = \frac{2\pi TN}{60} = \frac{2\pi \times 182{\cdot}3 \times 1200}{60} = 22\,900\,\text{W} \quad \text{or} \quad \underline{22{\cdot}9\,\text{kW}}$$

From Fig. 9.5,

$$\text{least axial force to hold clutch in engagement} = W + \mu N \cos\beta$$

where N is the normal force between the surfaces,

but $$N = W \operatorname{cosec}\beta$$

$$\therefore\ \text{axial force required} = W(1 + \mu\cot\beta)$$

$$= 1048(1 + 0{\cdot}3\cot 15^\circ) = \underline{2225\,\text{N}}$$

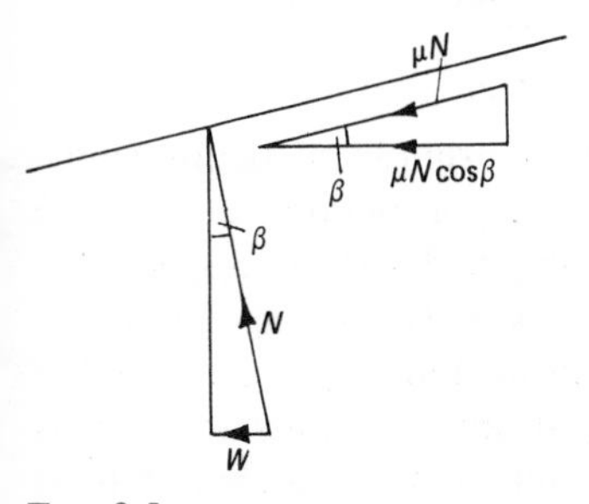

FIG. 9.5

Example 9.4 In a cone clutch, the contact surfaces have an effective diameter of 75 mm. The semi-angle of the cone is 15°. $\mu = 0{\cdot}3$. Find the torque required to produce slipping of the clutch if the axial force applied is 180 N.

This clutch is employed to connect an electric motor, running uniformly at 1000 rev/min, with a flywheel which is initially stationary. The flywheel has a mass of 13·5 kg and its radius of gyration is 150 mm. Calculate the time required for the flywheel to attain full speed and also the energy lost in the slipping of the clutch. (U. Lond.)

Solution

$$T = \mu WR \operatorname{cosec}\beta = 0{\cdot}3 \times 180 \times 0{\cdot}0375 \operatorname{cosec} 15^\circ = \underline{7{\cdot}824\ \text{N m}}$$

For the flywheel, $T = I\alpha$

$$\therefore\ 7{\cdot}824 = 13{\cdot}5 \times 0{\cdot}15^2 \times \alpha$$

$$\therefore\ \alpha = 25{\cdot}76\ \text{rad/s}^2$$

$$\therefore\ \text{time taken, } t = \left(\frac{2\pi}{60} \times 1000\right)\Big/ 25{\cdot}76 = \underline{4{\cdot}065\ \text{s}}$$

Let θ_m and θ_f be the angles turned through by the motor and flywheel respectively before slipping ceases.

Then
$$\theta_m = \left(\frac{2\pi}{60} \times 1000\right) \times 4{\cdot}065 = 425{\cdot}7\ \text{rad}$$

Since the flywheel accelerates uniformly from rest to 1000 rev/min,

$$\theta_f = \tfrac{1}{2}\theta_m = 212{\cdot}85\ \text{rad}$$

$$\text{Work done by motor} = T\theta_m$$

and
$$\text{work done on flywheel} = T\theta_f$$

$$\therefore\ \text{work lost in friction} = T(\theta_m - \theta_f) = 7{\cdot}824\,(425{\cdot}7 - 212{\cdot}85) = \underline{1665\ \text{J}}$$

Alternatively,

$$\text{work done by motor} = T\theta_m = 7{\cdot}824 \times 425{\cdot}7 = 3330\ \text{J}$$

$$\text{K.E. acquired by flywheel} = \tfrac{1}{2}I\omega^2$$

$$= \tfrac{1}{2} \times 13{\cdot}5 \times 0{\cdot}15^2 \times \left(\frac{2\pi}{60} \times 1000\right)^2 = 1665\ \text{J}$$

$$\therefore\ \text{work lost in friction} = 3330 - 1665 = \underline{1665\ \text{J}}$$

Example 9.5 An engine is coupled to a rotating drum by a friction clutch which has a single disc faced on each side with a ring of friction material. The inner and outer diameters of each ring are 240 mm and 320 mm respectively, $\mu = 0{\cdot}3$ and the axial pressure on the disc is 1·3 kN. The engine develops a constant torque of 40 N m, and its inertia is equivalent to that of a flywheel of mass 25 kg and radius of gyration 300 mm. The drum has a mass of 55 kg, its radius of gyration is 400 mm and the torque required to overcome friction is 7 N m.

If the clutch is engaged when the engine speed is 500 rev/min and the drum is at rest, find the speed when clutch slip ceases and the duration of slipping. Find also the total time taken for the drum to reach a speed of 500 rev/min.

(U. Lond.)

Solution

Assuming uniform wear,

$$T = \mu WR \times 2 = 0{\cdot}3 \times 1300 \times \left(\frac{0{\cdot}16 + 0{\cdot}12}{2}\right) \times 2 = 109{\cdot}2\ \text{N m}$$

$$\text{Net retarding torque on engine} = 109{\cdot}2 - 40 = 69{\cdot}2 \text{ N m}$$

$$\therefore\ 69{\cdot}2 = I_e \alpha_e = 25 \times 0{\cdot}3^2 \alpha_e$$

$$\therefore\ \alpha_e = 30{\cdot}75 \text{ rad/s}^2$$

$$\text{Net accelerating torque on drum} = 109{\cdot}2 - 7 = 102{\cdot}2 \text{ N m}$$

$$\therefore\ 102{\cdot}2 = I_d \alpha_d = 55 \times 0{\cdot}4^2 \alpha_d$$

$$\therefore\ \alpha_d = 11{\cdot}6 \text{ rad/s}^2$$

If N is the final steady speed in rev/min when slipping ceases and t is the time of duration of slipping, in seconds,

then
$$30{\cdot}75 = \frac{2\pi}{60} \frac{500 - N}{t}$$

and
$$11{\cdot}6 = \frac{2\pi}{60} \frac{N}{t}$$

$$\therefore\ N = \underline{137 \text{ rev/min}} \quad \text{and} \quad t = \underline{1{\cdot}24 \text{ s}}$$

After slipping has ceased,

$$\text{net accelerating torque on engine and drum} = 40 - 7 = 33 \text{ N m}$$

$$\therefore\ 33 = (25 \times 0{\cdot}3^2 + 55 \times 0{\cdot}4^2)\alpha$$

$$\therefore\ \alpha = 2{\cdot}99 \text{ rad/s}^2$$

If t' is the time in seconds for the system to reach 500 rev/min after slipping ceases, then

$$t' = \frac{2\pi}{60} \frac{500 - 137}{2{\cdot}99} = 12{\cdot}7 \text{ s}$$

$$\therefore\ \text{total time taken} = 1{\cdot}24 + 12{\cdot}7 = \underline{13{\cdot}94 \text{ s}}$$

Example 9.6 The disc brake on a vehicle consists of a pair of identical brake pads which can be clamped simultaneously to bear on opposite sides of a thin flat steel disc of diameter 250 mm which rotates between them. Each pad has a bearing surface whose shape is that of a 30° sector of an annulus with an inner radius of 70 mm and an outer radius of 120 mm. The pads are located such that the axis of the annulus from which they derive is coincident with that of the disc. The coefficient of friction between a pad and the disc is 0·4.

If the contact pressure is assumed to be constant over the pad surface, calculate the magnitude of this pressure and the normal force which must be applied to both pads simultaneously, in order to develop a braking torque on the disc of 9 N m.

Calculate also the lateral force exerted by the disc on the bearings which support it for the same braking torque. Hence deduce an 'effective radius' at which the brake pads may be considered to act. (U. Lond.)

Solution

Referring to Fig. 9.6,

$$\text{normal force on an elementary ring of radius } r = p \times \frac{\pi}{6} r\,\mathrm{d}r$$

Therefore the total axial force is given by

$$W = \frac{\pi}{6} p \int_{r_2}^{r_1} r\,\mathrm{d}r$$

$$= \frac{\pi}{12} p(r_1^2 - r_2^2) \qquad (1)$$

Since there are two pads,

$$\text{friction force on ring} = 2 \times \mu p \times \frac{\pi}{6} r\,\mathrm{d}r$$

$$\therefore \text{ friction torque on ring} = \frac{\mu p \pi}{3} r\,\mathrm{d}r \times r$$

Therefore total friction torque,

$$T = \frac{\mu p \pi}{3} \int_{r_2}^{r_1} r^2\,\mathrm{d}r$$

$$= \frac{\mu p \pi}{9}(r_1^3 - r_2^3)$$

$$= \frac{\mu \pi}{9}(r_1^3 - r_2^3) \times \frac{12W}{\pi(r_1^2 - r_2^2)}$$

$$= \frac{4}{3}\mu W\left(\frac{r_1^3 - r_2^3}{r_1^2 - r_2^2}\right)^* \qquad (2)$$

$$\therefore 9 = \frac{4}{3} \times 0{\cdot}4W\left(\frac{0{\cdot}12^3 - 0{\cdot}07^3}{0{\cdot}12^2 - 0{\cdot}07^2}\right)$$

from which $\underline{W = 115{\cdot}8\ \mathrm{N}}$

$$p = \frac{12W}{\pi(r_1^2 - r_2^2)} \qquad \text{from equation (1)}$$

$$= \frac{12 \times 115{\cdot}8}{\pi(0{\cdot}12^2 - 0{\cdot}07^2)} = \underline{46\,560\ \mathrm{N/m^2}}$$

*** This is the usual formula for the friction torque on a double-sided ring with uniform pressure, which is to be expected since friction force is independent of area.**

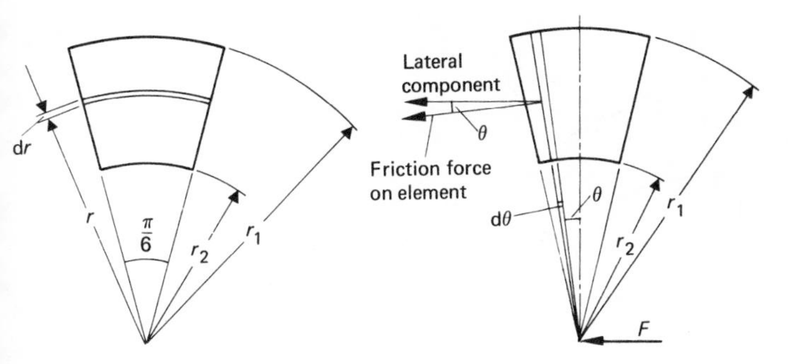

FIG. 9.6 FIG. 9.7

From Fig. 9.7,

$$\text{area of radial element} = \frac{d\theta}{2\pi} \times \pi(r_1^2 - r_2^2)$$

$$= \frac{d\theta}{2}(r_1^2 - r_2^2)$$

$$\text{Friction force on element} = \mu p \times \frac{d\theta}{2}(r_1^2 - r_2^2)$$

$$\text{Lateral component of force} = \mu p \times \frac{d\theta}{2}(r_1^2 - r_2^2)\cos\theta$$

$$\therefore \text{ total lateral force for two surfaces} = 2 \times \mu p \frac{(r_1^2 - r_2^2)}{2} \int_{-15^\circ}^{15^\circ} \cos\theta \, d\theta$$

$$= 0{\cdot}4 \times 46\,560 \times (0{\cdot}12^2 - 0{\cdot}07^2)[\sin\theta]_{-15^\circ}^{15^\circ}$$

i.e. lateral force on bearings,

$$F = \underline{91{\cdot}6\text{ N}}$$

If R is the effective radius at which this force acts, then

$$91{\cdot}6R = 9$$

i.e. $$R = 0{\cdot}0982\text{ m} \quad \text{or} \quad \underline{98{\cdot}2\text{ mm}}$$

Example 9.7 Figure 9.8 shows one of the four shoes of a centrifugal clutch. The contact surface is 160 mm radius and the friction force is μS, where S is the resultant of the normal forces on the surface and $\mu = 0{\cdot}25$. The centre of mass is 25 mm from the contact surface. The clutch is to commence engagement at 500 rev/min. If it is to transmit 20 kW at 750 rev/min, calculate the corresponding value of S and find the mass of each shoe and the pressure of the beam spring on the adjusting screw. If the spring has a stiffness of 170 kN/m, find the power transmitted at 750 rev/min when the shoes have worn 2 mm, if not adjusted. (I. Mech. E.)

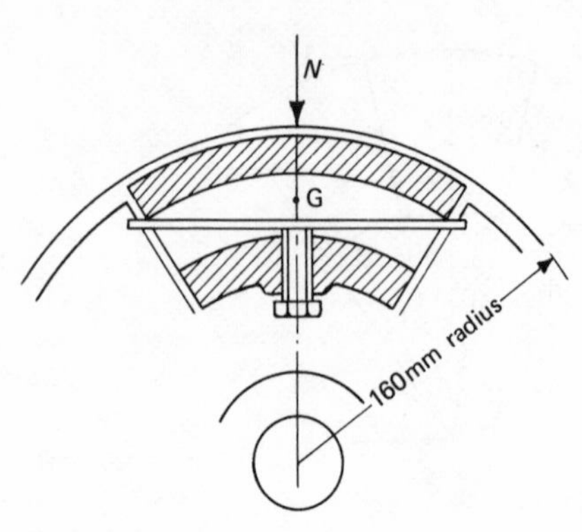

FIG. 9.8

Solution

$$\text{Radius of centre of mass of shoe} = 160 - 25 = 135\,\text{mm}$$

$$\therefore \text{ centrifugal force on each shoe, } F = m\omega^2 r$$

$$= m\left(\frac{2\pi}{60}N\right)^2 \times 0{\cdot}135$$

$$= 0{\cdot}001\,48mN^2\,\text{N}$$

At 750 rev/min, $T = \dfrac{\text{power} \times 60}{2\pi \times N} = \dfrac{20 \times 10^3 \times 60}{2\pi \times 750} = 255\,\text{N m}$

$$T = 4\mu SR \qquad \text{from equation 9.8}$$

i.e. $255 = 4 \times 0{\cdot}25 \times S \times 0{\cdot}16$

$$\therefore S = \underline{1594\,\text{N}}$$

When the shoe makes contact with the rim, the spring force remains constant as the speed rises and is therefore equal to the centrifugal force at 500 rev/min.

$$S = F - P$$

i.e. $1594 = 0{\cdot}001\,48m \times 750^2 - 0{\cdot}001\,48m \times 500^2$

$$\therefore m = \underline{3{\cdot}45\,\text{kg}}$$

Pressure on beam spring at 750 rev/min = spring force at 500 rev/min

$$= 0{\cdot}001\,48 \times 3{\cdot}45 \times 500^2$$

$$= 1275\,\text{N}$$

When shoe wears 2 mm,

$$\text{increase in spring force} = 170 \times 10^3 \times 0{\cdot}002 = 340\,\text{N}$$

$$\therefore \text{ new spring force} = 1275 + 340 = 1615\,\text{N}$$

$$\text{New centrifugal force} = 3{\cdot}55\left(\frac{2\pi}{60} \times 750\right)^2 \times 0{\cdot}137 = 3000\,\text{N}$$

$$\therefore\ S = 3000 - 1615 = 1385\,\text{N}$$

$$\therefore\ \text{power} = \frac{1385}{1590} \times 20 = \underline{17{\cdot}43\,\text{kW}}$$

Problems

1. An axial thrust of 50 kN is carried by a plain collar-type thrust bearing, having inner and outer diameters of 250 mm and 400 mm respectively. Assuming that μ between the thrust surfaces is 0·02, and that the local wear rate of these surfaces is proportional to the pressure and to the rubbing speed, determine the power absorbed in friction at a speed of 120 rev/min. (U. Lond.) (*Answer:* 2·04 kW)

2. Establish a formula for the maximum torque transmitted by a single-plate clutch of internal and external radii r_1 and r_2, if the limiting coefficient of friction is μ and the axial spring loading is P. Assume that the pressure intensity on the contact faces is uniform.

Apply this to determine the time required to accelerate a countershaft of rotating mass 500 kg and radius of gyration 200 mm to the full speed of 250 rev/min from rest through a single-plate clutch of internal and external radii 125 mm and 200 mm respectively, taking μ as 0·3 and the spring load as 600 N. (U. Glas.)
(*Answer:* 8·8 s)

3. A multi-plate clutch is to transmit 12 kW at 1500 rev/min. The inner and outer radii for the plates are to be 50 mm and 100 mm respectively. The maximum axial spring force is restricted to 1 kN. Calculate the necessary number of pairs of surfaces if $\mu = 0{\cdot}35$ assuming constant wear. What will be the necessary axial force?
(U. Lond.) (*Answer:* 3; 970 N)

4. A friction clutch is required to transmit 34·5 kW at 2000 rev/min. It is to be of single-plate disc type with both sides of the plate effective, the pressure being applied axially by means of springs and limited to 70 kN/m^2. If the outer diameter of the plate is to be 300 mm, find the required inner diameter of the clutch ring and the total force exerted by the springs. Assume the wear to be uniform. $\mu = 0{\cdot}3$. (U. Lond.)
(*Answer:* 200 mm; 2·2 kN)

5. A plate clutch consists of a flat driven plate gripped between a driving plate and presser plate so that there are two active driving surfaces, each having an inner diameter 200 mm and an outer diameter 350 mm. $\mu = 0{\cdot}4$. The working pressure is limited to 170 kN/m^2. Assuming the pressure is uniform, calculate the power which can be transmitted at 1000 rev/min. If the clutch becomes worn so that the intensity of pressure is inversely proportional to the radius, the total axial force on the presser plate remaining unaltered, calculate the power which can now be transmitted at 1000 rev/min and the greatest intensity of pressure on the friction surfaces. (U. Lond.)
(*Answer:* 130 kW; 127 kW; 234 kN/m^2)

6. A disc of radius R is rotating freely in its own horizontal plane about a fixed vertical axis with an angular velocity ω, when a second similar disc is placed on its upper surface. The second disc is at rest initially and its centre lies on the axis of rotation of the first disc. The surfaces in contact are rough. With what common angular velocity will the discs eventually rotate? Express the energy loss in the process as a fraction of the initial kinetic energy.

If the load of the upper disc is evenly distributed over the lower one, show that the time during which slipping occurs is given by $3\omega R/8\mu g$, where μ is the coefficient of friction between the surfaces in contact. (U. Lond.) (*Answer:* $\omega/2$; 0·5)

7. A machine is driven from a constant-speed shaft, rotating at 300 rev/min by means of a friction clutch. The moment of inertia of the rotating parts of the machine is 4·7 kg m^2. The clutch is of the disc type, both sides of the disc being effective in producing driving friction. The external and internal diameters of the discs are respectively 200 mm and 125 mm. The axial pressure applied to the disc is 70 kN/m^2 Assume that this pressure is uniformly distributed and that $\mu = 0{\cdot}25$.

If, when the machine is at rest, the clutch is suddenly engaged, what length of time will be required for the machine to attain its full speed? (U. Lond.) (*Answer:* 2·68 s)

8. A power shaft running at a steady speed of 175 rev/min drives a countershaft through a single-plate friction clutch of external and internal diameters 375 mm and 225 mm respectively. The masses on the countershaft have a radius of gyration of 250 mm and a total mass of 350 kg. The axial spring load operating the clutch is 450 N, and $\mu = 0{\cdot}3$. Assuming uniform acceleration, determine the time required to reach full speed from rest, and the work dissipated due to clutch slip during that time. (U. Lond.) (*Answer:* 9·9 s; 3·67 kJ)

9. A multi-plate clutch is used to connect two shafts in line. In this, one set of plates can slide axially in a shell attached to one shaft while the other set of plates can slide along the second shaft. Sketch the arrangement showing an operating mechanism for pressing the plates together. If the inner and outer diameters of the contact surfaces are 90 mm and 140 mm respectively and there are six contacts, find the axial thrust required to transmit 7·5 kW at 750 rev/min. Assume that $\mu = 0{\cdot}3$ and that the product of the contact pressure and radius is a constant over each surface. Find also the contact pressures at the inner and outer radii. (I. Mech. E.)

(*Answer:* 923 N; 130·5 kN/m^2; 83·9 kN/m^2)

10. An electric motor drives a co-axial rotor through a single-plate clutch which has two pairs of driving surfaces, each of 275 mm external and 200 mm internal diameter; the total spring load pressing the plates together is 500 N. The mass of the motor armature and shaft is 800 kg and its radius of gyration is 260 mm; the rotor has a mass of 1350 kg and its radius of gyration is 225 mm.

The motor is brought up to a speed of 1250 rev/min; the current is then switched off and the clutch suddenly engaged. Determine the final speed of the motor and rotor, and find the time taken to reach that speed and the kinetic energy lost during the period of slipping. How long would slipping continue if a constant torque of 50 N m were maintained on the armature shaft? Take $\mu = 0{\cdot}35$. (U. Lond.)

(*Answer:* 552 rev/min; 95 s; 259 kJ; 288·5 s)

11. A power shaft running at 180 rev/min drives a countershaft through a single-plate friction clutch of internal and external radii 75 mm and 150 mm respectively, μ being 0·3. The countershaft has a total moment of inertia of 8·4 kg m^2. If the time taken for the countershaft to attain full speed from rest is 6 s, determine the axial spring force in the clutch, and also the work dissipated due to clutch slip. Any formula for the clutch torque should be proved. Assume uniform wear. (U. Lond.) (*Answer:* 391 N; 1·49 kJ)

12. A multi-disc friction clutch has to be designed to transmit 75 kW from an engine running at 2000 rev/min. Assuming that the pressure distribution is uniform and limited to 150 kN/m^2, and that the inner and outer diameters of the lining are respectively 100 mm and 150 mm, determine the necessary end thrust and the necessary number of plates. Take $\mu = 0{\cdot}25$. If this clutch, after manufacture, is used to transmit the power from a larger engine to a rotor which has a mass of 1150 kg and a radius of gyration of 200 mm, determine the time required for this rotor to reach 1500 rev/min from standstill. Assume that the clutch is transmitting the maximum possible torque. (U. Lond.)

(*Answer:* 8 plates; 1·414 kN, since actual pressure = 144 kN/m^2; 20·2 s)

13. The rotating parts of a heavy grinding mill are carried on a vertical shaft running at 120 rev/min; the load of 2 t is carried by a conical bearing of 125 mm outer diameter and 50 mm inner diameter with an included angle of 120°. Assuming $\mu = 0{\cdot}075$ and that the intensity of pressure varies inversely as the radius, determine the power wasted in friction. (U. Lond.) (*Answer:* 935 W)

14. A motor drives a machine through a friction clutch which transmits 140 N m while slip occurs during engagement. For the motor, the rotor has a mass of 65 kg with radius of gyration 140 mm, and the inertia of the machine is equivalent to 25 kg with radius of gyration 75 mm. If the motor is running at 750 rev/min and the machine is at rest, find the speed after engaging the clutch and the time taken. Find also the energy absorbed in the clutch during engagement. Torques acting on the motor and machine other than at the clutch are to be neglected. (I. Mech. E.) (*Answer:* 675 rev/min; 0·071 s; 390 J)

15. A thrust of 30 kN along the axis of a shaft is taken by a pivot bearing consisting of the frustum of a cone. The outer and inner diameters are 200 mm and 100 mm respectively and the semi-angle of the cone is 60°. The shaft speed is 200 rev/min and $\mu = 0{\cdot}02$. Assuming that the intensity of pressure is uniform over the surface of the bearing, find, from first principles: (a) the magnitude of this pressure, in N/m^2, and (b) the power absorbed in friction. (U. Lond.)
(*Answer:* 1·272 MN/m^2; 1·13 kW)

16. A cone clutch is required to transmit 30 kW at 1200 rev/min. The mean diameter of the bearing surface is 250 mm and the cone angle is 25°. Assuming $\mu = 0{\cdot}3$ and a normal pressure of 140 kN/m^2, determine the axial width of the conical bearing surface and the axial load required. (U. Lond.) (*Answer:* 54·1 mm; 1·32 kN)

17. Two co-axial rotors A and B are connected by a single-plate clutch with two pairs of friction surfaces, each of 300 mm external and 220 mm internal diameter; the total spring load on the clutch pressing the plates together is 700 N. The masses and radii of gyration of A and B are 1100 kg, 200 mm, and 800 kg, 350 mm respectively. $\mu = 0{\cdot}3$.

The rotor A is given a speed of 1200 rev/min, while B is stationary, and the clutch is then suddenly engaged. Determine the time taken for A and B to reach the same speed, the magnitude of that speed and the amount of kinetic energy lost during the period of slipping of the clutch. Assume that the effective radius of the clutch is the arithmetic mean of the inner and the outer contact surface radii. (U. Lond.)
(*Answer:* 70 s; 372 rev/min; 240 kJ)

18. An induction motor has a centrifugal clutch fitted inside its belt pulley to allow it to start without load and to take the load automatically when well up to speed. The clutch consists of four 'shoes' which can move outwards radially under centrifugal forces against the inward pull of springs. These shoes then press against the inner drum surface of the pulley and take up the drive by friction.

Each shoe has a mass of 1·35 kg and the centre of mass is at 100 mm radius when the shoes are just touching the drum which has a radius of 125 mm. Each of the four spring-pulls is then 1·2 kN. The friction coefficient may be taken as 0·25 between shoe and drum.

Calculate the speed at which the shoes first touch the drum, and the torque and the power which can be transmitted at: (a) 1200 rev/min, and (b) 1450 rev/min.
(U. Glas.) (*Answer:* 900 rev/min; (a), 116 N m, 14·6 kW; (b), 239 N m, 36·3 kW)

19. A centrifugal clutch has four blocks which slide radially in a spider keyed to the driving shaft and make contact with the internal cylindrical surface of a drum keyed to the driven shaft. When the clutch is at rest, each block is pulled against a stop by a spring so as to leave a radial clearance of 6 mm between the block and the drum. The pull exerted by the spring is then 450 N and the centre of mass of the block is 200 mm from the axis of the clutch.

If the internal diameter of the drum is 500 mm, the mass of each block is 7 kg, the stiffness of each spring is 35 kN/m and the coefficient of friction between block and drum is 0·3, find the maximum power the clutch can transmit at 500 rev/min. (I. Mech. E.) (*Answer:* 51·7 kW)

20. A centrifugal friction clutch has a driving member consisting of a spider carrying four shoes which are kept from contact with the clutch case by means of the flat springs until increase of centrifugal force overcomes the resistance of the springs and power is transmitted by friction between the shoes and the case.

Determine the necessary mass of each shoe if 22·5 kW is to be transmitted at 750 rev/min with engagement beginning at 75 per cent of the running speed. The inside diameter of the drum is 300 mm and the radial distance of the centre of mass of each shoe from the shaft axis is 125 mm. Assume $\mu = 0{\cdot}25$. (U. Lond.)

(*Answer:* 5·66 kg)

10
Belt drives and band brakes

10.1 Ratio of belt tensions

Consider a flat belt partly wound round a pulley so that the angle of lap is θ, Fig. 10.1, and let T_1 and T_2 be the tensions in the belt when it is about to slip in the direction shown.

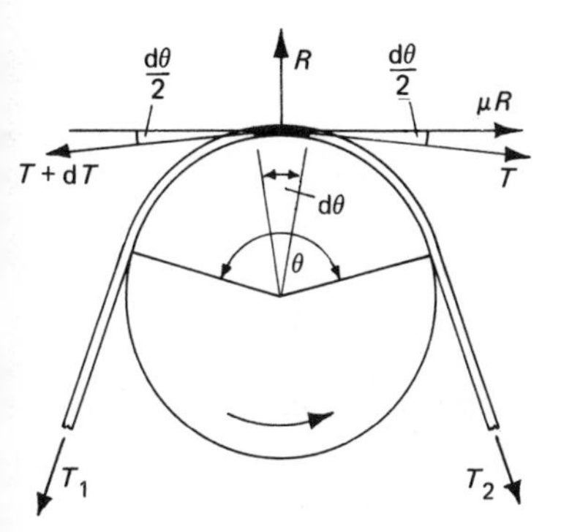

FIG. 10.1

If the tensions at the ends of an element subtending an angle $d\theta$ at the centre are T and $(T+dT)$ and the reaction between the belt and the pulley is R, then, resolving forces radially,

$$(T+dT)\frac{d\theta}{2}+T\frac{d\theta}{2}=R$$

Therefore, neglecting the second order of small quantities,

$$T\,d\theta = R \tag{10.1}$$

Resolving forces tangentially,

$$(T+dT)-T=\mu R$$

i.e.

$$dT=\mu R$$

$$=\mu T\,d\theta \qquad \text{from equation 10.1}$$

i.e.
$$\frac{dT}{T} = \mu \, d\theta$$

$$\therefore \int_{T_2}^{T_1} \frac{dT}{T} = \int_0^\theta \mu \, d\theta$$

i.e.
$$\ln \frac{T_1}{T_2} = \mu\theta$$

or
$$\frac{T_1}{T_2} = e^{\mu\theta} \tag{10.2}$$

If the belt is used to transmit power between two pulleys, Fig. 10.2, T_1 and T_2 are the tight and slack side tensions respectively. If the pulleys are of unequal diameters, the belt will slip first on the pulley having the smaller angle of lap, i.e. on the smaller pulley.

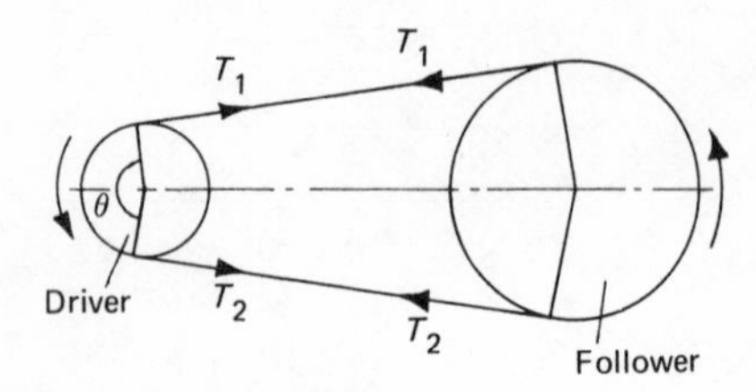

FIG. 10.2

If v is the speed of the belt in m/s and T_1 and T_2 are in newtons, then

$$\text{power transmitted} = (T_1 - T_2)v \text{ W} \tag{10.3}$$

$$= T_1\left(1 - \frac{1}{e^{\mu\theta}}\right)v \text{ W} \tag{10.4}$$

10.2 Modification for V-grooved pulley

For a V-grooved pulley, the normal force between the belt or rope and the pulley is increased since the radial component of this force must equal R. Thus, if the semi-angle of the groove is β, Fig. 10.3,

$$N = \frac{R}{2} \operatorname{cosec} \beta$$

$$\therefore \text{ frictional resistance } = 2\mu N$$

$$= \mu R \operatorname{cosec} \beta$$

The friction force is therefore increased in the ratio cosec β : 1, so that the V-grooved pulley is equivalent to a flat pulley having a coefficient of friction of μ cosec β.

Hence
$$\frac{T_1}{T_2} = e^{\mu\theta \operatorname{cosec} \beta} \tag{10.5}$$

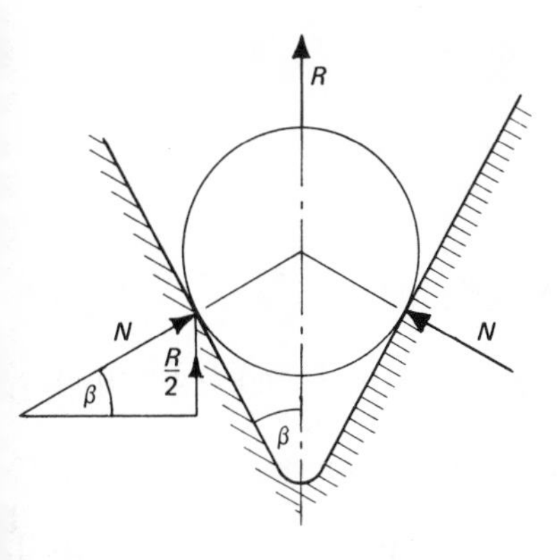

FIG. 10.3

10.3 Effect of centrifugal tension

Consider a belt, of mass m per unit length, wound round a pulley of radius r, Fig. 10.4. Let the speed of the belt be v and the centrifugal tension be T_c.

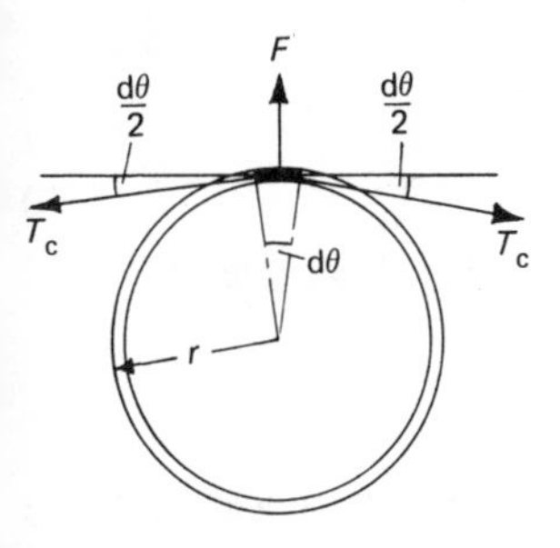

FIG. 10.4

If F is the centrifugal force acting on an element of the belt subtending an angle $d\theta$ at the centre, then, resolving forces radially,

$$F = 2T_c \frac{d\theta}{2}$$

i.e.

$$mr\, d\theta \frac{v^2}{r} = T_c d\theta$$

or

$$T_c = mv^2 \tag{10.6}$$

This is the tension caused by centrifugal force on the belt and is additional to the tension due to the transmission of power.

If allowance is made for this additional tension in determining the ratio of the belt tensions, equation 10.1 becomes

$$T\, d\theta = R + F = R + mr\, d\theta \frac{v^2}{r}$$

so that $$dT = \mu R = \mu(T - mv^2)d\theta = \mu(T - T_c)d\theta$$

i.e. $$\frac{dT}{T - T_c} = \mu d\theta$$

Integrating as before,

$$\frac{T_1 - T_c}{T_2 - T_c} = e^{\mu\theta} \quad \text{or} \quad e^{\mu\theta \operatorname{cosec} \beta} \tag{10.7}$$

$T_1 - T_c$ and $T_2 - T_c$ are the effective driving tensions and T_1 and T_2 are now the *total* tensions in the belt.

Allowing for centrifugal tension, equation 10.4 becomes

$$\text{power} = (T_1 - T_c)\left(1 - \frac{1}{e^{\mu\theta}}\right)v \tag{10.8}$$

If the belt velocity can be varied while the maximum belt tension T_1 remains fixed, the power transmitted will be a maximum when

$$\frac{d}{dv}\{(T_1 - T_c)v\} = 0$$

i.e. $$\frac{d}{dv}(T_1 v - mv^3) = 0$$

i.e. $$mv^2 = \tfrac{1}{3}T_1$$

or $$T_c = \tfrac{1}{3}T_1 \tag{10.9}$$

The maximum power is then obtained by substituting this value of T_c and the corresponding value of v in equation 10.8.

10.4 Initial tension

The belt is assembled with an initial tension, T_0. When power is being transmitted, the tension in the tight side increases from T_0 to T_1 and on the slack side decreases from T_0 to T_2. If the belt is assumed to obey Hooke's law and its length to remain constant, then the increase in length of the tight side is equal to the decrease in length of the slack side,

i.e. $$T_1 - T_0 = T_0 - T_2$$

since the lengths and cross-sectional areas of the belt are the same on each side.

Hence $$T_1 + T_2 = 2T_0 \tag{10.10}$$

Examples

Example 10.1 Two pulleys, one 450 mm diameter and the other 200 mm diameter, are on parallel shafts 1·95 m apart, and are connected by a crossed belt. Find the length of belt required and the angle of contact between the belt and each pulley.

What power can be transmitted by the belt when the larger pulley rotates at 200 rev/min, if the maximum permissible tension in the belt is 1 kN and the coefficient of friction between belt and pulley is 0·25? (U. Lond.)

Solution

From triangle ABC, Fig. 10.5,

$$\cos\frac{\phi}{2} = \frac{225 + 100}{1950} = 0{\cdot}1667$$

$$\therefore \frac{\phi}{2} = 80{\cdot}4°$$

∴ angle of lap for each pulley,

$$\theta = 360° - 2 \times 80{\cdot}4° = 199{\cdot}2° = \underline{3{\cdot}474 \text{ rad}}$$

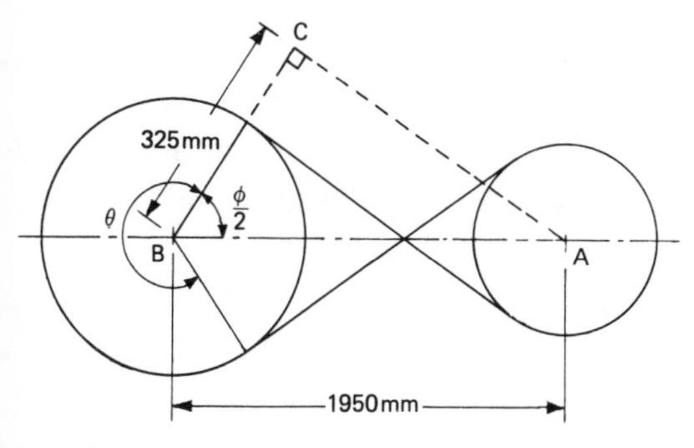

Fig. 10.5

$$\text{Length of belt} = 0{\cdot}225 \times 3{\cdot}474 + 0{\cdot}10 \times 3{\cdot}474 + 2 \times 1{\cdot}95 \sin 80{\cdot}4°$$

$$= \underline{4{\cdot}975 \text{ m}}$$

$$T_1 = 1000 \text{ N}$$

$$\therefore T_2 = \frac{1000}{e^{0{\cdot}25 \times 3{\cdot}474}} = \frac{1000}{2{\cdot}384} = 419{\cdot}3 \text{ N}$$

$$\therefore \text{power} = (T_1 - T_2)v \qquad \text{from equation 10.3}$$

$$= (1000 - 419{\cdot}3) \times \frac{2\pi}{60} \times 200 \times 0{\cdot}225 = 2740 \text{ W} \quad \text{or} \quad \underline{2{\cdot}74 \text{ kW}}$$

Example 10.2 A belt drive consists of two V-belts in parallel, on grooved pulleys of the same size. The angle of the groove is 30°. The cross-sectional area of each belt is 750 mm^2 and $\mu = 0{\cdot}12$. The density of the belt material is 1·2 Mg/m^3 and the maximum safe stress in the material is 7 MN/m^2.

Calculate the power that can be transmitted between pulleys of 300 mm diameter rotating at 1500 rev/min. Find also the shaft speed in rev/min at which the power transmitted would be a maximum. (U. Lond.)

Solution

$$\theta = \pi \text{ rad}$$

$$v = \frac{2\pi}{60} \times 1500 \times 0{\cdot}15 = 23{\cdot}6 \text{ m/s}$$

$$m = 1{\cdot}2 \times 10^3 \times 1 \times 750 \times 10^{-6} = 0{\cdot}9 \text{ kg/m}$$

$$T_c = mv^2 \qquad \text{from equation 10.6}$$

$$= 0{\cdot}9 \times 23{\cdot}6^2 = 502 \text{ N}$$

$$T_1 = 7 \times 10^6 \times 750 \times 10^{-6} = 5250 \text{ N}$$

$$e^{\mu\theta \operatorname{cosec} \beta} = e^{0\cdot 12\pi \operatorname{cosec} 15^\circ} = 4{\cdot}291$$

$$\frac{T_1 - T_c}{T_2 - T_c} = e^{\mu\theta \operatorname{cosec} \beta} \qquad \text{from equation 10.7}$$

i.e.
$$\frac{5250 - 502}{T_2 - 502} = 4{\cdot}291$$

from which $T_2 = 1610$ N

$$\therefore \text{power} = (T_1 - T_2)v \times 2 \qquad \text{from equation 10.3}$$

$$= (5250 - 1610) \times 23{\cdot}6 \times 2 = 172\,000 \text{ W} \quad \text{or} \quad \underline{172 \text{ kW}}$$

For maximum power,

$$T_c = \tfrac{1}{3}T_1 = 1750 \text{ N}$$

$$\therefore 1750 = 0{\cdot}9v^2$$

$$\therefore v = 44{\cdot}1 \text{ m/s}$$

$$\therefore N = 1500 \times 44{\cdot}1/23{\cdot}6 = \underline{2800 \text{ rev/min}}$$

Example 10.3 An open-belt drive connects two pulleys, 1·2 and 0·5 m diameter, on parallel shafts 3·6 m apart. The belt has a mass of 0·9 kg/m length, and the maximum tension in it is not to exceed 2 kN.

The 1·2 m pulley, which is the driver, runs at 200 rev/min. Due to belt slip on one of the pulleys, the velocity of the driven shaft is only 450 rev/min. Calculate the torque on each of the two shafts, the power transmitted and the power lost in friction. $\mu = 0{\cdot}3$.

What is the efficiency of the drive? (U. Lond.)

Solution

From Fig. 10.6,
$$\cos\frac{\theta}{2} = \frac{0{\cdot}6 - 0{\cdot}25}{3{\cdot}6} = 0{\cdot}0972$$

$$\therefore \frac{\theta}{2} = 84{\cdot}42^\circ = 1{\cdot}472 \text{ rad}$$

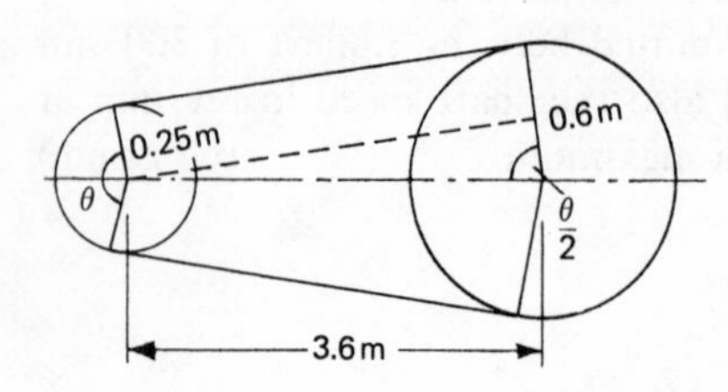

Fig. 10.6

∴ angle of lap on the smaller pulley,

$$\theta = 2{\cdot}944 \text{ rad}$$

The belt speed is that corresponding to the peripheral speed of the larger (driving) pulley.

$$\therefore \; v = 200 \times \frac{2\pi}{60} \times 0{\cdot}6 = 12{\cdot}57 \text{ m/s}$$

$$T_c = mv^2 = 0{\cdot}9 \times 12{\cdot}57^2 = 142{\cdot}3 \text{ N}$$

$$\frac{T_1 - T_c}{T_2 - T_c} = e^{\mu\theta}$$

i.e.

$$\frac{2000 - 142{\cdot}3}{T_2 - 142{\cdot}3} = e^{0{\cdot}3 \times 2{\cdot}944} = 2{\cdot}419$$

from which

$$T_2 = 910 \text{ N}$$

$$\therefore \text{ torque on driver} = (2000 - 910) \times 0{\cdot}6 = \underline{654 \text{ N m}}$$

and

$$\text{torque on follower} = (2000 - 910) \times 0{\cdot}25 = \underline{272{\cdot}5 \text{ N m}}$$

$$\text{Power of driver} = \frac{(2000 - 910) \times 12{\cdot}57}{1000} = \underline{13{\cdot}7 \text{ kW}}$$

If there were no slip,

$$\text{speed of follower would be } \frac{48}{20} \times 200 = 480 \text{ rev/min}$$

$$\therefore \text{ power transmitted to follower} = 13{\cdot}7 \times \frac{450}{480} = \underline{12{\cdot}85 \text{ kW}}$$

$$\therefore \text{ power lost in friction} = 13{\cdot}7 - 12{\cdot}85 = \underline{0{\cdot}85 \text{ kW}}$$

$$\text{Efficiency of drive} = \frac{12{\cdot}85}{13{\cdot}7} \times 100 = \underline{93{\cdot}8 \text{ per cent}}$$

Example 10.4 A flat belt is to be designed to transmit 110 kW at a belt speed of 25 m/s between two pulleys of diameters 250 mm and 400 mm, having a pulley centre distance of 1 m. The allowable belt stress is 8·5 MN/m^2, and belts are available having a thickness-to-width ratio of 0·1 and a material density of 1100 kg/m^3. Given that the coefficient of friction between the belt and pulleys is 0·3, determine the minimum required belt width.

What would be the necessary installation force between the pulley bearings, and what will be the force between the pulley bearings when the full power is being transmitted? (U. Lond.)

Solution

As in Example 10.3,

$$\cos\frac{\theta}{2} = \frac{200 - 125}{1000} = 0{\cdot}075$$

Therefore $$\theta = 171{\cdot}4^\circ = = 2{\cdot}99\,\text{rad}$$

$$e^{\mu\theta} = e^{0{\cdot}3\times 2{\cdot}99} = 2{\cdot}453$$

If the width of the belt is w, the thickness is $0{\cdot}1w$, and so

$$T_1 = \sigma a = 8{\cdot}5\times 10^6\times w\times 0{\cdot}1w = 85\times 10^4 w^2\,\text{N}$$

$$T_c = mv^2 = 1100\times w\times 0{\cdot}1w\times 25^2 = 6{\cdot}875\times 10^4 w^2\,\text{N}$$

$$\text{Power} = (T_1 - T_c)\left(1-\frac{1}{e^{\mu\theta}}\right)v \qquad \text{from equation 10.8}$$

i.e. $$110\times 10^3 = (85-6{\cdot}875)10^4 w^2\left(1-\frac{1}{2{\cdot}453}\right)\times 25$$

from which $$w = 0{\cdot}0975\,\text{m} \quad \text{or} \quad \underline{97{\cdot}5\,\text{mm}}$$

$$T_1 = 85\times 10^4\times 0{\cdot}0975^2 = 8080\,\text{N}$$

$$T_c = 6{\cdot}875\times 10^4\times 0{\cdot}0975^2 = 653\,\text{N}$$

$$\frac{T_1 - T_c}{T_2 - T_c} = e^{\mu\theta} \qquad \text{from equation 10.7}$$

i.e. $$\frac{8080-653}{T_2-653} = 2{\cdot}453$$

from which $$T_2 = 3681\,\text{N}$$

$$T_0 = \frac{T_1 + T_2}{2} \qquad \text{from equation 10.10}$$

$$= \frac{8080+3681}{2} = 5880\,\text{N}$$

If the inclination of the belt to the centre line of the drive is ϕ, then

$$\phi = 90^\circ - \frac{171{\cdot}4^\circ}{2} = 4{\cdot}3^\circ$$

Initial force between bearings $= 2\,T_0\cos\phi = 2\times 5880\cos 4{\cdot}3^\circ = \underline{11\,727\,\text{N}}$

Force between bearings when transmitting power

$$= \{(T_1 - T_c) + (T_2 - T_c)\}\cos\phi$$

$$= \{(8080-653)+(3681-653)\}\cos 4{\cdot}3^\circ$$

$$= \underline{10\,425\,\text{N}}$$

Example 10.5 A small air compressor is belt-driven from a lay shaft in a workshop, the pulley on the compressor being 300 mm diameter, and the angle of lap of the belt is 165°. When the belt is moved from the loose to the fast pulley, it slips for 8 s until the compressor attains its constant speed of 300 rev/min. The flywheel of the compressor has a moment of inertia of 4 kg m^2 and the friction requires a constant torque of 4 N m. If the coefficient of friction is 0·28 during the accelerating period, find the tensions in both reaches of the belt, and also the distance that the belt slips and the energy lost in that time due to the belt slip. (U. Glas.)

Solution

While slip is taking place, ratio of belt tensions,

$$\frac{T_1}{T_2} = e^{0{\cdot}28(165\times\pi/180)} = 2{\cdot}24 \qquad (1)$$

$$\text{Acceleration of compressor} = \left(300\times\frac{2\pi}{60}\right)\Big/8 = \frac{5\pi}{4}\,\text{rad/s}^2$$

$$\text{Net torque on compressor} = (T_1 - T_2)\times 0{\cdot}15 - 4$$

$$\therefore\ (T_1 - T_2)\times 0{\cdot}15 - 4 = 4\times\frac{5\pi}{4}$$

i.e.

$$T_1 - T_2 = 131{\cdot}4\,\text{N} \qquad (2)$$

Therefore, from equations (1) and (2),

$$T_1 = \underline{237\,\text{N}} \quad \text{and} \quad T_2 = \underline{106\,\text{N}}$$

$$\text{Belt velocity} = \left(300\times\frac{2\pi}{60}\right)\times 0{\cdot}15 = 4{\cdot}71\,\text{m/s}$$

$$\therefore\ \text{distance moved in 8 s} = 8\times 4{\cdot}71 = 37{\cdot}68\,\text{m}$$

Since the pulley accelerates uniformly until its circumferential speed is 4·71 m/s,

$$\text{distance moved by a point on the circumference} = \tfrac{1}{2}\times 37{\cdot}68 = 18{\cdot}84\,\text{m}$$

$$\therefore\ \text{slip of belt relative to pulley} = 37{\cdot}68 - 18{\cdot}84 = \underline{18{\cdot}84\,\text{m}}$$

$$\text{Energy lost due to slip} = (T_1 - T_2)\times\text{distance slipped}$$
$$= 131{\cdot}4\times 18{\cdot}84 = \underline{2470\,\text{J}}$$

Example 10.6 Two parallel horizontal shafts, whose centre lines are 4·8 m apart, one being vertically above the other, are connected by an open belt drive. The pulley on the upper shaft is 1·05 m diameter, that on the lower shaft is 1·5 m diameter. The belt is 150 mm wide and the initial tension in it when stationary and when no torque is being transmitted is 3 kN. The belt has a mass of 1·5 kg/m length; the gravitational force on it may be neglected but centrifugal force must be taken into account. The material of the belt may be assumed to obey Hooke's law, and the free lengths of the belt between pulleys may be assumed to be straight. The coefficient of friction between the belt and either pulley is 0·3.

Calculate: (a) the pressure in N/m^2 between the belt and the upper pulley when the belt and pulleys are stationary and no torque is being transmitted, (b) the tension in the belt and the pressure between the belt and the upper pulley if the upper shaft rotates at 400 rev/min and there is no resisting torque on the lower shaft, hence no power being transmitted, and (c) the greatest tension in the belt if the upper shaft rotates at 400 rev/min and the maximum possible power is being transmitted to the lower shaft. (U. Lond.)

Solution

(a) Let the pressure on an element subtending an angle $d\theta$ at the centre, Fig. 10.7, be p N/m². Then, resolving forces radially,

$$p \times 0{\cdot}525\, d\theta \times 0{\cdot}15 = 2 \times 3000 \frac{d\theta}{2}$$

$$\therefore\ p = \underline{38\,100\ \text{N/m}^2}$$

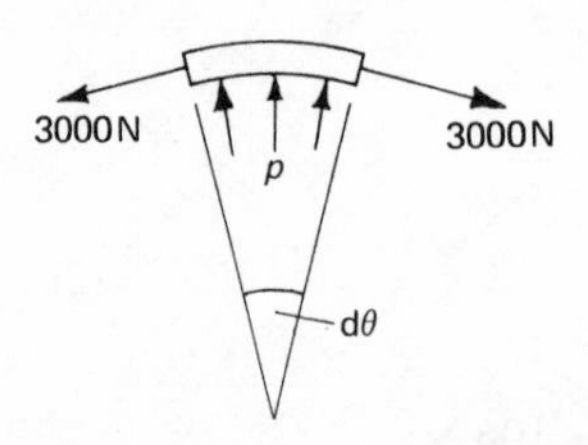

FIG. 10.7

(b) $$T_c = mv^2 = 1{\cdot}5 \times \left(\frac{2\pi}{60} \times 400 \times 0{\cdot}525\right)^2 = 725\ \text{N}$$

The total length of the belt remains constant, and since the material obeys Hooke's law, the tension remains constant at 3000 N. Part of this force is now due to centrifugal tension, however, and the reaction between the belt and the pulley is reduced.

$$\text{Effective tension} = 3000 - 725 = 2275\ \text{N}$$

$$\therefore\ p = \frac{2275}{3000} \times 38\,100 = \underline{28\,900\ \text{N/m}^2}$$

(c) $$\cos\frac{\theta}{2} = \frac{0{\cdot}75 - 0{\cdot}525}{4{\cdot}8} = 0{\cdot}0468 \qquad \text{(as in Example 10.3)}$$

$$\therefore\ \theta = 174{\cdot}63^\circ = 3{\cdot}05\ \text{rad}$$

$$T_1 + T_2 = 2T_0 = 6000\ \text{N} \qquad (1)$$

Also $$\frac{T_1 - 725}{T_2 - 725} = e^{0{\cdot}3 \times 3{\cdot}05} = 2{\cdot}5 \qquad (2)$$

Therefore, from equations (1) and (2),

$$T_1 = \underline{3970\ \text{N}}$$

Example 10.7 Figure 10.8 shows a belt drive fitted with a gravity idler. The driver rotates anticlockwise at 360 rev/min and the coefficient of friction between belt and pulley is 0·3. Determine the initial belt tension and the power transmitted. Neglect any sag in the belt. (U. Lond.)

Solution
Referring to Fig. 10.9, let

P = normal reaction at O between idler and lever

R = resultant of initial belt tensions, T_0, at idler = $\sqrt{2}\,T_0$

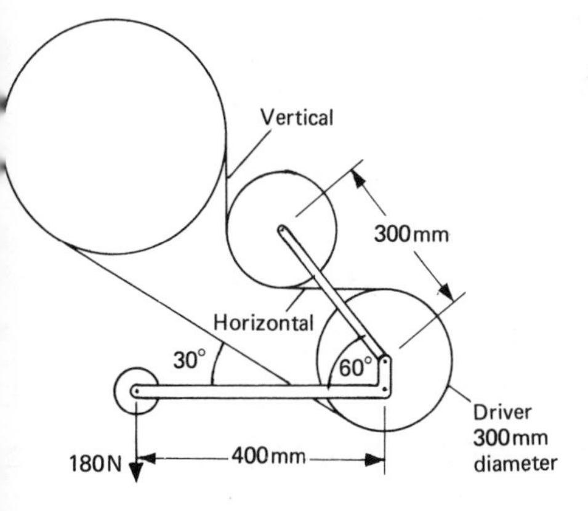

FIG. 10.8

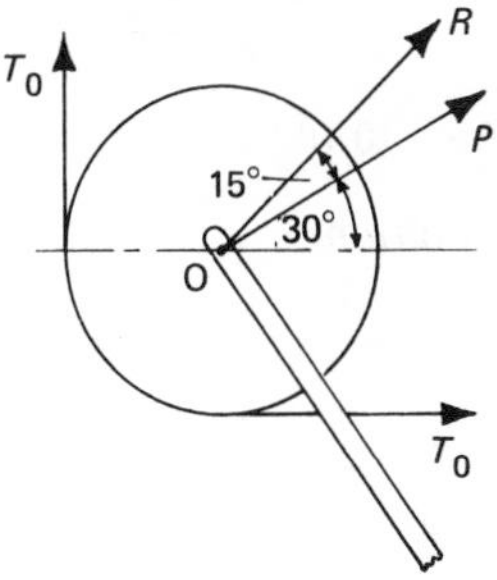

FIG. 10.9

Taking moments about lever pivot,

$$P \times 300 = 180 \times 400$$

$$\therefore\ P = 240\text{ N}$$

$$\therefore\ R = 240 \sec 15^\circ = 249\text{ N}$$

$$\therefore\ T_0 = 249/\sqrt{2} = \underline{176\text{ N}}$$

$$\text{Angle of lap on driving pulley} = 210^\circ = 3{\cdot}66\text{ rad}$$

$$\therefore\ \frac{T_1}{T_2} = e^{0{\cdot}3 \times 3{\cdot}66} = 3{\cdot}0$$

Due to the action of the idler, *the slack side tension remains constant at 176 N*.

$$\therefore\ T_1 = 176 \times 3 = 528\text{ N}$$

$$\therefore\ \text{power} = (T_1 - T_2)v \qquad \text{from equation 10.3}$$

$$= (528 - 176) \times \frac{2\pi}{60} \times 360 \times 0{\cdot}15 = \underline{1990\text{ W}}$$

Example 10.8 Figure 10.10 shows the layout of a band brake applied to the brake drum of a hoist, where the braking force P is applied at one end of a lever which is pivoted on a fixed fulcrum at F. The drum diameter is 1 m, the arc of contact 225° and $\mu = 0{\cdot}3$.

Calculate the force P to give a braking torque of 400 N m if the drum is rotating: (a) clockwise, and (b) anticlockwise. Comment on the large difference between the answers to (a) and (b). (U. Lond.)

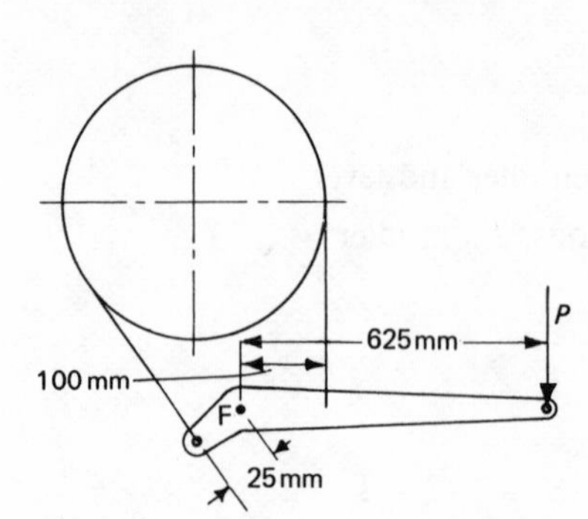

FIG. 10.10

Solution

$$(T_1 - T_2) \times 0{\cdot}5 = 4000$$

$$\therefore\ T_1 - T_2 = 8000\ \text{N m} \qquad (1)$$

Also
$$\frac{T_1}{T_2} = e^{0\cdot 3(225 \times \pi/180)} = 3{\cdot}248 \qquad (2)$$

Therefore, from equations (1) and (2),

$$T_1 = 11\,560\ \text{N} \quad \text{and} \quad T_2 = 3560\ \text{N}$$

The tight side tension T_1 is on the side which is being pulled on to the pulley.
(a) Clockwise rotation, Fig. 10.11.
Taking moments about F,

$$P \times 625 + 11\,560 \times 25 = 3560 \times 100$$

$$\therefore\ P = \underline{107{\cdot}2\ \text{N}}$$

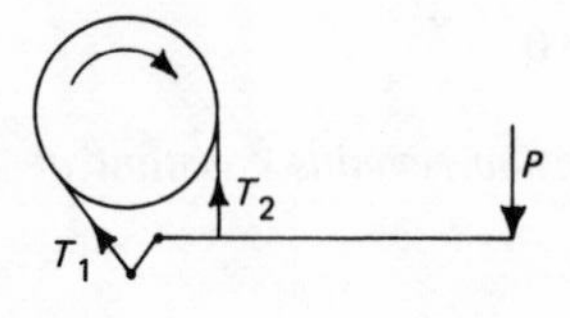

FIG. 10.11

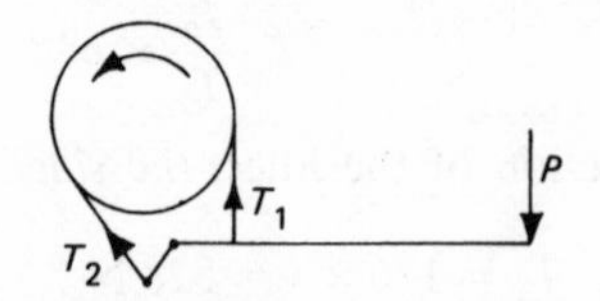

FIG. 10.12

(b) Anticlockwise rotation, Fig. 10.12.
Taking moments about F,

$$P \times 625 + 3560 \times 25 = 11\,560 \times 100$$

$$\therefore\ P = \underline{1710\ \text{N}}$$

In case (a) the couple about the fulcrum due to belt friction is acting in the same direction as that due to P, whereas, in case (b) it is acting in the opposite direction.

Example 10.9 The essential features of a transmission dynamometer are shown in Fig. 10.13. A is the driving pulley which runs at 500 rev/min. B and C are jockey pulleys mounted on a horizontal beam pivoted at D, about which point the complete beam is balanced when at rest. E is the driven pulley, and all portions of the belt between the pulleys are vertical. A, B and C are each 300 mm diameter, and the thickness and mass of the belt are to be neglected. DF is 750 mm.

Find: (a) the value of the mass m necessary to maintain the beam in a horizontal position when 4 kW is being transmitted, and (b) the value of m when the belt just begins to slip on A, μ being 0·2 and the maximum tension in the belt 1100 N. (U. Lond.)

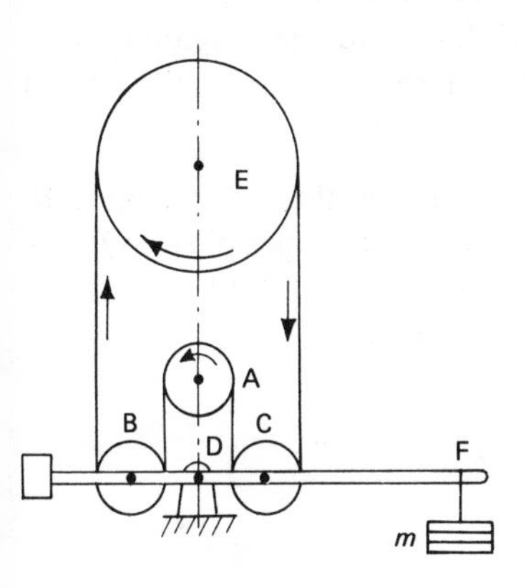

FIG. 10.13

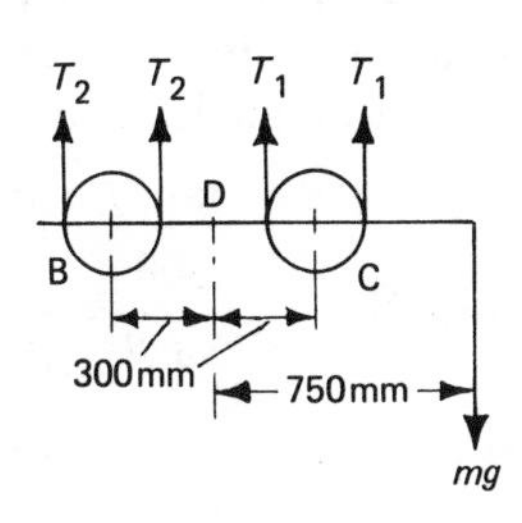

FIG. 10.14

Solution

(a)

$$\text{Diameter of E} = 0{\cdot}9\text{ m}$$

$$\therefore \text{ speed of E} = 500/3 \text{ rev/min}$$

$$\text{Torque on E} = \frac{\text{power} \times 60}{2\pi N}$$

i.e.

$$(T_1 - T_2) \times 0{\cdot}45 = \frac{4 \times 10^3 \times 60}{2\pi \times 500/3}$$

$$\therefore T_1 - T_2 = 509 \text{ N}$$

Taking moments about D, Fig. 10.14,

$$mg \times 750 + 2T_2 \times 300 = 2T_1 \times 300$$

i.e.

$$mg = \tfrac{4}{5}(T_1 - T_2)$$

$$\therefore m = \tfrac{4}{5} \times 509/9{\cdot}81 = \underline{41{\cdot}5 \text{ kg}}$$

(b) When the belt is about to slip,

$$\frac{T_1}{T_2} = e^{0 \cdot 2\pi} = 1{\cdot}875$$

$$T_1 = 1100\,\text{N}$$

$$\therefore\; T_2 = 1100/1{\cdot}875 = 586{\cdot}5\,\text{N}$$

$$\therefore\; m = \frac{4(1100 - 586{\cdot}5)}{5 \quad 9{\cdot}81} = \underline{41{\cdot}9\,\text{kg}}$$

Example 10.10 For the simple brake shown in Fig. 10.15, find a relationship between the braking torque and the applied force P N, if the coefficient of friction between the brake drum and brake block is 0·35.

What is the braking torque if $P = 500$ N? Also find the magnitude and direction of the resultant force at each of the hinges A and B. (U. Lond.)

Solution

The reaction, R, between the block and the drum passes through the point B, Fig. 10.16, and at the point of intersection with the drum periphery, Q, it is inclined at the friction angle, ϕ, to the radius at that point, OQ.

$$\phi = \tan^{-1}\mu = \tan^{-1}0{\cdot}35 = 19{\cdot}3°$$

From triangle OQB,
$$\frac{300}{\sin\theta} = \frac{375}{\sin(180° - 19{\cdot}3°)}$$

$$\therefore\; \theta = 15{\cdot}33°$$

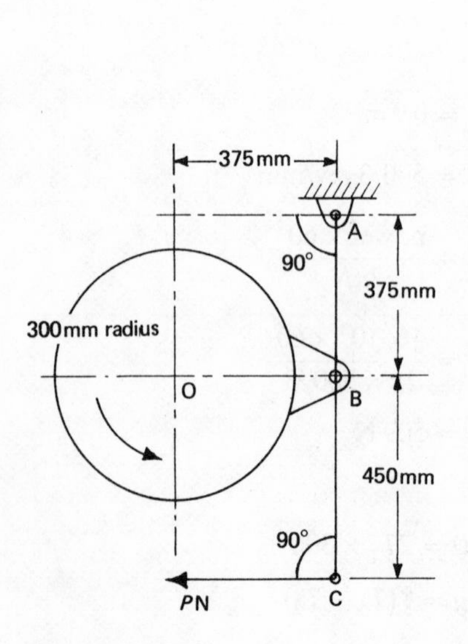

Fig. 10.15

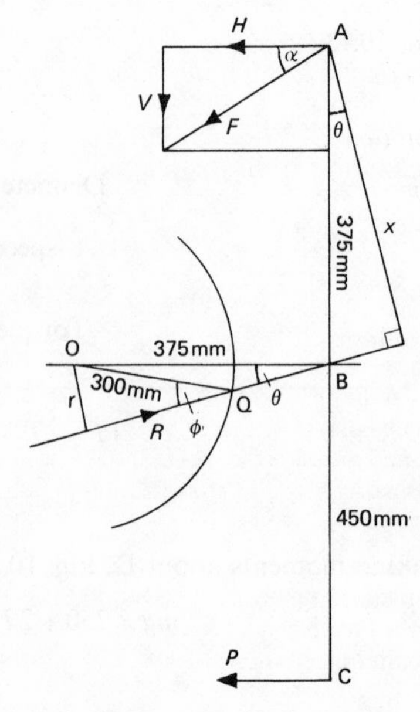

Fig. 10.16

Taking moments about A,

$$P \times 825 = R \times x$$

$$\therefore \quad R = \frac{825}{375\cos\theta}$$

$$\therefore \text{ friction torque, } T = Rr$$

$$= \frac{825P}{375\cos\theta} \times 375\sin\theta$$

$$= 825P \times \tan 15{\cdot}33^\circ$$

$$= \underline{226P \text{ N mm}}$$

When $P = 500$ N,

$$T = \frac{226 \times 500}{1000} = \underline{113 \text{ N m}}$$

Resultant force at hinge B,

$$R = \frac{825 \times 500}{375\cos 15{\cdot}33^\circ} = \underline{1142 \text{ N}}$$

Inclination to horizontal,

$$\theta = \underline{15{\cdot}33^\circ}$$

Horizontal component of reaction at A,

$$H = R\cos\theta - P = 1142\cos 15{\cdot}33^\circ - 500 = 600 \text{ N}$$

Vertical component of reaction at A,

$$V = R\sin\theta = 1142\sin 15{\cdot}33^\circ = 302 \text{ N}$$

Therefore resultant reaction at A,

$$F = \sqrt{(H^2 + V^2)} = \sqrt{(600^2 + 302^2)} = \underline{671{\cdot}7 \text{ N}}$$

Inclination to horizontal,

$$\alpha = \tan^{-1}\frac{V}{H} = \tan^{-1}\frac{302}{600} = \underline{26{\cdot}68^\circ}$$

Problems

1. A ship is dragged through a lock by means of a capstan and rope. The capstan, which has a diameter of 500 mm, turns at 30 rev/min. The rope makes 3 complete turns around the capstan, μ being 0·25, and at the free end of the rope a pull of 100 N is applied. Find: (a) the pull on the ship, and (b) the power required to drive the capstan. (U. Lond.) (*Answer:* 11·1 kN; 8·66 kW)

2. A leather belt, 125 mm wide and 6 mm thick, transmits power from a pulley 750 mm diameter which runs at 500 rev/min. The angle of lap is 150° and $\mu = 0{\cdot}3$. If the mass of 1 m^3 of leather is 1 Mg and the stress in the belt is not to exceed 2·75 MN/m^2, find the maximum power which can be transmitted. (I. Mech. E.) (*Answer:* 18·94 kW)

3. In a belt drive, the angle of lap of the belt on the small pulley is 150°. With a belt speed of 20 m/s and a tension in the tight side of the belt of 1·35 kN, the greatest power which can be transmitted without slip is 10 kW. What increase of power would be obtained for the same belt speed and maximum tension by using an idler pulley so as to increase the angle of lap to 210°? Take into account the centrifugal effect, the mass of the belt being 0·75 kg/m. (U. Lond.) (*Answer:* 2·5 kW)

4. A pulley is driven by a flat belt, the angle of lap being 120°. The belt is 100 mm wide by 6 mm thick and has a mass of 1 Mg/m^3. If $\mu = 0{\cdot}3$ and the maximum stress in the belt is not to exceed 1·5 MN/m^2, find the greatest power which the belt can transmit and the corresponding speed of the belt. (U. Lond.) (*Answer:* 6·26 kW; 22·36 m/s)

5. Power is transmitted between two shafts, 4·5 m apart, by an open wire rope passing round two pulleys, of 3 m and 2 m diameter respectively, the groove angle being 40°. If the rope has a mass of 4 kg/m, and the maximum working tension is 20 kN, determine the maximum power that the rope can transmit, and the corresponding speed of the smaller pulley. $\mu = 0{\cdot}2$. (U. Lond.) (*Answer:* 445 kW; 390 rev/min)

6. Power is transmitted from an electric motor to a machine tool by an open belt drive. The effective diameter of the pulley on the motor shaft is 150 mm, while that on the machine tool is 200 mm, with a centre distance of 600 mm. If the motor speed is 1440 rev/min and the maximum permissible belt tension is 900 N, then the maximum power transmissible is 6 kW.

It is necessary that the power transmissible be increased to 6·75 kW, using the same pulleys, centre distance and motor speed. The belt is treated with a special preparation that increases its coefficient of friction by 10 per cent of its existing value, and in addition a jockey pulley may be fitted.

Determine: (a) the existing coefficient of friction, and (b) the new angle of lap. (U. Lond.) (*Answer:* 0·29; 195°)

7. A vertical open-belt drive connects two pulleys, A and B, the centres of which are 4 m apart. The belt has a mass of 1·15 kg/m. Pulley A is 1 m diameter, has radius of gyration of 420 mm and a mass of 25 kg. Pulley B is 0·5 m diameter, has radius of gyration of 225 mm and a mass of 18 kg. When at rest the tension in the belt is 700 N. Assuming that the belt obeys Hooke's law, determine the tensions in the two portions of the belt between the pulleys when 1·5 kW is being transmitted, the speed of A being 180 rev/min. Neglect belt stretch over the pulleys. Find also the kinetic energy of the belt and pulleys under these conditions. (U. Lond.)
(*Answer:* 779·6 N; 620·4 N; 530 J; 1430 J)

8. A belt drive consists of a V-belt working on a grooved pulley, with an angle of lap of 160°. The cross-sectional area of the belt is 650 mm^2, the groove angle is 30° and $\mu = 0{\cdot}1$. The density of the belt material is 1 Mg/m^3 and its maximum safe stress is 8 MN/m^2 of cross-section.

Derive an expression for the ratio of the tensions on the two sides of the drive when the belt is about to slip.

Calculate the power that can be transmitted at a belt speed of 25 m/s. (U. Lond.)
(*Answer:* 79 kW)

9. The following particulars apply to one pulley of a rope drive between two parallel shafts:

Effective diameter of pulley	1·5 m
Minimum angle of lap	180°
Mass of rope per metre run	0·45 kg
Total angle of groove	45°
Maximum permitted load per rope	650 N
Coefficient of friction	0·25

(a) Find the power transmitted per rope at a pulley speed of 200 rev/min, if centrifugal tension may be neglected.
(b) Find the pulley speed when centrifugal tension accounts for half the permitted load in the rope, and the power which can be transmitted at that speed. (U. Lond.)
(*Answer:* 8·9 kW; 342 rev/min; 7·62 kW)

10. Power is transmitted from a shaft rotating at 250 rev/min by 10 ropes running in grooves in the periphery of a wheel of effective diameter 1·65 m (to the centre line of the rope). The groove angle is 50° and the arc of contact round the wheel rim is 180°. The maximum permissible load in each rope is 900 N and its mass is 0·55 kg/m.

If the coefficient of friction between the rope and wheel surface is 0·3, what power can be transmitted under the above conditions? (U. Lond.) (*Answer:* 124 kW)

11. A rope drive is required to transmit 35 kW at 160 rev/min. The grooved pulley has a mean diameter to the rope centre of 1·2 m and the groove angles are 45°. Taking μ as 0·25 and the arc of contact of the ropes as 190°, determine the number of ropes required if the greatest pull in each rope is limited to 700 N. (U. Lond.) (*Answer:* 6 ropes)

12. A small generator is driven by means of a V-belt which has a total angle of 60° between the faces of the V. The angle of lap on the pulley is 120° and the mean radius of the belt as it passes round the pulley is 50 mm. If $\mu = 0{\cdot}2$ and the mass of the belt is 0·45 kg/m, find the tension in each side of the belt when 750 W is being transmitted at a pulley speed of 1800 rev/min. (U. Lond.) (*Answer:* 180 N; 100·5 N)

13. A 4-to-1 speed reduction drive between two parallel shafts at 2 m centres is provided by means of five parallel V-belts running on suitable pulleys mounted on the shafts. The effective diameter of the driving pulley is 350 mm and the driving shaft rotates at 740 rev/min. The included angle of each pulley groove is 40°, each V-belt has a mass of 0·45 kg/m and the coefficient of friction between belt and groove is 0·28.

Determine what power can be transmitted by the drive, if the tension in each belt is not to exceed 800 N. (U. Lond.) (*Answer:* 42·9 kW)

14. Derive the expression $T_1/T_2 = e^{\mu\theta}$ for a vertical belt drive, where T_1 and T_2 are the tight and slack side tensions respectively when slipping is about to commence, μ is the effective coefficient of friction and θ is the angle of lap.

Starting from the above equation, and allowing for centrifugal tension, calculate: (a) the belt speed at which the maximum power will be transmitted by a V-belt of mass 0·7 kg/m, the maximum belt tension being limited to 2 kN, and (b) the power transmitted at this speed.

The angle of lap on the pulley is 170°, $\mu = 0{\cdot}17$ and the belt is running in a groove of total angle 45°. (U. Lond.) (*Answer:* 30·86 m/s; 30·15 kW)

15. In a belt drive, power is being transmitted from one pulley to another by a flat belt. The smaller pulley has a diameter of 0·3 m and is running at 150 rev/min. The angle of lap on this pulley is 160° and the coefficient of friction between the belt and pulley is 0·25. The belt is on the point of slipping when 3 kW is being transmitted.

Calculate which of the following two alternatives would be the more effective in increasing the power which can be transmitted: (a) increasing the coefficient of friction to 0·27, or (b) increasing the initial static tension in the belt by 10%. (U. Lond.)
(*Answers:* 3·21 kW; 3·31 kW; (b) more effective)

16. An electric motor running at 1400 rev/min transmits power by 3 V-belts, each of 320 mm^2 cross-sectional area, the total angle of groove being 45°. The density of the belt material is 1·65 Mg/m^3 and the maximum allowable working stress in the belts is 2 MN/m^2. $\mu = 0{\cdot}2$. The angle of lap on the motor pulley is 145°. Calculate the maximum power which can be transmitted and the corresponding diameter of the motor pulley. (U. Glas.) (*Answer:* 18·9 kW; 275 mm)

17. The drive from an electric motor to a line shaft is by 3 V-belts in parallel. The mean diameter of the pulley on the motor is 150 mm and the motor and shaft speeds are 1600 rev/min and 400 rev/min respectively. The shafts are 1 m apart. The belt has a mass of 1·4 Mg/m^3, the cross-sectional area of each belt is 800 mm^2, the angle between the sides of the V-groove is 30°, and $\mu = 0{\cdot}13$.

If the tensile stress in the belt material is not to exceed 8 MN/m^2, find the power that can be transmitted by the drive. (U. Lond.) (*Answer:* 173·6 kW)

18. The winding drum of a crane has an effective diameter of 450 mm and carries a brake wheel of 1·2 m diameter. The angle of lap of the brake band is 270°, and $\mu = 0{\cdot}33$. Find the maximum load which the brake can sustain on the winding drum, if the safe working tension in the brake band is 15 kN.

Derive any formula used, and make a neat diagram of the brake and operating lever. (U. Lond.) (*Answer:* 31·5 kN)

19. A band brake is fitted to the circumference of a wheel of 0·9 m diameter, the angle of contact being 220°. The band is fixed at one end and a pull of 180 N can be applied at the other end. The mass of the wheel is 550 kg, and its radius of gyration is 375 mm. $\mu = 0{\cdot}2$.

If the wheel is rotating at 600 rev/min and the brake is then fully applied, find the time required for the wheel to come to rest. (U. Lond.) (*Answer:* 52 s)

20. A cylinder, whose axis is horizontal, rests in contact with a vertical wall and is supported by a flexible belt as shown in Fig. 10.17. The belt is secured at A and D, the portions AB and CD being horizontal and vertical respectively. The cylinder weighs 200 N and has a radius of 0·25 m. The coefficients of friction between cylinder and belt and between cylinder and wall are 0·25 and 0·1 respectively.

Determine the magnitude of the torque required to rotate the cylinder: (a) in a clockwise direction, and (b) in an anticlockwise direction. (C.E.I.)
(*Answer:* 21 N m; 27·4 N m)

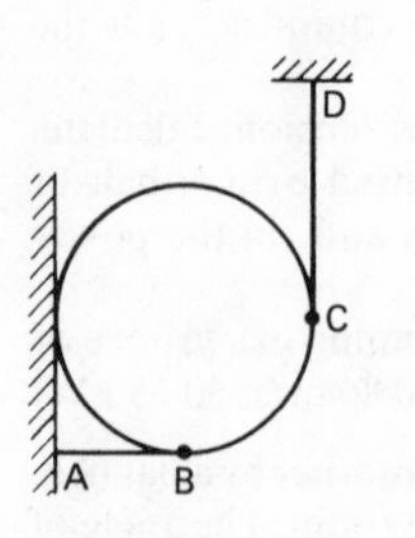

FIG. 10.17

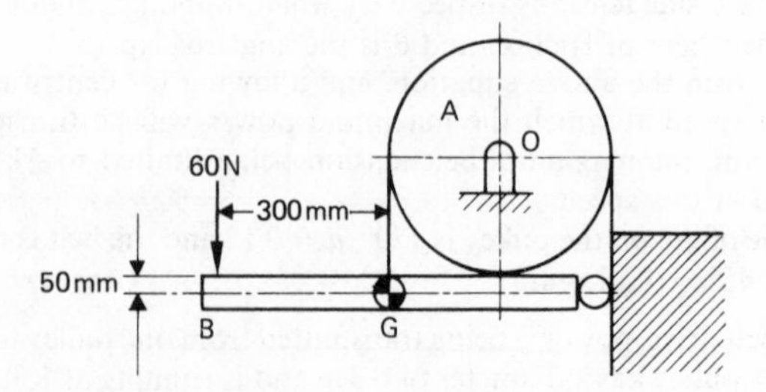

FIG. 10.18

21. A braking mechanism shown in Fig. 10.18 includes a uniform circular disc, A, which can rotate about its centre, which is fixed at O. The disc has a diameter of 250 mm and a point on its circumference below its centre is in contact with a horizontal bar, B, of mass 2·5 kg. A vertical force of 60 N acts on one end of the bar while the other end of the bar presses a light, smooth, circular pin against a vertical abutment. At its centre of mass, G, the bar is attached to one end of a light flexible belt, which passes over the disc A and is attached at its other end to the vertical abutment. The centre of the pin and the centre of mass of the bar are 50 mm below the contact point between the disc and the bar.

If the coefficient of friction between the disc and the bar is 0·4, and that between the disc and the belt is 0·3, what is the magnitude of the moment applied about the centre of the disc when the disc is on the point of turning anticlockwise? (U. Lond.)

(*Answer:* 61·55 N m)

22. The band brake indicated in Fig. 10.19 is applied to a shaft carrying a flywheel of 360 kg mass, with a radius of gyration of 450 mm, and running at 360 rev/min.

Find: (a) the torque applied due to a hand load of 90 N, given that $\mu = 0{\cdot}2$, and (b) the number of turns of the wheel before it is brought to rest. (U. Lond.)

(*Answer:* 58·4 N m; 141 revolutions)

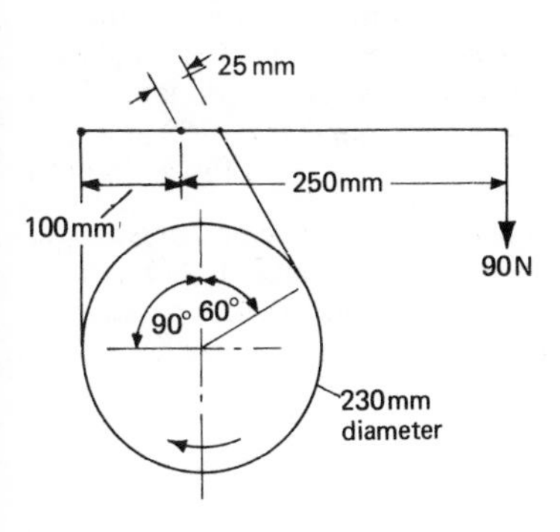

Fig. 10.19

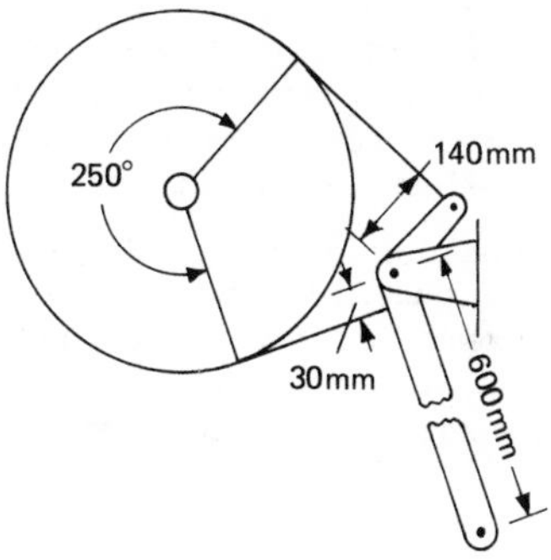

Fig. 10.20

23. The hand-brake shown in Fig. 10.20 is used on a wall crane. The brake drum is 600 mm diameter, the lifting drum is 400 mm diameter and carries a load of 4 kN. If $\mu = 0{\cdot}3$, find the least force at the end of the 0·6 m lever to support the load. (U. Lond.)

(*Answer:* 47·4 N)

24. A brake consists of a flexible band on the periphery of a wheel of 750 mm diameter. One end of the band is attached to a fixed pin and the other is subjected to a pull of 225 N. The angle of lap is 270° and $\mu = 0{\cdot}2$.

Sketch the arrangement, showing the direction of rotation of the wheel, and find the value of the maximum braking torque. If the wheel and the parts attached to it have a moment of inertia of 65 kg m^2, find the time to come to rest from an initial speed of 600 rev/min. (U. Lond.) (*Answer:* 132·2 N m; 31 s)

25. The application of a band brake to a circular drum is shown in Fig. 10.21. The two ends of the flat belt are connected to the cranked lever which pivots about B. The angle of lap is 210° and the coefficient of friction is 0·20. The braking force F is applied at A as shown.

Find the magnitude of F necessary to provide a braking torque of 30 N m on the drum when the latter is rotating: (a) clockwise, and (b) anticlockwise. Explain the difference in your answers. (C.E.I.) (*Answer:* 406·7 N; 106·7 N)

26. Figure 10.22 shows a band brake for use on a winch, consisting of a drum 450 mm diameter around a portion of the circumference of which is wrapped a band APQB the ends of which are attached to the brake lever as shown at points A and B. The lever can move around the fixed point O.

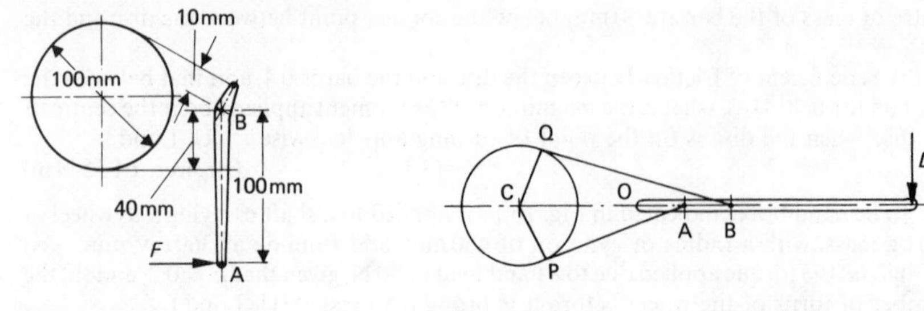

FIG. 10.21 FIG. 10.22

Determine the load to be applied to the lever 600 mm from O in order to apply a torque of 6·75 kN m to the drum, the rotation being clockwise. $\mu = 0{\cdot}1$. The band may be considered to be inextensible, so that the linkage does not depart appreciably from the position shown when under load. Lengths OC = 375 mm, OA = 150 mm and OB = 300 mm. (U. Lond.) (*Answer:* 500 N)

27. Figure 10.23 shows the driving wheel, D, of a machine tool. The belt passes over two idler pulleys, P, which are freely pinned to the light horizontal bar, AO, pinned at O. Each pulley, P, weighs 80 N and has a diameter of 0·3 m. The belt also passes over the driven wheel above. The coefficient of friction between the belt and the wheels is 0·3 and the angle of lap of the driving wheel is 180°.

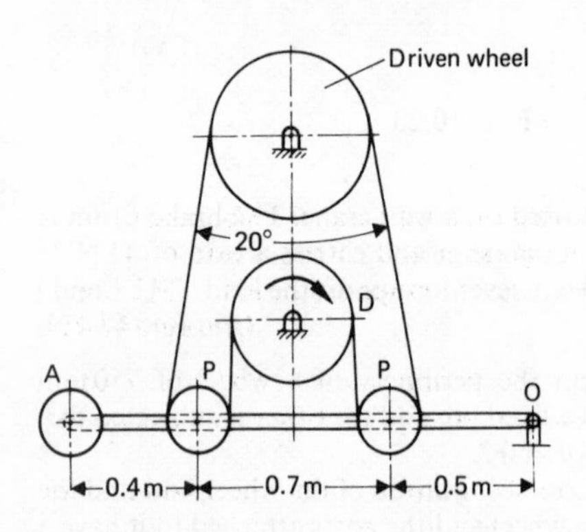

FIG. 10.23

Neglecting the mass of the belt and all bearing friction, estimate the mass required to be attached at A to the bar such that 2·25 kW is transmitted with a peripheral belt speed of 15 m/s. (U. Lond.) (*Answer:* 34·68 kg)

11
Spur gearing

11.1 Definitions

When two gears are in mesh, the larger is termed the *wheel* or *spur* and the smaller the *pinion*. Quantities relating to the wheel and pinion are denoted by capital and small letters respectively.

Pitch circle diameters (D, d): the diameters of discs which would transmit the same velocity ratio by friction as the gear wheels. If T and t are the numbers of teeth and Ω and ω are the angular velocities of the wheel and pinion respectively, then, since the circumferential velocity of both pitch circles is the same,

$$\frac{\Omega}{\omega} = \frac{d}{D} = \frac{t}{T}$$

The pitch circle radii are denoted by R and r. The *pitch point* is the point of contact of two pitch circles.

Circular pitch (p): the distance between a point on one tooth and the corresponding point on an adjacent tooth, Fig. 11.1, measured along the pitch circle,

i.e. $$p = \frac{\pi D}{T} = \frac{\pi d}{t}$$

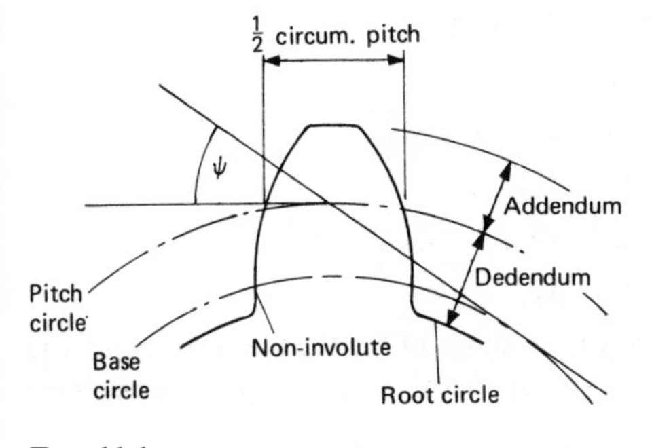

FIG. 11.1

Diametral pitch (P): the number of teeth per millimetre of p.c.d.,

i.e.
$$P = \frac{T}{D} = \frac{t}{d} = \frac{\pi}{p}$$

Module (m): the number of millimetres of p.c.d. per tooth,

i.e.
$$m = \frac{D}{T} = \frac{d}{t} = \frac{1}{P} = \frac{p}{\pi}$$

Base circle: the circle from which the involute curves* forming the tooth profiles are drawn.

Addendum: the radial height of a tooth above the pitch circle. The radii of the addendum circles are denoted by R_a and r_a.

Dedendum: the radial depth of a tooth below the pitch circle.

Working depth: the sum of the addenda of two mating teeth.

Pressure angle or angle of obliquity (ψ): the angle between the common normal to two teeth in contact and the common tangent to the pitch circles.

Standard proportions: the proportions recommended by the British Standards Institution in B.S. 436–1940 are:

$$\text{addendum} = 1/P = m$$
$$\text{dedendum} = 1{\cdot}25/P = 1{\cdot}25m$$
$$\text{working depth} = 2/P = 2m$$
$$\text{pressure angle} = 20^\circ$$

Standard modules are : $\frac{1}{2} \times \frac{1}{2} - 6\frac{1}{2}, 7 \times 1 - 16, 18, 20, 25, 30, 35, 40, 45, 50.$

11.2 Condition for transmission of constant velocity ratio

In Fig. 11.2, O_1 and O_2 are the centres of the pinion and wheel respectively, I_1I_2 is the common normal at the point of contact, C, and O_1I_1 and O_2I_2 are the perpendiculars from O_1 and O_2 respectively to the common normal.

Let v_{C_1} be the velocity of point C on the pinion and v_{C_2} be the velocity of point C on the wheel. If the teeth are to remain in contact, the components of these velocities along the common normal I_1I_2 must be equal,

i.e.
$$v_{C_1} \cos\alpha = v_{C_2} \cos\beta$$

i.e.
$$\omega\, O_1C \cos\alpha = \Omega\, O_2C \cos\beta$$

i.e.
$$\omega\, O_1I_1 = \Omega\, O_2I_2$$

$$\therefore \quad \frac{\omega}{\Omega} = \frac{O_2I_2}{O_1I_1} = \frac{O_2P}{O_1P}$$

If ω/Ω is to be constant, P must be the pitch point for the two wheels, i.e. the common normal at the point of contact must pass through the pitch point. This

* See *Practical Geometry and Engineering Graphics*, W. Abbott, Blackie (1972), p. 36.

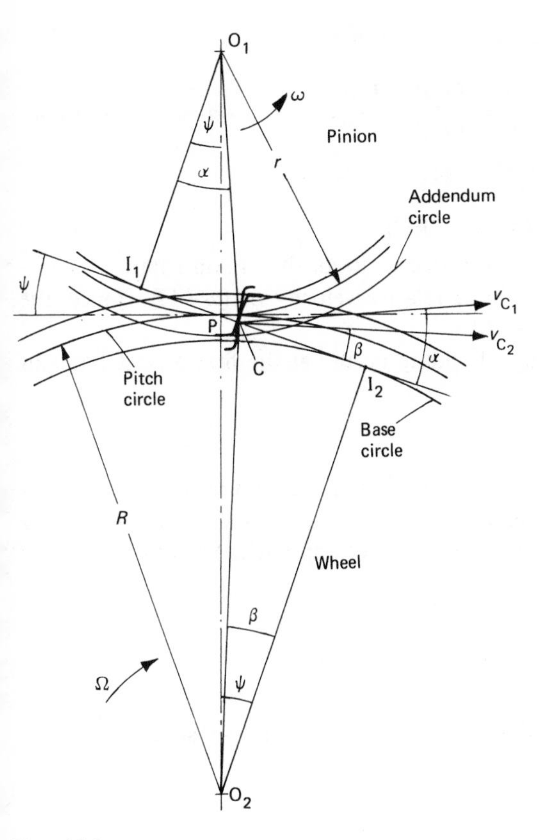

FIG. 11.2

condition is fulfilled by teeth of involute form, provided that the base circles from which the profiles are generated are tangential to the common normal. Since all points of contact lie on the common normal, it is called the *line of contact*. The force between two mating teeth acts along this line in the absence of friction, so that it is also called the *pressure line*.

11.3 Velocity of sliding

The velocity of sliding is the velocity of one tooth relative to its mating tooth along the common tangent at the point of contact. From Fig. 11.2, the tangential components of velocities v_{C_1} and v_{C_2} at the point C are $v_{C_1} \sin\alpha$ and $v_{C_2} \sin\beta$. Hence

$$
\begin{aligned}
\text{velocity of sliding at C} &= v_{C_1} \sin\alpha - v_{C_2} \sin\beta \\
&= \omega\, O_1C \sin\alpha - \Omega\, O_2C \sin\beta \\
&= \omega\, I_1C - \Omega\, I_2C \\
&= \omega(I_1P + PC) - \Omega(I_2P - PC)
\end{aligned}
$$

However, from Section 11.2,

$$\frac{\omega}{\Omega} = \frac{O_2P}{O_1P} = \frac{I_2P}{I_1P} \qquad \text{from similar triangles}$$

Therefore $\omega\, I_1P = \Omega\, I_2P$

Hence velocity of sliding at C = $(\omega + \Omega)$PC (11.1)

Alternatively, if the wheel is regarded as fixed, the pinion rotates instantaneously about the point P with a relative velocity of $(\omega + \Omega)$. Hence, the velocity of sliding at C = $(\omega + \Omega)$PC.

Thus the maximum velocity of sliding occurs at the first or last point of contact.

11.4 Path of contact

Assuming the pinion to be the driver, the first and last points of contact are A and B, Fig. 11.3, where the addenda circles cut the common normal. The path of contact is AB, which is divided into the path of approach, AP, and the path of recess, BP.

$$AP = AI_2 - PI_2 = \sqrt{(R_a^2 - R^2\cos^2\psi)} - R\sin\psi \qquad (11.2)$$

$$BP = BI_1 - PI_1 = \sqrt{(r_a^2 - r^2\cos^2\psi)} - r\sin\psi \qquad (11.3)$$

and

$$AB = AP + BP$$

$$= \sqrt{(R_a^2 - R^2\cos^2\psi)} + \sqrt{(r_a^2 - r^2\cos^2\psi)} - (R + r)\sin\psi \qquad (11.4)$$

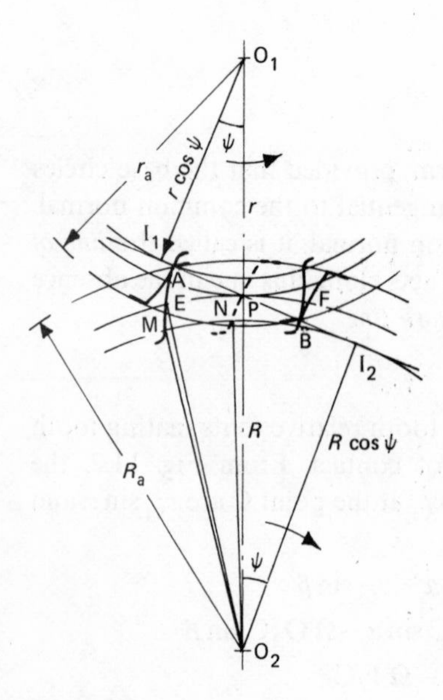

Fig. 11.3

11.5 Arc of contact and contact ratio

The arc of contact is the arc of the pitch circle EF, Fig. 11.3, between the positions of a tooth at the first and last points of contact with its mating tooth. It is divided into the arc of approach, EP, and the arc of recess, PF.

$$\angle \mathrm{MO_2N} = \angle \mathrm{EO_2P}$$

i.e.

$$\frac{\text{arc MN}}{R\cos\psi} = \frac{\text{arc EP}}{R}$$

$$\therefore \text{arc EP} = \frac{\text{arc MN}}{\cos\psi}$$

But, from the construction of the involute curve, arc MN = line AP.

$$\therefore \text{arc of approach} = \frac{\text{path of approach}}{\cos\psi} \tag{11.5}$$

Similarly,

$$\text{arc of recess} = \frac{\text{path of recess}}{\cos\psi} \tag{11.6}$$

$$\therefore \text{arc of contact} = \frac{\text{path of contact}}{\cos\psi} \tag{11.7}$$

Number of pairs of teeth in contact, or *contact ratio* $= \dfrac{\text{arc of contact}}{\text{circular pitch}}$ (11.8)

The maximum and minimum pairs of teeth in contact are the nearest whole numbers above and below this value.

11.6 Interference

For correct tooth action, the points of contact on two mating teeth must lie on the involute profiles. If the addendum of one tooth is too large, however, contact may occur between the tip of that tooth and the non-involute portion of the mating tooth between the base circle and the dedendum circle, Fig. 11.1. This causes undercutting of the mating tooth and interference is said to occur.

For no interference between the teeth, the first and last points of contact must lie between the points of tangency, I_1 and I_2, Fig. 11.3, i.e. the addendum circles must cut the common tangent to the base circles between I_1 and I_2. The limiting case occurs when the addendum circles pass through these points and, since the limiting addendum for the pinion is larger than that for the wheel, it is usually interference between the tips of the wheel teeth and the flanks of the pinion teeth which has to be prevented. This limits either the maximum wheel addendum or the minimum number of teeth on the pinion.

For no undercutting of the pinion teeth, the maximum permissible addendum radius of the wheel is given by

$$\begin{aligned} R_a &= \mathrm{O_2I_1} \\ &= \sqrt{(\mathrm{O_2I_2^2} + \mathrm{I_1I_2^2})} \\ &= \sqrt{\{R^2\cos^2\psi + (r\sin\psi + R\sin\psi)^2\}} \end{aligned}$$

i.e. $$R_a = \sqrt{(R^2 + 2Rr\sin^2\psi + r^2\sin^2\psi)} \quad (11.9)$$

$$\therefore \text{ maximum wheel addendum} = \sqrt{(R^2 + 2Rr\sin^2\psi + r^2\sin^2\psi)} - R \quad (11.10)$$

If the standard addendum, m, is used, then, for no interference,

$$m = \frac{2r}{t} \leqslant \sqrt{(R^2 + 2Rr\sin^2\psi + r^2\sin^2\psi)} - R$$

i.e. $$\frac{1}{t} \leqslant \tfrac{1}{2}\sqrt{(G^2 + 2G\sin^2\psi + \sin^2\psi)} - \frac{G}{2}$$

where G is the gear ratio, R/r.

$$\therefore t \geqslant \frac{2}{\sqrt{\{G^2 + \sin^2\psi(1 + 2G)\}} - G} \quad (11.11)$$

11.7 Rack and pinion

A rack may be regarded as a wheel of infinite radius; the teeth are straight-sided and normal to the line of contact. Referring to Fig. 11.4, and assuming the pinion to be the driver,

$$\text{path of approach, AP} = a\operatorname{cosec}\psi$$

where a = rack addendum, and

$$\text{path of recess, PB} = \text{BI} - \text{PI} = \sqrt{(r_a^2 - r^2\cos^2\psi)} - r\sin\psi$$

$$\therefore \text{ path of contact, AB} = \sqrt{(r_a^2 - r^2\cos^2\psi)} - r\sin\psi + a\operatorname{cosec}\psi \quad (11.12)$$

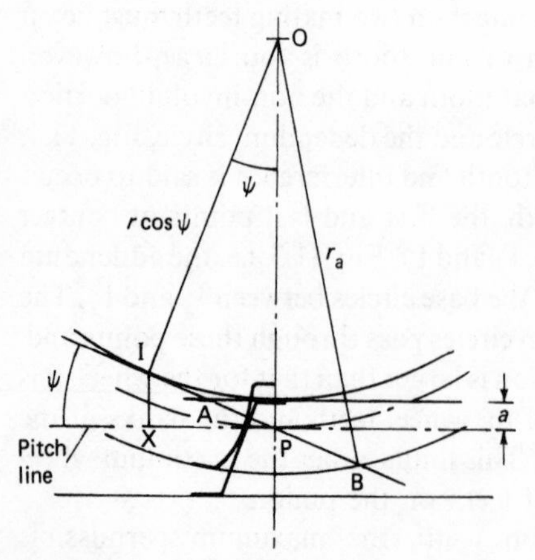

Fig. 11.4

For no interference between the rack tips and the pinion flanks,

$$\text{maximum rack addendum} = \text{IX} = \text{IP}\sin\psi = r\sin^2\psi \quad (11.13)$$

$$\text{Maximum path of contact} = \text{IB} = \sqrt{(r_a^2 - r^2\cos^2\psi)} \quad (11.14)$$

If the standard addendum, $1/P$, is used for the rack, then

$$m = \frac{2r}{t} \leqslant r\sin^2\psi \qquad (11.15)$$

$$\therefore\ t \geqslant 2\operatorname{cosec}^2\psi \qquad (11.16)$$

If $\psi = 14{\cdot}5°$,* $\qquad t \geqslant 32$

If $\psi = 20°$, $\qquad t \geqslant 18$

Examples

Example 11.1 Two gear wheels of 4·5 module have 24 and 33 teeth respectively. The pressure angle is 20° and each wheel has a standard addendum of 1 module. Find the length of the arc of contact and the maximum sliding velocity if the speed of the smaller wheel is 120 rev/min. (U. Lond.)

Solution

$$r = \frac{mt}{2} = \frac{4{\cdot}5 \times 24}{2} = 54\ \text{mm}$$

$$R = \frac{mT}{2} = \frac{4{\cdot}5 \times 33}{2} = 74{\cdot}25\ \text{mm}$$

and $\qquad$ addendum $= 4{\cdot}5$ mm

$$\therefore\ r_a = 58{\cdot}5\ \text{mm} \quad \text{and} \quad R_a = 78{\cdot}75\ \text{mm}$$

Assuming the pinion to be the driver,

$$\text{path of approach} = \sqrt{(R_a^2 - R^2\cos^2\psi)} - R\sin\psi \qquad \text{from equation 11.2}$$

$$= \sqrt{(78{\cdot}75^2 - 74{\cdot}25^2\cos^2 20°)} - 74{\cdot}25\sin 20° = 11{\cdot}14\ \text{mm}$$

$$\text{path of recess} = \sqrt{(r_a^2 - r^2\cos^2\psi)} - r\sin\psi \qquad \text{from equation 11.3}$$

$$= \sqrt{(58{\cdot}5^2 - 54^2\cos^2 20°)} - 54\sin 20° = 10{\cdot}67\ \text{mm}$$

$$\therefore\ \text{path of contact} = 11{\cdot}14 + 10{\cdot}67 = 21{\cdot}81\ \text{mm}$$

$$\therefore\ \text{arc of contact} = \frac{\text{path of contact}}{\cos\psi} \qquad \text{from equation 11.7}$$

$$= 21{\cdot}81/\cos 20° = \underline{23{\cdot}18\ \text{mm}}$$

$$\text{Velocity of sliding} = (\omega + \Omega)\text{PC} \qquad \text{from equation 11.1}$$

The maximum value of PC is the path of approach, i.e. 11·14 mm,

$$\omega = 120 \times \frac{2\pi}{60} = 12{\cdot}57\ \text{rad/s}$$

$$\Omega = 12{\cdot}57 \times \frac{24}{33} = 9{\cdot}14\ \text{rad/s}$$

$$\therefore\ \text{maximum velocity of sliding} = (12{\cdot}57 + 9{\cdot}14) \times 11{\cdot}14 = \underline{242\ \text{mm/s}}$$

* American standard (Brown and Sharpe).

Example 11.2 When in mesh, two gear wheels with 95 and 20 teeth of involute form rotate in opposite directions and operate with a pressure angle of 20°, a module of 4 mm and a contact ratio of 1·5. The arc of recess is 1·2 times the arc of approach. The smaller wheel runs at 2000 rev/min and transmits 4 kW to the larger wheel.

Determine: (a) the addenda of the two wheels, (b) the greatest speed of sliding between mating teeth, and (c) the greatest force between a pair of mating teeth, if the effects of friction between the teeth are neglected. (U. Lond.)

Solution

(a)
$$R = \frac{mT}{2} = \frac{4 \times 95}{2} = 190 \text{ mm}$$
$$r = \frac{mt}{2} = \frac{4 \times 20}{2} = 40 \text{ mm}$$
$$p = \pi m = \pi \times 4 = 12{\cdot}57 \text{ mm}$$
$$\text{Contact ratio} = \frac{\text{arc of contact}}{p} \qquad \text{from equation 11.8}$$
$$\therefore \text{ arc of contact} = 1{\cdot}5 \times 12{\cdot}57 = 18{\cdot}85 \text{ mm}$$
$$\text{Arc of recess} = 1{\cdot}2 \times \text{arc of approach} \qquad (1)$$

and
$$\text{arc of contact} = \text{arc of approach} + \text{arc of recess} \qquad (2)$$

From equations (1) and (2),
$$\text{arc of approach} = 8{\cdot}568 \text{ mm}$$

and
$$\text{arc of recess} = 10{\cdot}28 \text{ mm}$$

From equation 11.5,
$$\text{path of approach} = \text{arc of approach} \times \cos\psi$$
$$= 8{\cdot}568 \times \cos 20° = 8{\cdot}051 \text{ mm}$$

and
$$\text{path of recess} = 10{\cdot}28 \times \cos 20° = 9{\cdot}660 \text{ mm}$$

Hence
$$8{\cdot}051 = \sqrt{(R_a^2 - R^2\cos^2\psi)} - R\sin\psi \qquad \text{from equation 11.2}$$
$$= \sqrt{(R_a^2 - 190^2\cos^2 20°)} - 190\sin 20°$$

from which
$$R_a = 192{\cdot}9 \text{ mm}$$
$$\therefore \text{ addendum for wheel} = 192{\cdot}9 - 190 = 2{\cdot}9 \text{ mm}$$

Similarly,
$$9{\cdot}660 = \sqrt{(r_a^2 - r^2\cos^2\psi)} - r\sin\psi \quad \text{from equation 11.3}$$
$$= \sqrt{(r_a^2 - 40^2\cos^2 20°)} - 40\sin 20°$$

from which
$$r_a = 44{\cdot}24 \text{ mm}$$
$$\therefore \text{ addendum for pinion} = 44{\cdot}24 - 40 = \underline{4{\cdot}24 \text{ mm}}$$

(b)
$$\text{Speed of pinion, } \omega = \frac{2\pi}{60} \times 2000 = 209{\cdot}44 \text{ rad/s}$$
$$\text{Speed of wheel, } \Omega = 209{\cdot}44 \times \frac{20}{95} = 44{\cdot}09 \text{ rad/s}$$

Furthest point of contact from pitch point = path of recess = 9·66 mm

$\therefore$ greatest speed of sliding $= (\omega + \Omega) \times PC$ from equation 11.1

$$= (209{\cdot}44 + 44{\cdot}09) \times 9{\cdot}66$$

$$= 2450\ \text{mm/s} \quad \text{or} \quad \underline{2{\cdot}45\ \text{m/s}}$$

(c) Since the contact ratio is 1·5, the least number of pairs of teeth in contact is 1.

$$\text{Pinion torque} = \frac{\text{power}}{\omega} = \frac{4 \times 10^3}{209{\cdot}44} = 19{\cdot}1\ \text{N m}$$

$$\therefore \text{normal force between teeth} = \frac{19{\cdot}1}{\text{base circle radius of pinion}}$$

$$= \frac{19{\cdot}1}{0{\cdot}04 \cos 20^\circ} = \underline{508\ \text{N}}$$

Example 11.3 The following particulars of a single reduction spur gear are given: gear ratio = 10 to 1; distance between centres = 660 mm, approximately; pinion transmits 500 kW at 1800 rev/min; involute teeth of standard proportions (addendum = m) with a pressure angle of 22·5°; permissible normal pressure between teeth = 175 N per mm of width.

Find: (a) the nearest standard module if no interference is to occur, (b) the number of teeth in each wheel, (c) the necessary width of the pinion, and (d) the loads on the bearings of the wheels due to the power transmitted.

(U. Lond.)

Solution

(a) For no interference, the maximum addendum radius for the larger wheel is given by

$$R_a = \sqrt{(R^2 + 2Rr \sin^2 \psi + r^2 \sin^2 \psi)} \quad \text{from equation 11.9}$$

but $R + r = 660$ mm and $R/r = 10$,

from which $R = 600$ mm and $r = 60$ mm

$$\therefore R_a = \sqrt{(600^2 + 2 \times 600 \times 60 \sin^2 22{\cdot}5^\circ + 60^2 \sin^2 22{\cdot}5^\circ)} = 609.1\ \text{mm}$$

$\therefore$ maximum addendum = 9·1 mm = m

$\therefore$ nearest standard module is <u>nine</u>

(b) Number of teeth on wheel = $(2 \times 600)/9 = 133{\cdot}3$

and number of teeth on pinion = $(2 \times 60)/9 = 13{\cdot}33$

Therefore for a gear ratio of 10 : 1, tooth numbers are <u>130</u> and <u>13.</u>

Corrected value of $R = (133 \times 9)/2 = 585$ mm

and corrected value of $r = (13 \times 9)/2 = 58{\cdot}5$ mm

(c) $$\text{Torque transmitted by pinion} = \frac{500 \times 10^3 \times 60}{2\pi \times 1800} = 2650\ \text{N m}$$

Assuming the whole of the torque to be transmitted by one pair of teeth,

normal force between teeth along line of contact

$$= \frac{2650}{\text{base circle radius of pinion}}$$

$$= \frac{2650}{0{\cdot}0585 \cos 22{\cdot}5^\circ} = 49\,000 \text{ N}$$

$\therefore$ necessary width of pinion $= 49\,000/175 = \underline{280 \text{ mm}}$

(d) Load on each bearing $= 49\,000/2 = \underline{24\,500 \text{ N}}$

Example 11.4 Two parallel shafts are connected by spur gearing through an intermediate wheel or idler as shown in Fig. 11.5 (a). The lines joining the centres of the wheels are at right angles and the number of teeth on the idler is 40, 6·5 module. Neglecting friction and the weight, find the force on the idler shaft when 4 kW is being transmitted at an idler speed of 360 rev/min, the direction of rotation being: (a) clockwise, and (b) anticlockwise. The teeth are 20° involute. What will be the base circle diameter of the idler? (U. Lond.)

Solution

$$\text{p.c.d. of idler} = 40 \times 6{\cdot}5 = 260 \text{ mm}$$

$$\text{Torque transmitted by idler} = \frac{4 \times 10^3 \times 60}{2\pi \times 360} = 106{\cdot}2 \text{ N m}$$

$$\therefore \text{tangential force at p.c.d. of idler} = \frac{106{\cdot}2}{0{\cdot}13} = 816 \text{ N}$$

$$\therefore \text{force along common normal} = \frac{816}{\cos 20^\circ} = 869 \text{ N}$$

Assuming A to be the driver, the driving and resisting forces acting on the idler for clockwise rotation of the idler are as shown in Fig. 11.5 (b), and the resultant force on the shaft is given by

$$R = 2P \cos 65^\circ = 2 \times 869 \times 0{\cdot}4226 = \underline{740 \text{ N}}$$

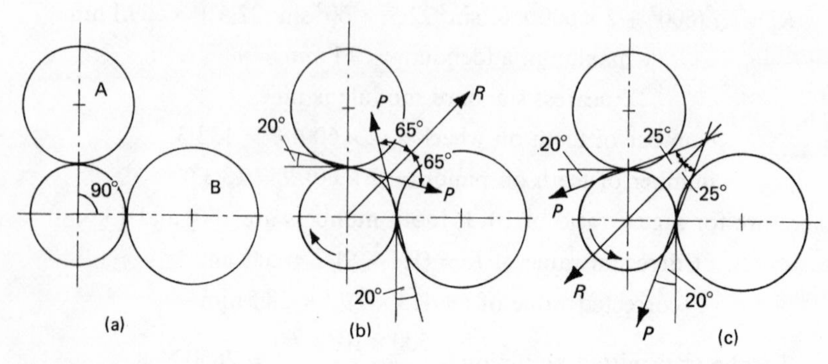

Fig. 11.5

For anticlockwise rotation of the idler, the forces acting upon it are as shown in Fig. 11.5 (c) and

$$R = 2P\cos 25^\circ = 2\times 869\times 0{\cdot}9063 = \underline{1575\ \text{N}}$$

$$\text{Base circle diameter of idler} = 2\times 0{\cdot}13\cos 20^\circ = \underline{245\ \text{mm}}$$

Example 11.5 A pinion of 20 involute teeth and 125 mm p.c.d. drives a rack. The addendum of both pinion and rack is 6 mm. What is the least pressure angle which can be used to avoid undercutting? With this pressure angle, find the length of the arc of contact and the minimum number of teeth in contact at a time. (U. Lond.)

Solution

$$r = 62{\cdot}5\ \text{mm} \quad \text{and} \quad r_a = 62{\cdot}5 + 6 = 68{\cdot}5\ \text{mm}$$

For no interference, $\text{rack addendum} \leqslant r\sin^2\psi$ from equation 11.13

$$\therefore \sin^2\psi \geqslant \frac{6}{62{\cdot}5}$$

i.e. $$\sin\psi \geqslant 0{\cdot}31$$

$$\therefore \text{least pressure angle, } \psi = \underline{18{\cdot}05^\circ}$$

$$\text{Length of path of contact} = \sqrt{(r_a^2 - r^2\cos^2\psi)} \qquad \text{from equation 11.14}$$

$$= \sqrt{(68{\cdot}5^2 - 62{\cdot}5^2\cos^2 18{\cdot}05^\circ)} = 34{\cdot}1\ \text{mm}$$

$$\therefore \text{length of arc of contact} = 34{\cdot}1/\cos 18{\cdot}05^\circ \qquad \text{from equation 11.7}$$

$$= \underline{35{\cdot}9\ \text{mm}}$$

$$\therefore \text{number of pairs of teeth in contact} = \frac{\text{arc of contact}}{\text{circular pitch}} \qquad \text{from equation 11.8}$$

$$= \frac{35{\cdot}9}{125\pi/20} = 1{\cdot}83$$

$\therefore$ minimum number of teeth in contact is <u>one pair</u>

Example 11.6 What is the effect of increasing the centre distance of a pair of spur wheels having involute teeth upon the following: (a) effective addenda, (b) working depth, (c) arc of contact, and (d) backlash?

A pinion with 25 teeth is cut with a cutter of 2·5 module and 14·5° pressure angle at the standard distance. It is made to mesh with a rack having straight-sided teeth at an angle of 70° to the pitch line. The pinion rotates at 120 rev/min. Determine the speed of the rack. (U. Lond.)

Solution

Increasing the centre distance of a pair of spur wheels results in a decrease in the effective addenda, the working depth and the arc of contact. The backlash is increased.

$$\text{p.c.d. of pinion} = 25\times 2{\cdot}5 = 62{\cdot}5\ \text{mm}$$

and $$\text{base circle diameter} = 62{\cdot}5\cos 14{\cdot}5^\circ = 60{\cdot}5\ \text{mm}$$

When made to mesh with the rack, the new pressure angle becomes 20°, Fig. 11.6. The base circle of the pinion remains unaltered but the new p.c.d. becomes

$$60{\cdot}5/\cos 20° = 64{\cdot}4 \text{ mm}$$

$$\text{Speed of rack} = \text{pitch circle velocity of pinion}$$

$$= \frac{2\pi}{60} \times 120 \times \frac{64{\cdot}4}{2}$$

$$= \underline{405 \text{ mm/s}}$$

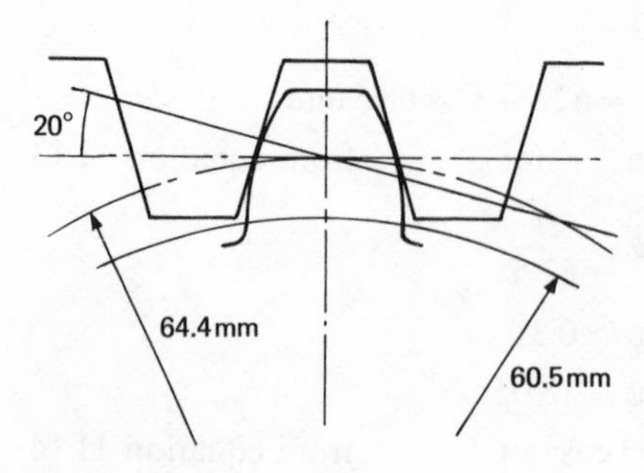

FIG. 11.6

Example 11.7 A gear wheel has a mass of 900 kg and has a radius of gyration of 0·6 m. It has 300 teeth of 5 module. If, owing to incorrect tooth form, the speed fluctuates ±0·04 per cent during the time of contact, find: (a) the variation in torque, and (b) the variation in tooth pressure, when the wheel rotates at 300 rev/min. Assume constant acceleration and deceleration.

(U. Lond.)

Solution

$$\text{Change in angular velocity during time of contact} = \frac{0{\cdot}04}{100} \times \left(\frac{2\pi}{60} \times 300\right)$$

$$= 0{\cdot}004\pi \text{ rad/s}$$

Assuming only one pair of teeth is in contact,

$$\text{time of contact} = \frac{1}{300} \times \frac{60}{300} = \frac{1}{1500} \text{ s}$$

Assuming constant acceleration and deceleration,

$$\text{angular acceleration or deceleration} = \frac{0{\cdot}004\,\pi}{1/1500} = 6\pi \text{ rad/s}^2$$

$$\therefore \text{variation in torque} = \pm I\alpha$$

$$= \pm 900 \times 0{\cdot}6^2 \times 6\pi = \underline{\pm 6110 \text{ N m}}$$

$$\text{p.c.d. of wheel} = 300 \times 5 = 1500 \text{ mm}$$

$$\therefore \text{variation in tooth pressure} = \pm 6110/0{\cdot}75 = \underline{8140 \text{ N}}$$

Example 11.8 A pinion of 100 mm p.c.d. drives a wheel of 250 mm p.c.d. The teeth are of involute form, with an angle of obliquity 20°. If the addendum on each wheel is 5 mm, the pinion speed is 2000 rev/min and $\mu = 0{\cdot}1$, find for the first point of contact: (a) the sliding velocity between the teeth, and (b) the transmission efficiency. (U. Lond.)

Solution

$$r = 50 \text{ mm}, \quad r_a = 55 \text{ mm}, \quad R = 125 \text{ mm}, \quad R_a = 130 \text{ mm}$$

(a) At the first point of contact, C, Fig. 11.7,

$$\text{velocity of sliding} = (\omega + \Omega)\,\text{PC} \quad \text{from equation 11.1}$$

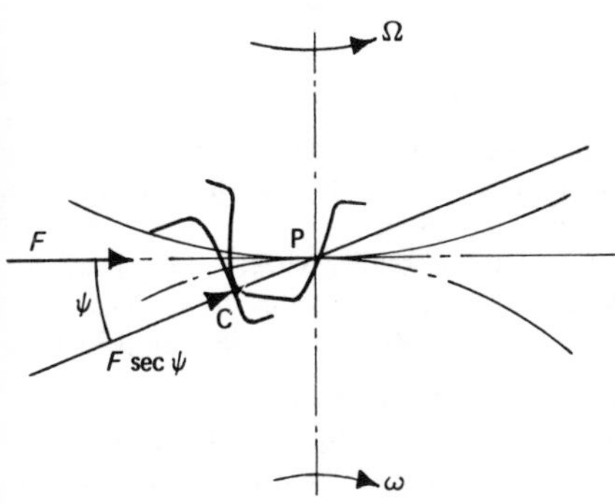

FIG. 11.7

But PC is the path of approach and, since the pinion is the driver,

$$\text{PC} = \sqrt{(R_a^2 - R^2\cos^2\psi)} - R\sin\psi \qquad \text{from equation 11.2}$$

$$= \sqrt{(130^2 - 125^2\cos^2 20°)} - 125\sin 20° = 12{\cdot}63 \text{ mm}$$

$$\omega = \frac{2\pi}{60} \times 2000 = \frac{200\pi}{3} \text{ rad/s}$$

and

$$\Omega = \frac{\omega}{2{\cdot}5} = \frac{80\pi}{3} \text{ rad/s}$$

$$\therefore \text{ velocity of sliding} = \left(\frac{200\pi}{3} + \frac{80\pi}{3}\right) \times 12{\cdot}63$$

$$= 3700 \text{ mm/s} \quad \text{or} \quad \underline{3{\cdot}7 \text{ m/s}}$$

(b) At the first point of contact, let the pitch circles rotate through a distance δ. Then the angle turned through by the pinion is δ/r and the time taken is $\delta/\omega r$.

$$\text{Distance of sliding in this time} = \text{velocity of sliding} \times \text{time}$$

$$= \text{velocity of sliding} \times \frac{\delta}{\omega r}$$

$$= \frac{3700\delta}{(200\pi/3) \times 50}$$

$$= 0{\cdot}353\delta \text{ mm}$$

If F is the tangential force between the teeth, the normal force along the line of contact is $F \sec \psi$.

$$\therefore \text{ friction force at point of contact} = \mu F \sec \psi$$

$$\therefore \text{ work lost in friction} = \mu F \sec \psi \times 0{\cdot}353\delta$$

$$\text{Work done by pinion in this time} = F\delta$$

$$\therefore \text{ instantaneous transmission efficiency} = \frac{F\delta - \mu F \sec \psi \times 0{\cdot}353\delta}{F\delta}$$

$$= 1 - 0{\cdot}1 \times \sec 20^\circ \times 0{\cdot}353$$

$$= 0{\cdot}962 \quad \text{or} \quad 96{\cdot}2 \text{ per cent}$$

Problems

1. Two spur wheels with involute teeth gear together externally. The pitch circle diameters are 100 mm and 250 mm and the smaller wheel has 16 teeth. The pressure angle is 20° and the addendum is 0·3 of the circular pitch.

Determine: (a) the length of the arc of approach if the smaller wheel is the driver, and (b) the maximum velocity of sliding between the teeth when the driver rotates at 150 rev/min. (U. Lond.) (*Answer:* 15·98 mm; 0·33 m/s)

2. For a pair of gear wheels of involute teeth, show diagrammatically the two pitch circles in contact, the base circles, the addendum circles and the line along which the point of tooth contact moves. If the numbers of teeth in the wheels are 20 and 29, with a pressure angle of 20°, module equal to 6·5 and standard addendum m, calculate the length of the path of contact. Find also the angle turned by each wheel while any one pair of teeth is in contact. (I. Mech. E.) (*Answer:* 30·73 mm; 19·88°; 28·8°)

3. A gear wheel having 20 teeth of involute form of 6·5 module and an angle of obliquity of 20° drives another wheel of the same dimensions. Calculate the length of the arc of contact if the addendum is 6·5 mm.

If the addendum were altered so that the arc of contact was the maximum possible, what would be the length of this arc and the addendum required for this? (U. Lond.) (*Answer:* 31·8 mm; 47·2 mm; 10·52 mm)

4. What condition must be satisfied if two wheels in gear are to have a constant velocity ratio? Show that this condition is satisfied by teeth of involute profile.

A single reduction gear transmits 120 kW, the p.c.d. of the pinion being 230 mm and its speed 650 rev/min. If the pinion is supported by bearings on each side, calculate the total load on these bearings due to the power transmitted. The wheel teeth are of involute profile with a pressure angle of 20°. If the pressure angle is reduced to 15°, what is the percentage reduction in bearing load? (U. Lond.) (*Answer:* 16·3 kN; 2·7)

5. A gear wheel has involute teeth, the pressure angle being 20°. There are 18 teeth, module 12, and the wheel rotates at 250 rev/min. If this wheel meshes with one having 30 teeth, find: (a) the length of the line of contact, and (b) the maximum rubbing velocity. The addendum for each wheel is equal to the module. (U. Lond.) (*Answer:* 56·4 mm: 1·226 m/s)

6. Two equal gear wheels of 360 mm p.c.d. and 6 module are in mesh. The teeth are of involute form, of 20° angle of obliquity. Determine the maximum addendum necessary if there are always to be at least two pairs of teeth in contact. If the gear wheels rotate at 110 rev/min and 6 kW is being transmitted, find the normal force between the teeth, assuming that the total force is divided equally between the two pairs of teeth. Neglect friction. (U. Lond.) (*Answer:* 6·8 mm; 1·54 kN)

7. Two mating gear wheels have 20 and 40 involute teeth of 12 module and 20° obliquity, and the addendum on each wheel is to be made of such a length that the line of contact on each side of the pitch point has half the maximum possible length.

Determine the addendum height for each gear wheel and the length of the line of contact.

If the smaller wheel rotates at 250 rev/min, find: (a) the velocity of the point of contact along the surface of each tooth at the instant when the tip of a tooth on the smaller wheel is in contact, and (b) the velocity of sliding at this instant. (U. Lond.)
(*Answer:* 19·5 mm; 7·78 mm; 61·6 mm; 2·15 m/s; 0·537 m/s; 1·614 m/s)

8. A pinion having 20 involute teeth of 6 module rotates at 200 rev/min and transmits 1·5 kW to a gear wheel having 50 teeth; the addendum on both wheels is $\frac{1}{4}$ of the circular pitch and the angle of obliquity is 20°.

Find: (a) the lengths of the path of approach and the arc of approach, and (b) the normal force between the teeth at an instant when there is only one pair of teeth in contact. (U. Lond.) (*Answer:* 12·5 mm; 13·3 mm; 1·27 kN)

9. Two gear wheels have teeth of 5 module, angle of obliquity 20°, the numbers of teeth being 15 and 20. The addendum is the same for each wheel and is as large as possible while avoiding interference.

Find: (a) the length of the addendum, (b) the contact ratio, and (c) the sliding velocity at the first point of contact when the smaller wheel is driving at 3500 rev/min. (U. Lond.) (*Answer:* 5·7 mm; 1·69; 8·21 m/s)

10. Two involute gear wheels operate with a pressure angle of 20° and rotate in opposite directions. The smaller wheel has 20 teeth of 5 mm module and runs at 2000 rev/min. The larger wheel has 80 teeth and is the driven member.

If the contact ratio is 1·4, and if the arc of recess is 1·2 times as large as the arc of approach, determine: (a) the addenda of the teeth on both wheels, (b) the greatest speed of sliding between the teeth, (c) the centre distance of the gears, and (d) the working depth. (U. Lond.) (*Answer:* 3·4 mm; 4·89 mm; 2·95 m/s; 250 mm; 8·29 mm)

11. Two gear wheels of 3 module and of 20° pressure angle, each having 100 teeth, are in mesh together. Determine the minimum addendum length necessary if the contact ratio is to be 2·0. How much greater is the addendum length than that of a standard tooth of the same pitch?

If these gears transmit 7·5 kW at 1000 rev/min and the tooth load is shared uniformly between those pairs of teeth in contact, determine the maximum and minimum normal loads between each pair of teeth. (U. Lond.)
(*Answer:* 3·27 mm; 0·27 mm; 254 N and 169 N, since number of pairs of teeth in contact varies between 2 and 3)

12. A pair of spur gears with involute tooth form have a module of m and equal addenda of m. Show that involute interference will occur if the number of teeth on the pinion is less than

$$\frac{2}{\sqrt{\{G^2 + (1 + 2G)\sin^2\psi\}} - G}$$

where ψ is the pressure angle and G is the gear ratio.

A pair of gears is required to give a ratio of 2·5 : 1. A module of 3 is to be used and a pressure angle of 20°. Determine suitable numbers of teeth for the wheel and pinion and the exact centre distance. (U. Lond.) (*Answer:* 16; 40; 84 mm)

13. A rack and pinion are in gear, the teeth being of involute form with pressure angle ψ and standard addendum m. Make a sketch showing the pitch line and circle of contact, the addendum line and circle, the line of pressure and the pinion base circle. For no contact inside the base circle, show that the number of pinion teeth must exceed $2 \operatorname{cosec}^2 \psi$. For $\psi = 20°$, the least number of teeth and module $m = 3$, find the length of the path of tooth contact. (I. Mech. E.) (*Answer:* 18 teeth; 16·1 mm)

14. A pinion of 250 mm p.c.d. drives a rack. The addendum height for both pinion and rack is 12·5 mm and the teeth, of involute form, have a pressure angle of 20°.
(a) Show that interference does not occur.
(b) Find the minimum number of teeth on the pinion to ensure continuity of contact. (U. Lond.)
(*Answer*: maximum permissible rack addendum = 14·625 mm; 12)

15. A pinion with 24 involute teeth of 150 mm p.c.d. drives a rack. The addendum of the pinion and rack is 6·25 mm. What is the least pressure angle which can be used if undercutting of the teeth is to be avoided? Using this pressure angle, find the length of the arc of contact and the minimum number of teeth in contact at one time. (U. Lond.) (*Answer:* 16·78°; 39·71 mm; 2 pairs)

16. Explain what is meant by the term 'interference' as applied to toothed gearing with involute profiles.

A pinion with 25 teeth, 5 module, is required to mesh with a rack, the teeth being of involute form with a pressure angle of 15°. Find the largest addendum that can be used on the rack if interference is to be avoided.

If the addendum is the same for both rack and pinion, find the length of the arc of contact in terms of the circular pitch of the teeth. (I. Mech. E.)
(*Answer:* 4·187 mm; 1·82p)

17. A pinion has involute teeth of standard proportions; pressure angle 20° and addendum equal to the module. What is the smallest number of teeth that may be used if undercutting of the flanks is to be avoided when the pinion gears with a rack?

How may the proportions be modified so as to allow a pinion with a smaller number of teeth to be used? (I. Mech. E.) (*Answer:* 18; by increasing the pressure angle)

18. A pinion of 100 mm p.c.d. drives a rack. The teeth are of involute form, the angle of obliquity is 20° and the total depth of the teeth on each wheel is to be 17·5 mm, the root clearance being 1 mm.

Determine the lengths of the addendum and root for both pinion and rack, the addendum of the rack being as large as possible consistent with correct tooth action.

Calculate the length of the path of approach. (U. Lond.)
(*Answer:* 10·65 mm; 6·85 mm; 5·85 mm; 11·65 mm; 17·1 mm)

12

Gear trains

12.1 Simple trains

In the simple trains shown in Fig. 12.1, let

N_A and N_B = speeds of A and B

T_A and T_B = numbers of teeth on A and B

r_A and r_B = pitch circle radii of A and B

In train (a),
$$\frac{N_A}{N_B} = -\frac{T_B}{T_A} = -\frac{r_B}{r_A} \tag{12.1}$$

In train (b),
$$\frac{N_A}{N_B} = +\frac{T_B}{T_A} = +\frac{r_B}{r_A} \tag{12.2}$$

In train (c),
$$\frac{N_A}{N_B} = -\frac{T_B}{T_A} = -\frac{r_B}{r_A} \tag{12.3}$$

The negative signs signify that A and B rotate in opposite directions.

The idler, C, Fig. 12.1 (b) and (c), does not affect the velocity ratio of A to B but decides the direction of B.

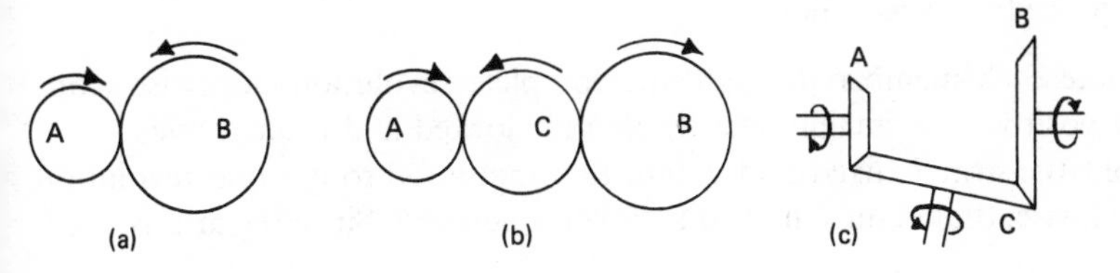

Fig. 12.1

12.2 Compound trains

In the compound trains shown in Fig. 12.2, wheels C and D are rigidly connected together to form a compound wheel which rotates on the intermediate shaft, XX.

Then
$$\frac{N_A}{N_B} = \frac{T_C}{T_A}\frac{T_B}{T_D} \tag{12.4}$$

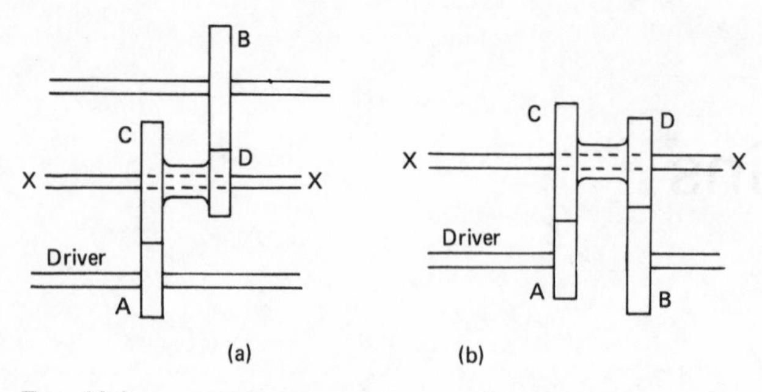

FIG. 12.2

or
$$\frac{\text{speed of driver}}{\text{speed of follower}} = \frac{\text{product of teeth on followers}}{\text{product of teeth on drivers}}$$

In the *reverted* or *co-axial* train, Fig. 12.2(b), the equation

$$r_A + r_C = r_B + r_D$$

must also be satisfied,

i.e.
$$\frac{m_1 T_A}{2} + \frac{m_1 T_C}{2} = \frac{m_2 T_B}{2} + \frac{m_2 T_D}{2} \tag{12.5}$$

where m_1 and m_2 are the modules of A and B respectively.

The tooth numbers to satisfy equations 12.4 and 12.5 are usually found by trial.

12.3 Epicyclic trains

Referring to the epicyclic train shown in Fig. 12.3, S is the sun wheel, A is the annulus, having internal teeth, and P is a planet wheel which can rotate freely on a pin attached to the arm L. The arm L rotates freely about the axis of S.

Suppose it is required to find the velocity ratio of S to L when A is held fixed. The procedure is as follows:

(a) Rotate each member through one complete revolution clockwise (considered positive), i.e. imagine the whole gear locked and rotated once.
(b) Hold the arm L fixed and rotate the annulus through one revolution anticlockwise, thus returning it to its former position. Since the arm is fixed,

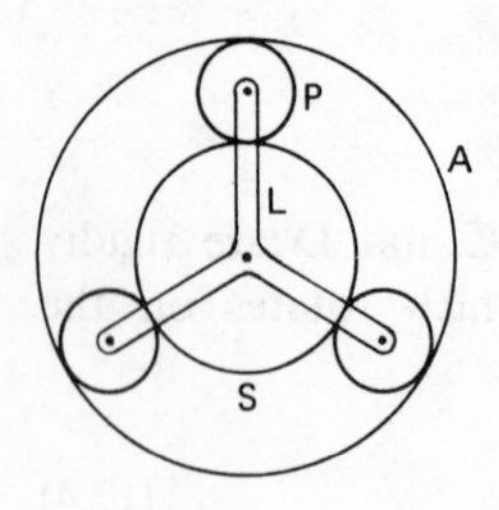

FIG. 12.3

wheels A, P and S form a simple train and the revolutions of P and S due to the rotation of A are $-T_A/T_P$ and $+T_A/T_S$ respectively.
(c) Add the corresponding rotations of each member in operations (a) and (b) to obtain the resultant motion.

These operations are set out in tabular form thus:

	L	A	P	S
(a) Turn whole gear clockwise 1 revolution	+1	+1	+1	+1
(b) Hold arm L, turn A anticlockwise 1 revolution	0	−1	$-\frac{T_A}{T_P}$	$+\frac{T_A}{T_S}$
(c) Resulting motion[= (a) + (b)]	+1	0	$1-\frac{T_A}{T_P}$	$1+\frac{T_A}{T_S}$

The last line of the table gives the relative motion of the arm, sun wheel and planet wheel when the annulus is fixed. It is always the fixed wheel which is given −1 revolution in line (b).

Where all the members are rotating, a modification of the method is necessary. The whole gear is given $+a$ revolutions in line (a). In line (b), the arm is held fixed and *any* wheel is given $+b$ revolutions. The motion of the other wheels is then found as before. The resulting motion found by the addition of lines (a) and (b) is in terms of the constants a and b, which are then evaluated from the known speeds of two of the members.

Thus, in tabular form:

	L	A	P	S
(a) Give whole gear $+a$ revolutions	$+a$	$+a$	$+a$	$+a$
(b) Hold arm L, give A $+b$ revolutions	0	$+b$	$+\frac{T_A}{T_P}b$	$-\frac{T_A}{T_S}b$
(c) Resulting motion	$+a$	$a+b$	$a+\frac{T_A}{T_P}b$	$a-\frac{T_A}{T_S}b$

12.4 Torques on gear trains

In the gear units shown in Fig. 12.4, let C_A, C_B and C_C be respectively the applied input torque, the resisting torque on the output shaft and the torque

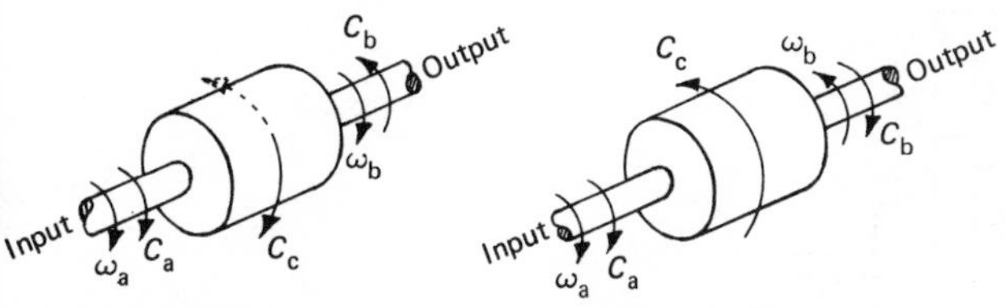

(a) Input and output shafts rotating in the same direction

(b) Input and output shafts rotating in opposite directions

FIG. 12.4

to hold the casing fixed. If there is to be no acceleration of the system, the net torque applied to the unit *about any one axis* must be zero,

i.e. $$C_A + C_B + C_C = 0 \tag{12.6}$$

Also, if there is no acceleration, the kinetic energy of the system remains constant so that the net work done per second is zero,

i.e. $$C_A\omega_A + C_B\omega_B = 0 \tag{12.7}$$

The appropriate signs must be given to both torques and speeds in equations 12.6 and 12.7. In Fig. 12.4 (a), the direction of the fixing torque C_C will depend on the relative magnitudes of C_A and C_B.

If, allowing for friction, the efficiency of the unit is η, then

$$\eta = \frac{\text{output power}}{\text{input power}}$$

so that equation 12.7 becomes

$$\eta C_A\omega_A + C_B\omega_B = 0 \tag{12.8}$$

If the casing is not fixed, C_C represents either an input or output torque and equation 12.7 becomes

$$C_A\omega_A + C_B\omega_B + C_C\omega_C = 0 \tag{12.9}$$

12.5 Compound epicyclic trains

A compound train consists of two or more co-axial simple trains with members forming part of two consecutive trains.

Consider the compound train shown in Fig. 12.5, where the annulus A_1 of the train A_1S_1L also forms the arm of the train $A_2S_2A_1$. Let A_2 be fixed.

First obtain the speed ratio of L to S_1 when A_1 is fixed, thus:

	L	A_1	S_1
(a) Give whole train +1 revolution	+1	+1	+1
(b) Hold L and give A_1 −1 revolution	0	−1	$+\frac{T_{A_1}}{T_{S_1}}$
(c) Add [(a)+(b)]	1	0	$1+\frac{T_{A_1}}{T_{S_1}}$

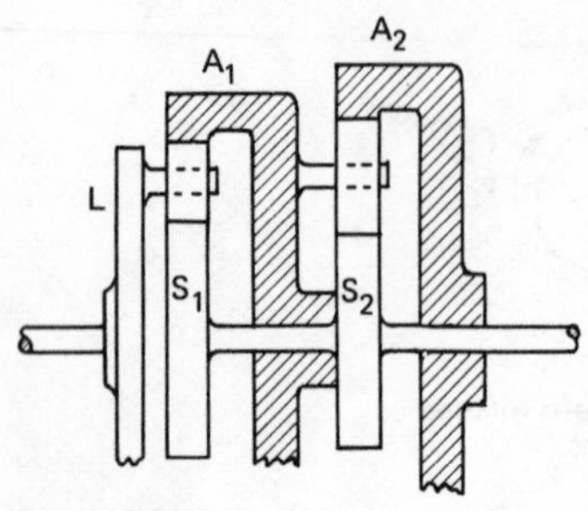

FIG. 12.5

Thus $$\frac{N_L}{N_{S_1}} = \frac{1}{1 + T_{A_1}/T_{S_1}} \quad \text{when } A_1 \text{ is fixed} \tag{12.10}$$

Then consider the whole train with A_1 as the arm, using the above result in the second line of the table when A_1 is fixed.

	L	A_1	S_1S_2	A_2
(a) Give whole train +1 revolution	+1	+1	+1	+1
(b) Hold A_1 and give A_2 −1 revolution	$\frac{T_{A_2}/T_{S_2}{}^*}{1+T_{A_1}/T_{S_1}}$	0	$+\frac{T_{A_2}}{T_{S_2}}$	−1
(c) Add [(a)+(b)]	$1+\frac{T_{A_2}/T_{S_2}}{1+T_{A_1}/T_{S_1}}$	1	$1+\frac{T_{A_2}}{T_{S_2}}$	0

* From equation 12.10.

Thus $$\frac{N_L}{N_{S_1}} = \left(1 + \frac{T_{A_2}/T_{S_2}}{1 + T_{A_1}/T_{S_1}}\right) \bigg/ (1 + T_{A_2}/T_{S_2})$$

The advantage of this type of gear over a simple train is that if A_2 is released and A_1 is fixed, a different ratio N_L/N_{S_1} is obtained.

Examples

Example 12.1 Two shafts, A and B, in the same straight line are geared together through an intermediate parallel shaft, C. The wheels connecting A and C have a module of 2, those connecting C and B a module of 3·5. The speed of B is to be about, but less than, 1/10 that of A. If the two pinions have each 24 teeth, find suitable numbers of teeth for the wheels, the actual ratio and the corresponding distance of shaft C from A. (I. Mech. E.)

Solution

In Fig. 12.6, $$\frac{N_A}{N_B} \geqslant 10$$

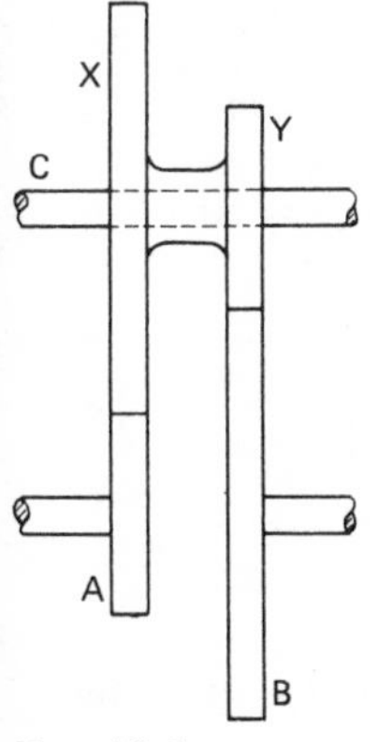

FIG. 12.6

and $$T_A = T_Y = 24$$

i.e. $$\frac{T_X}{T_A}\frac{T_B}{T_Y} \geqslant 10$$

i.e. $$T_X T_B \geqslant 5760 \qquad (1)$$

$$r_A + r_X = r_B + r_Y$$

i.e. $$\frac{2T_A}{2} + \frac{2T_X}{2} = \frac{3{\cdot}5T_B}{2} + \frac{3{\cdot}5T_Y}{2}$$

or $$T_X - 1{\cdot}75T_B = 18 \qquad (2)$$

Therefore, from equations (1) and (2),

$$T_X = \underline{116} \quad \text{and} \quad T_B = \underline{56}$$

$$\frac{N_A}{N_B} = \frac{116 \times 56}{24 \times 24} = \underline{11{\cdot}28}$$

$$\text{Centre distance} = r_A + r_X = \frac{2 \times 24}{2} + \frac{2 \times 116}{2} = \underline{140\,\text{mm}}$$

Example 12.2 A gear box is to be arranged for four speeds in approximate geometrical progression, one of which is to be a direct drive. The driving shaft transmits 30 kW at 2400 rev/min, the speed of the layshaft is approximately 1600 rev/min, and the speed of the driven shaft in lowest gear is to be approximately 400 rev/min. The distance between the axes of the driving shaft and the layshaft is 180 mm and all teeth are 6 module.
(a) Find the necessary numbers of teeth in each pair of gears.
(b) Find the torque on the driven shaft and the torque on the gear-box frame in lowest gear, neglecting losses. (U. Lond.)

Solution
(a) Referring to Fig. 12.7,

$$T_A + T_B = T_C + T_D = T_E + T_F = T_G + T_H = 180 \times 2 \times \frac{1}{m}$$

$$= \frac{180 \times 2}{6} = 60$$

Also $$\frac{T_B}{T_A} = \frac{T_K}{T_J} = \frac{2400}{1600} = \frac{3}{2}$$

$$\therefore\ T_A = T_J = \underline{24} \quad \text{and} \quad T_B = T_K = \underline{36}$$

$$\frac{T_D}{T_C} = \frac{400}{1600} = \frac{1}{4}$$

Therefore the common ratio r of the geometrical progression is given by

$$\tfrac{1}{4}r^3 = \tfrac{3}{2}$$

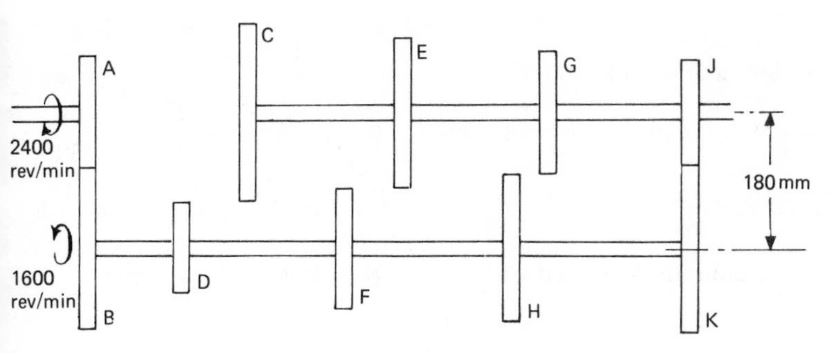

FIG. 12.7

from which $$r = \sqrt[3]{6} = 1{\cdot}817$$

$$\therefore \quad \frac{T_F}{T_E} = 1{\cdot}817 \times \frac{1}{4} = 0{\cdot}454$$

and $$\frac{T_H}{T_G} = 1{\cdot}817^2 \times \frac{1}{4} = 0{\cdot}825$$

$$\therefore \quad \underline{T_E = 41} \quad \text{and} \quad \underline{T_F = 19}$$

$$\underline{T_G = 33} \quad \text{and} \quad \underline{T_H = 27}$$

(b) Torque on driving shaft $= \dfrac{\text{power}}{\omega} = \dfrac{30 \times 10^3 \times 60}{2\pi \times 2400} = 119{\cdot}4\ \text{N m}$

and torque on driven shaft $= \dfrac{2400}{400} \times 119{\cdot}4 = \underline{716{\cdot}4\ \text{N m}}$

The torques on the unit correspond to those in Fig. 12.4(a). Therefore if C_C is the torque on the frame,

$$119{\cdot}4 + C_C - 716{\cdot}4 = 0 \qquad \text{from equation 12.6}$$

$$\therefore \quad \underline{C_C = 597\ \text{N m}} \quad \text{(in same direction as driving torque)}$$

Example 12.3 An epicyclic train has a sun wheel with 30 teeth and two planet wheels of 50 teeth, the latter meshing with the internal teeth of a fixed annulus. The input shaft, carrying the sun wheel, transmits 4 kW at 300 rev/min. The output shaft is connected to an arm which carries the planet wheels.

What is the speed of the output shaft and the torque transmitted if the overall efficiency is 95 per cent?

If the annulus is rotated independently, what should be its speed in order to make the output shaft rotate at 10 rev/min? (U. Lond.)

Solution

Referring to Fig. 12.3, $$r_A = 2r_P + r_S$$

i.e. $$T_A = 2T_P + T_S$$

$$= 2 \times 50 + 30 = 130$$

	L	A	P	S
(a) Give whole gear $+a$ revolutions	$+a$	$+a$	$+a$	$+a$
(b) Hold arm L and give A $+b$ revolutions	0	$+b$	$+\frac{130}{50}b$	$-\frac{130}{30}b$
(c) Add [(a)+(b)]	$+a$	$a+b$	$a+\frac{13}{5}b$	$a-\frac{13}{3}b$

When the annulus A is fixed and the sun S rotates at 300 rev/min,

$$N_A = a+b = 0$$

and $$N_S = a-\tfrac{13}{3}b = 300$$

Hence $$a = 900/16 \quad \text{and} \quad b = -900/16$$

$\therefore$ speed of output shaft L

$$= a = 900/16$$

$$= \underline{+56{\cdot}26 \text{ rev/min}} \quad \text{(in same direction as input shaft)}$$

$$\text{Output power} = 0{\cdot}95 \times 4 \times 10^3 = \frac{2\pi N_L C_L}{60}$$

$$\text{Therefore torque transmitted, } C_L = \frac{0{\cdot}95 \times 4 \times 10^3 \times 60}{2\pi \times 56{\cdot}25} = \underline{645 \text{ N m}}$$

When the output shaft L rotates at 10 rev/min and the sun S rotates at 300 rev/min,

$$N_S = a-\tfrac{13}{3}b = 300$$

and $$N_L = a = 10$$

$$\therefore\ b = -870/13$$

$\therefore$ speed of annulus A $= a+b$

$$= 10-870/13$$

$$= \underline{-56{\cdot}9 \text{ rev/min}} \quad \text{(in opposite direction to input shaft)}$$

Example 12.4 Figure 12.8 shows an epicyclic gear train in which the wheel D is held stationary by the shaft A and the arm B is rotated at 200 rev/min. The

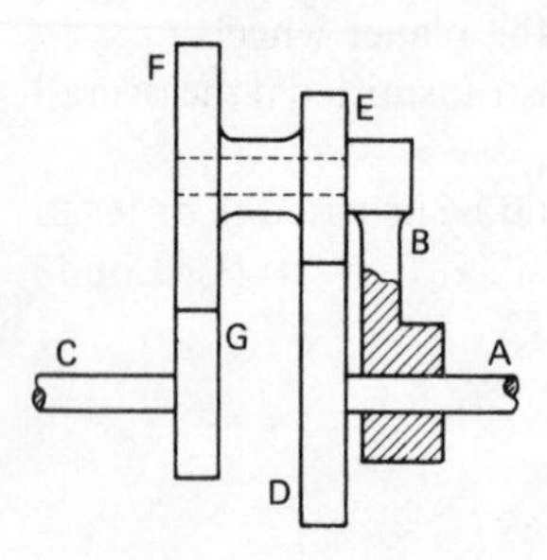

FIG. 12.8

wheels E and F are fixed together and rotate freely on the pin carried by the arm. The wheel G is rigidly attached to the shaft C.

Find the speed of the shaft C, stating the direction of rotation relative to that of B.

The numbers of teeth are as follows: E, 20; F, 40; G, 30.

If the gearing transmits 7·5 kW, what will be the torque required to hold the shaft A stationary if all frictional losses are neglected? (U. Lond.)

Solution

$$r_D + r_E = r_G + r_F$$

$$\therefore \; T_D + T_E = T_G + T_F \quad \text{(assuming pitches are equal)}$$

i.e.

$$T_D + 20 = 30 + 40$$

$$\therefore \; T_D = 50$$

	B	C, G	F, E	A, D
(a) Give whole gear +1 revolution	+1	+1	+1	+1
(b) Hold arm B and give A −1 revolution	0	$-\frac{50 \times 40}{20 \times 30}$	$+\frac{50}{20}$	−1
(c) Add [(a)+(b)]	+1	$-\frac{7}{3}$	–	0
(d) Multiply (c) by 200	+200	−466·7	–	0

$$\therefore \text{ speed of C} = \underline{-466{\cdot}7 \text{ rev/min}} \quad \text{(in opposite direction to B)}$$

$$\text{Applied input torque, } C_B = \frac{\text{power}}{\omega} = \frac{7{\cdot}5 \times 10^3 \times 60}{2\pi \times 200} = 358 \text{ N m}$$

Since net work done is zero,

$$C_C N_C + C_B N_B = 0$$

$$\therefore \; C_C = -\frac{N_B}{N_C} C_B = -\left(-\frac{3}{7}\right) \times 358$$

$$= +153{\cdot}5 \text{ N m} \quad \text{(in the same direction as } C_B\text{)}$$

Since there is no acceleration,

$$C_A + C_B + C_C = 0$$

i.e.

$$C_A + 358 + 153{\cdot}5 = 0$$

$$\therefore \; C_A = \underline{-511{\cdot}5 \text{ N m}} \quad \text{(in opposite direction to } C_B\text{)}$$

Example 12.5 In the gear drive shown in Fig. 12.9, the casing is fixed. Gear wheel B rotates freely on shaft Q and carries a pin on which rotates the wheels C and D, which are fixed together. Wheels F and G are fixed together and rotate freely on shaft Q. The pitch of the teeth is the same for all wheels, and the numbers of teeth are shown for all wheels except E, which is fixed on shaft Q.

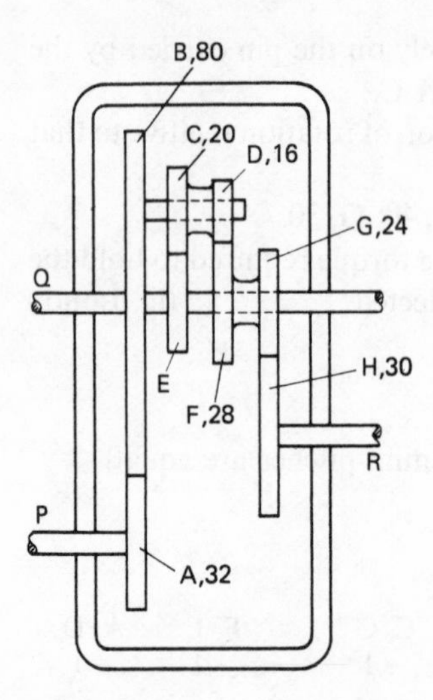

FIG. 12.9

If shaft P rotates at 200 rev/min, find the speed and direction of rotation of shaft Q for the two cases: (a) shaft R is fixed, and (b) shaft R rotates at 100 rev/min in the same direction as P. (U. Lond.)

Solution

$$T_E = 28 + 16 - 20 = 24$$

The wheels A and H are not part of the epicyclic train and must therefore be treated separately.

$$N_B = -\frac{T_A}{T_B} N_A = -\frac{32}{80} \times 200 = -80$$

$$N_G = -\frac{T_H}{T_G} N_H = -\frac{30}{24} \times N_H = -\frac{5}{4} N_H$$

Wheel B is the arm of the epicyclic train.

	B	C, D	E, Q	F, G
(a) Give whole gear $+a$ revolutions	$+a$	$+a$	$+a$	$+a$
(b) Hold arm and give F $+b$ revolutions	0	$-\frac{28}{16}b$	$+\frac{28 \times 20}{16 \times 24}b$	$+b$
(c) Add [(a) + (b)]	$+a$	$a-\frac{7}{4}b$	$a+\frac{35}{24}b$	$a+b$

(a) When R is fixed,

$$a = -80$$

and $$a+b=0$$
$$\therefore\ b=80$$

$$\therefore\ N_Q = -80+\frac{35}{24}\times 80 = \underline{+36{\cdot}7\ \text{rev/min}} \quad \text{(in same direction as P)}$$

(b) When R rotates at 100 rev/min,

$$N_G = -\frac{5}{4}\times 100 = -125\ \text{rev/min}$$

$$a = -80$$
and $$a+b = -125$$
$$\therefore\ b = -45$$

$$\therefore\ N_Q = -80-\frac{35}{24}\times 45 = \underline{-145{\cdot}6\ \text{rev/min}} \quad \text{(in opposite direction to P)}$$

Example 12.6 In the gear train shown in Fig. 12.10, the wheel C is fixed, X is the driving shaft and the compound wheel BD can revolve on a spindle which can turn freely about the axis of X and Y.

Show that: (a) if the ratio of the tooth numbers T_B/T_D is greater than T_C/T_E, the wheel E will rotate in the same direction as wheel A, and (b) if the ratio T_B/T_D is less than T_C/T_E, the direction of E is reversed.

If the numbers of teeth on wheels A, B, C, D and E are respectively 17, 60, 75, 19 and 25, and 7·5 kW is put into the shaft X at 500 rev/min, what is the output torque on the shaft Y, and what are the forces (tangential to the pitch cones) at the contact points between wheels D and E and between B and C, if the mean diameters of B and D are 300 mm and 95 mm? (U. Lond.)

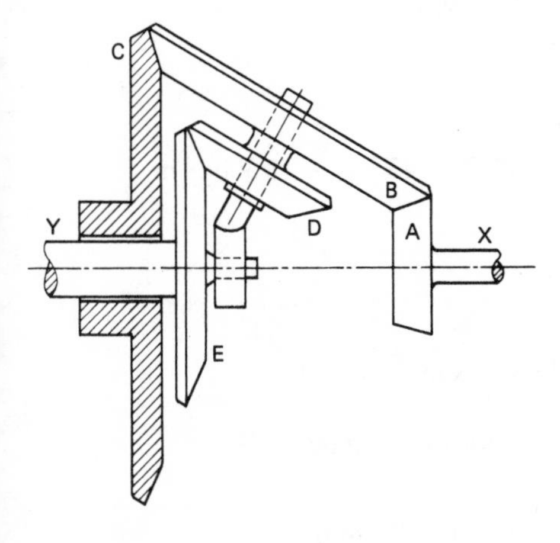

Fig. 12.10

Solution

	Arm	E, Y	C	B, D	A, X
(a) Give whole gear +1 revolution	+1	+1	+1	+1	+1
(b) Hold arm and give C −1 revolution	0	$-\dfrac{T_C}{T_B}\dfrac{T_D}{T_E}$	−1	—	$+\dfrac{T_C}{T_A}$
(c) Add [(a)+(b)]	+1	$1-\dfrac{T_C}{T_B}\dfrac{T_D}{T_E}$	0	—	$1+\dfrac{T_C}{T_A}$

If $1-\dfrac{T_C}{T_B}\dfrac{T_D}{T_E}$ is positive, E is of same sign as A,

i.e.
$$1 > \frac{T_C}{T_B}\frac{T_D}{T_E}$$

i.e.
$$\underline{\frac{T_B}{T_D} > \frac{T_C}{T_E}}$$

If $1-\dfrac{T_C}{T_B}\dfrac{T_D}{T_E}$ is negative, E is of opposite sign to A,

i.e.
$$1 < \frac{T_C}{T_B}\frac{T_D}{T_E}$$

i.e.
$$\underline{\frac{T_B}{T_D} < \frac{T_C}{T_E}}$$

$$\frac{N_Y}{N_X} = \left(1-\frac{T_C}{T_B}\frac{T_D}{T_E}\right)\Bigg/\left(1+\frac{T_C}{T_A}\right) = \left(1-\frac{75}{60}\frac{19}{25}\right)\Bigg/\left(1+\frac{75}{17}\right) = \frac{1}{108{\cdot}2}$$

$$\text{Input torque, } C_X = \frac{7{\cdot}5\times10^3\times60}{2\pi\times500} = 143{\cdot}3\ \text{N m}$$

Since net work done is zero,

$$C_Y N_Y + C_X N_X = 0$$

i.e.
$$\text{output torque, } C_Y = -108{\cdot}2\times143{\cdot}3 = \underline{-15\,500\ \text{N m}}$$

Since there is no acceleration,

$$C_X + C_C + C_Y = 0$$

i.e.
$$143{\cdot}3 + C_C - 15\,500 = 0$$

$$\therefore\ C_C = 15\,357\ \text{N m}$$

$$\text{Mean diameter of C} = \frac{T_C}{T_B}\times\text{mean diameter of B}$$

$$= \frac{75}{60}\times300 = 375\ \text{mm}$$

Since whole torque C_C is taken by wheel C,

$$\text{force between C and B} = \frac{C_C}{r_C} = \frac{15\,500 \times 2}{0.375}\,\text{N} = \underline{82.6\,\text{kN}}$$

$$\text{Mean diameter of E} = \frac{T_E}{T_D} \times \text{mean diameter of D}$$

$$= \frac{25}{19} \times 95 = 125\,\text{mm}$$

Since whole torque C_E is taken by wheel E,

$$\text{force between E and D} = \frac{C_E}{r_E} = \frac{15\,500 \times 2}{0.125}\,\text{N} = \underline{248\,\text{kN}}$$

Example 12.7 In the train of gear wheels shown in Fig. 12.11, the shaft X runs at 2400 rev/min and drives the bevel wheel B by the pinion A. The compound wheel EC revolves freely on a spindle rigidly attached to B. Wheel C gears with D, and E with F, both wheels D and F being concentric with wheel B. Wheel F is held stationary and wheel K, solid with D, drives L on shaft Z. There are three sets of the wheels CE carried on spindles equally spaced on B. The tooth numbers are: A, 18; B, 60; C = D = 22; E, 23; F, 21; K, 19; L, 64.
(a) Find the speed of shaft Z.
(b) If the gear is to be modified by simply substituting different wheels of the same pitch for E and F, so that the speed of Z is to be reversed in direction and is to be between 35 rev/min and 45 rev/min, find the number of teeth on the new wheels E and F. (U. Lond., *modified*)

Solution

The wheels A and L are not part of the epicyclic train and must therefore be treated separately.

$$N_B = \frac{T_A}{T_B} N_X = \frac{18}{60} \times 2400 = 720\,\text{rev/min}$$

Also
$$N_Z = -N_K \times \frac{19}{64}$$

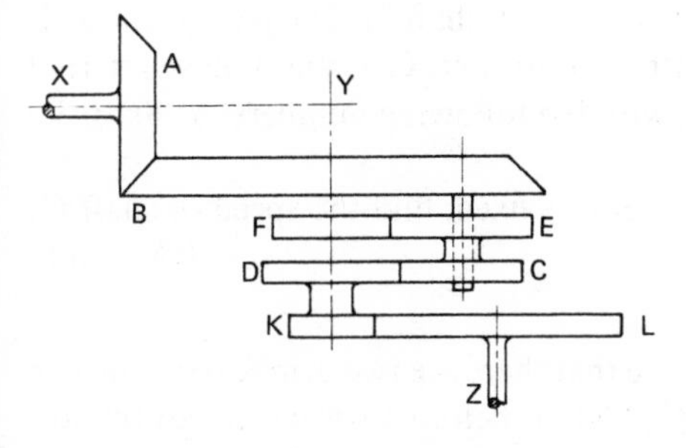

FIG. 12.11

	B	E, C	F	D, K
(a) Give whole gear +1 revolution	+1	+1	+1	+1
(b) Hold arm B and give F −1 revolution	0	$+\frac{T_F}{T_E}$	−1	$-\frac{T_F}{T_E}\frac{22}{22}$
(c) Add [(a)+(b)]	+1	$1+\frac{T_F}{T_E}$	0	$1-\frac{T_F}{T_E}$
(d) Multiply (c) by 720	+720	—	—	$720\left(1-\frac{T_F}{T_E}\right)$

(a) When $T_F = 21$ and $T_E = 23$,

$$N_K = 720\left(1-\frac{21}{23}\right) = 62{\cdot}6\,\text{rev/min}$$

$$\therefore\; N_Z = -62{\cdot}6 \times \frac{19}{64} = \underline{-18{\cdot}6\,\text{rev/min}} \quad \text{(opposite in direction to B)}$$

(b) $$N_Z = -720\left(1-\frac{T_F}{T_E}\right)\times\frac{19}{64} = n \quad \text{where} \quad 35 < n < 45$$

i.e. $$\frac{T_F}{T_E} = 1 + 0{\cdot}004\,68\,n \tag{1}$$

Also $$T_F + T_E = 23 + 21 = 44 \tag{2}$$

Therefore, from equations (1) and (2),

$$T_E = \frac{44}{2 + 0{\cdot}004\,68\,n}$$

When $n = 35$, $$T_E = \frac{44}{2+0{\cdot}164} = 20{\cdot}3$$

When $n = 45$, $$T_E = \frac{44}{2+0{\cdot}211} = 19{\cdot}1$$

$$\therefore\; T_E = \underline{20} \quad \text{and} \quad T_F = \underline{24}$$

Example 12.8 A compound epicyclic gear is shown in Fig. 12.12. C and D form a compound wheel which rotates freely on shaft G. The planet wheels B and E rotate on pins fixed on arms attached to shaft G. C and F have internal teeth; the others have external teeth, with the following numbers: A, 40; B, 30; D, 50; E, 20.

If A rotates at 500 rev/min and wheel F is fixed, find the speed of shaft G. (U. Lond.)

Solution

This gear is compound only in the sense that there are two arms carried by *one* shaft. The problem is solved by a single table as before, since when the arms are held fixed in the second operation, a simple train results. The speeds of the

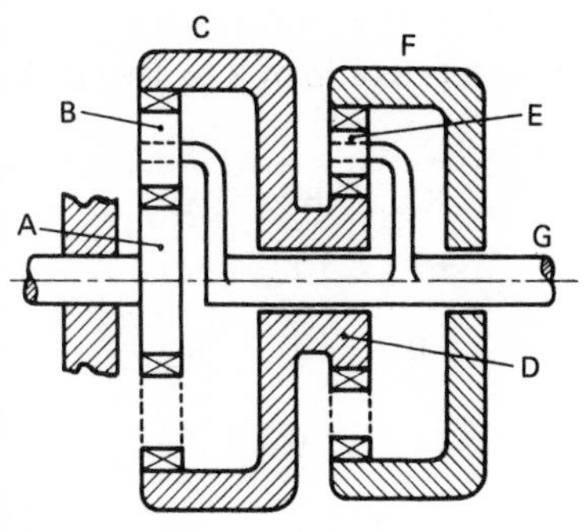

FIG. 12.12

planets B and E are irrelevant to the solution and may therefore be omitted from the table.

$$r_F = r_D + 2r_E$$

i.e. $$T_F = 50 + 2 \times 20 = 90$$

Similarly $$T_C = 100$$

	G	A	F	C, D
(a) Give whole gear +1 revolution	+1	+1	+1	+1
(b) Hold shaft G and give F −1 revolution	0	$-\frac{90}{50} \times \frac{100}{40}$	−1	$+\frac{90}{50}$
(c) Add [(a)+(b)]	+1	−3·5	0	+2·8
(d) Multiply (c) by $-\frac{500}{3 \cdot 5}$	−143	500	—	—

$\therefore$ speed of G = $\underline{-143 \text{ rev/min}}$ (in opposite direction to A)

Example 12.9 In the epicyclic gear shown in Fig. 12.13, the sun shaft P is driven in a clockwise direction at 1200 rev/min while the annulus A is driven in an anticlockwise direction at 600 rev/min. Determine the speeds and directions of rotation of the annulus B and of the shaft R.

The numbers of teeth in the various gears are as follows: S, 30; A, 100; T, 25; B, 75. (U. Lond.)

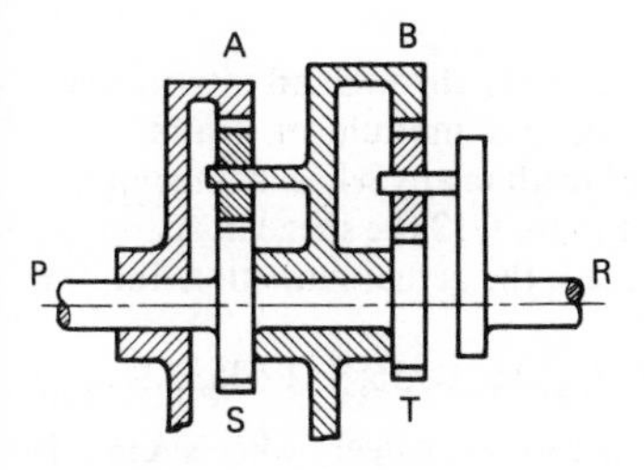

FIG. 12.13

Solution

Consider first the train R, B, T in order to find the speed ratio of R to T when B is fixed.

	R	B	T
(a) Give whole train +1 revolution	+1	+1	+1
(b) Hold arm R and give B −1 revolution	0	−1	$+\frac{75}{25}$
(c) Add [(a)+(b)]	+1	0	+4

Therefore, when B is fixed,

$$N_R = \tfrac{1}{4}N_T \tag{1}$$

The whole gear is now considered with B as the arm and the speed ratio of R to T already found is used to complete the table. Since no members are fixed, the whole train is given an initial rotation of $+a$ revolutions.

	B	A	P, S, T	R
(a) Give whole train $+a$ revolutions	$+a$	$+a$	$+a$	$+a$
(b) Hold arm B and give A $+b$ revolutions	0	$+b$	$-\frac{100}{30}b$	$-\frac{1}{4}\times\frac{100}{30}b$*
(c) Add [(a)+(b)]	a	$a+b$	$a-\frac{10}{3}b$	$a-\frac{5}{6}b$

* From equation (1).

Hence $\quad N_P = a - \frac{10}{3}b = 1200$

and $\quad N_A = a + b = -600$

$$\therefore\ a = -184{\cdot}6 \quad \text{and} \quad b = -415{\cdot}4$$

$$\therefore\ N_B = a = \underline{-184{\cdot}6\ \text{rev/min}} \quad \text{(anticlockwise)}$$

and $\quad N_R = a - \frac{5}{6}b = \underline{+161{\cdot}8\ \text{rev/min}} \quad$ (clockwise)

Problems

1. Two shafts A and B in the same line are geared together through an intermediate parallel shaft C. The wheels connecting A and C have a module of 4 and those connecting C and B a module of 9, the least number of teeth in any wheel being not less than 15. The speed of B is to be about, but not greater than, 1/12 the speed of A and the ratio at each reduction is the same. Find suitable wheels, the actual reduction and the distance of shaft C from A and B. (I. Mech. E.)

(*Answer:* A, 36; C_1, 126; C_2, 16; B, 56; 1 : 12·25; 324 mm)

2. Two parallel shafts X and Y are to be connected by toothed wheels; wheels A and B form a compound pair which can slide along, but rotate with, shaft X; wheels C and D are rigidly attached to shaft Y, and the compound pair may be moved so that A engages with C, or B engages with D.

Shaft X rotates at 640 rev/min and the speeds of shaft Y are to be 340 rev/min exactly, and 240 rev/min as nearly as possible. Using a module of 12 for all wheels, find the minimum distance between the shaft axes, suitable tooth numbers for the wheels, and the lower speed of Y. (U. Lond.)

(*Answer:* 294 mm; A, 13; B, 17; C, 36; D, 32; 231·1 rev/min)

3. In the arrangement of gears shown diagrammatically in Fig. 12.14, the driving shaft X runs at 1000 rev/min; the compound gears D and F rotate with the driven shaft Z, and can be moved so as to engage D with C or F with E. All wheels are to have a module of 6, wheels A and C have equal tooth numbers, and the distance between the lay shaft (Y) axis and the axis of X and Z is fixed at 135 mm.

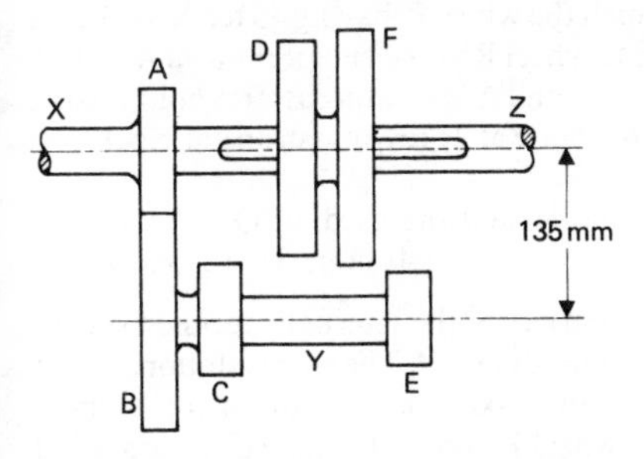

FIG. 12.14

If shaft Z is to run at 640 rev/min exactly, and 480 rev/min as nearly as possible, determine suitable numbers of teeth for all wheels and the actual lower speed of Z. (U. Lond.) (*Answer:* A, 20; B, 25; C, 20; D, 25; E, 17; F, 28; 486 rev/min)

4. The first and third shafts of a double reduction spur gear box are in line and a total reduction of approximately 10:1 is required. The module of the high speed pair is to be 5, that of the low speed pair is to be 8, and no wheel is to have fewer than 20 teeth. Find suitable values of the centre distance between the first and the second shafts and the numbers of teeth on the wheels so as to satisfy the above conditions. What is the actual gear ratio? (I. Mech. E.)

(*Answer:* 340 mm; A, 32; B, 104; C, 20; D, 65; 1:9·5)

5. An epicyclic gear has a fixed annular wheel C concentric with a sun wheel A. A planet wheel B gears with A and C and can rotate freely on a pin carried by an arm D which rotates about an axis co-axial with that of A and C. If T_1 and T_2 are the numbers of teeth on A and C respectively, show that the ratio of the speeds of D to A is $T_1/(T_1+T_2)$. If the least number of teeth on any wheel is 18 and $T_1+T_2 = 120$, find the greatest and least speeds of D when the wheel A rotates at 500 rev/min. (U. Lond.)

(*Answer:* 175; 75 rev/min)

6. In the epicyclic gear shown in Fig. 12.15, the pinion B and the internal wheels E and F are mounted independently on the spindle, O, while C and D form a compound wheel which rotates on the pin P attached to the arm A. The wheels B, C and D have 15, 30 and 25 teeth, all of the same pitch.

(a) If wheel E is fixed, what is the ratio of the speed of F to that of B?

(b) If wheel B is fixed, what are the ratios of the speeds of E and F to that of A?

(I. Mech. E.) (*Answer:* 1:56; 6:5; 33:28)

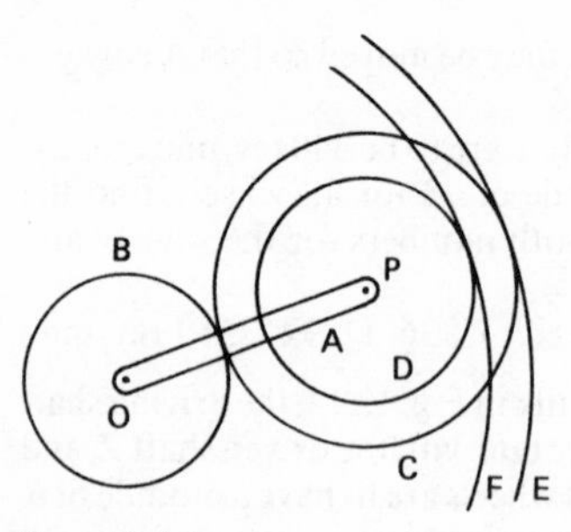

FIG. 12.15

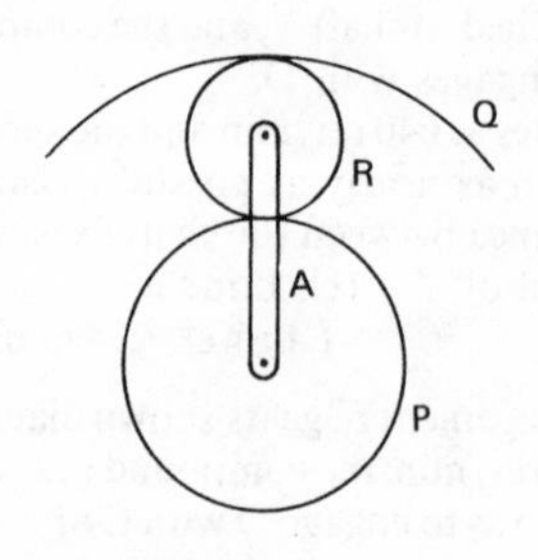

FIG. 12.16

7. Figure 12.16 shows an epicyclic gear in which the wheel P, having 45 teeth of 15 mm pitch, is geared with Q through the intermediate wheel R at the end of the arm A. When P is rotating at 63 rev/min in a clockwise direction, and A is rotating at 9 rev/min, also in a clockwise direction, the wheel Q is required to rotate at 21 rev/min in an anticlockwise direction.

Find the necessary numbers of teeth in Q and R, and the p.c.d. of Q. (U. Lond.)
(*Answer:* 81; 18; 386·9 mm)

8. In the epicyclic gear shown in Fig. 12.16, the p.c.d. of the internally toothed ring Q is to be about 250 mm and the teeth are to have a module of 4. The ring is stationary but the spider A carrying the planet wheels R is to make one revolution for every 5 revolutions of the driving spindle carrying the wheel P and in the same direction. Find suitable numbers of teeth for all the wheels. (U. Lond.) (*Answer:* 64; 16; 24)

9. If, in the epicyclic gear train shown in Fig. 12.17, gear A rotates at 1000 rev/min clockwise, while E rotates at 500 rev/min anticlockwise, determine the speed and direction of rotation of the annulus D, and of the shaft F. All gears are of the same pitch, and the number of teeth in A is 30, in B is 20 and in E is 80. (U. Lond.)
(*Answer:* 372 rev/min anticlockwise; 40 rev/min clockwise)

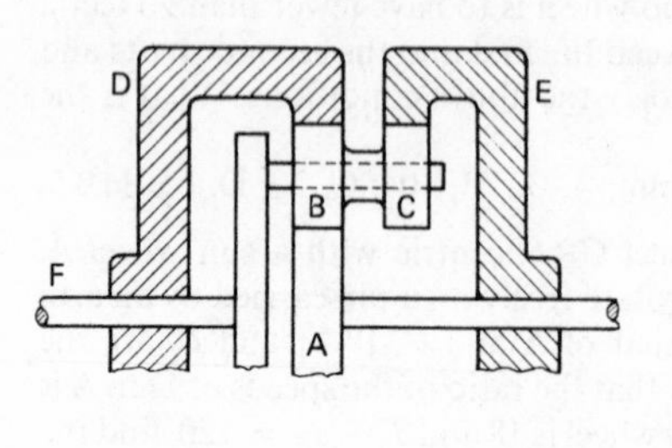

FIG. 12.17

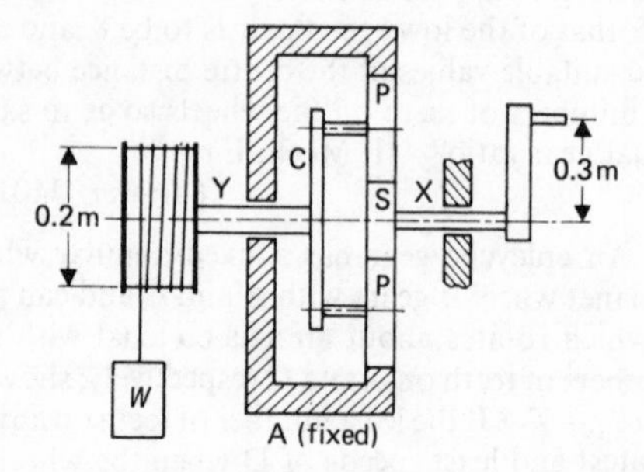

FIG. 12.18

10. The manually-operated hoist shown in Fig. 12.18 is based on an epicyclic gear train. The sun wheel, S, is rigidly connected by shaft X to the handle, and the planet carrier, C, is rigidly connected by shaft Y to the winding drum. The annulus, A, is fixed.

If the annulus is to have 90 teeth, determine the number of teeth on the sun and planet wheels so that the tangential force on the handle is approximately 25 times less than the load being lifted. (U. Lond.) (*Answer:* 12; 39)

11. In an epicyclic gear of the sun and planet type, the p.c.d. of the internally toothed ring is to be as nearly as possible 220 mm, and the teeth are to have a module of 4. When the ring is stationary, the spider, which carries three planets of equal size, is to make one revolution for every five revolutions of the driving spindle carrying the sun wheel. Determine suitable numbers of teeth for all the wheels and the exact diameter of the pitch circle of the ring.

If a torque of 12 N m is applied to the spindle carrying the sun wheel, what torque will be required to keep the ring stationary? (U. Lond.)

(*Answer:* 56, 21, 14; 224 mm; 48 N m)

12. An epicyclic gear consists of two sun wheels, S_1 and S_2 with 24 and 28 teeth respectively, engaged with a compound planet wheel with 26 and 22 teeth. S_1 is keyed to the driven shaft and S_2 is a fixed wheel co-axial with the driven shaft. The planet wheel is carried on an arm from the driving shaft. Find the velocity ratio of the gear. If 750 W is transmitted, the output speed being 100 rev/min, what torque is required to hold S_2? (U. Lond.) (*Answer:* $-2{\cdot}64:1$; 98·7 N m)

13. An epicyclic gear consists of a sun wheel which has 24 teeth, planet wheels which have 28 teeth, and an internally toothed annulus which is held stationary. Neglecting friction, find the torque required to hold the annulus at rest when 9 kW is being transmitted, the sun wheel rotating at 700 rev/min.

If the teeth have a module of 4, what is the diameter of the circle traced out by the centres of the planet wheels? (U. Lond.) (*Answer:* 409 N m; 208 mm)

14. In the epicyclic train shown in Fig. 12.19, shaft A rotates at 1000 rev/min in a clockwise direction while shaft B is driven at 500 rev/min in an anticlockwise direction. The torque input at shaft B is 50 N m. Determine: (a) the speed and direction of rotation of the shaft C, and (b) the torques in shafts A and C, stating whether these torques are inputs to the system or outputs from the system.

All teeth are of the same module and the numbers of teeth in the various gears are as follows: D, 16; E, 24; F, 43; H, 35; K, 30. (U. Lond.)

(*Answer:* 2060 rev/min clockwise; A, 22·9 N m input; C, 23·25 N m output)

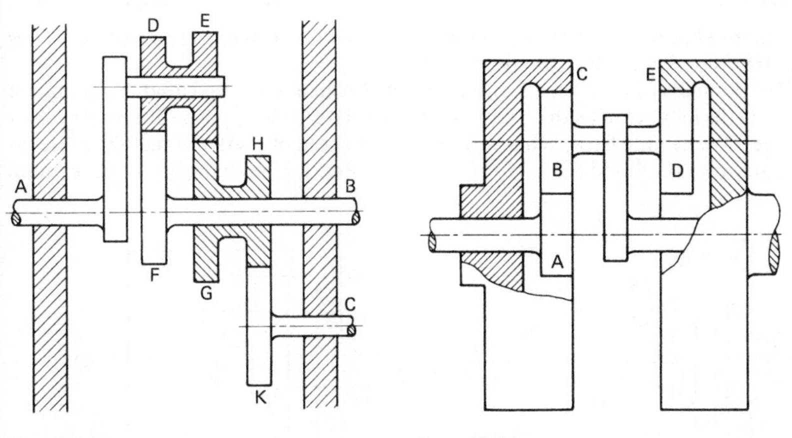

FIG. 12.19

FIG. 12.20

15. In the epicyclic gear shown in Fig. 12.20, the driving wheel A has 14 teeth and the fixed annular wheel C, 100 teeth; the ratio of tooth numbers in wheels E and D is 98 : 41.

If 2 kW at 1200 rev/min is supplied to the wheel A, find the speed and direction of rotation of E and the fixing torque required at C. (U. Lond.)

(*Answer:* 3·99 rev/min in the same direction as A; 4·77 kN m)

16. In the epicyclic speed reducing gear shown in Fig. 12.21, the input shaft A runs at 12 000 rev/min and the annular wheel B is fixed. Find the speed of the output shaft Z and the speed of the planet wheels relative to the spindle on which they are mounted. The numbers of teeth in the wheels are: A, 15; C, 41; C_1, 25; B, 81.

If there are three planet systems and the teeth have 20° involute profiles, find the tangential and radial forces at the tooth contacts of C and C_1 when the gear transmits 2 MW. (I. Mech. E.)

(*Answer:* 1218 rev/min; – 3945 rev/min; 9·9 kN; 3·6 kN; 16·24 kN; 5·91 kN)

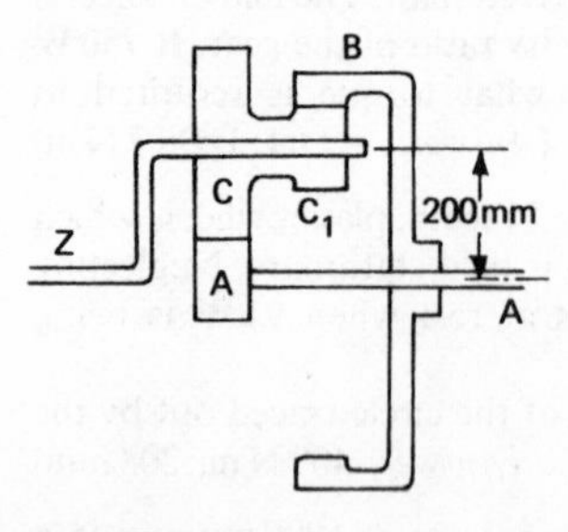

Fig. 12.21

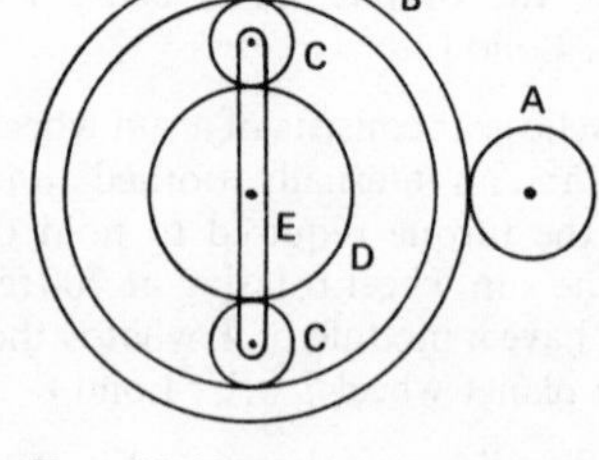

Fig. 12.22

17. In the epicyclic gear shown in Fig. 12.22, the gear B has 120 teeth externally and 100 teeth internally. The driver A has 20 teeth and the arm E is connected to the driven shaft. Gear D has 60 teeth. If A revolves at + 100 rev/min and D revolves at + 27 rev/min, find the speed of the arm E.

If D is now fixed and A transmits a torque of 10 N m at + 100 rev/min, what will be the available torque at the arm E, assuming 96 per cent efficiency of transmission? (U. Lond.)

(*Answer:* – 7/24 rev/min; 92 N m)

18. In the Humpage gear shown in Fig. 12.23, the wheel C is fixed and shaft X rotates at 650 rev/min. Determine the speed of shaft Y.

If the numbers of teeth on all wheels except E remain unaltered, find the minimum change in the number of teeth on E to cause Y to rotate in the opposite direction to shaft X, wheel C remaining fixed. What will then be the speed of shaft Y if shaft X rotates at 650 rev/min? (U. Lond.)

(*Answer:* 9·75 rev/min; 2; – 1·05 rev/min)

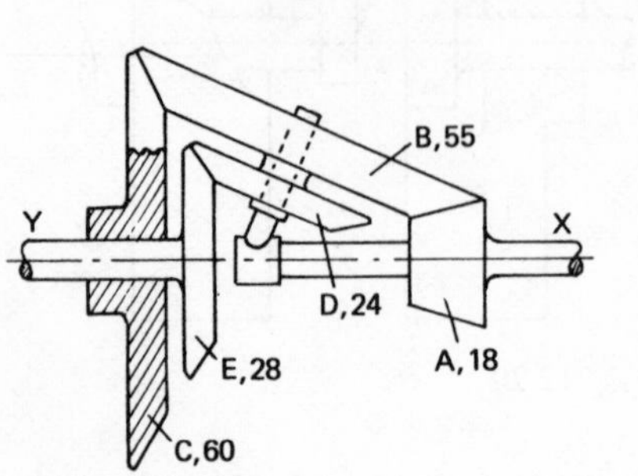

Fig. 12.23

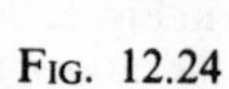

F
C
D
E
B
X
X
A

Fig. 12.24

19. An epicyclic gear consists of bevel wheels arranged as shown in Fig. 12.24. The driving pinion A has 20 teeth and meshes with the wheel B which has 25 teeth. The wheels B and C are fixed together and turn freely on the shaft F. The shaft F can rotate radially about the main axis XX. The wheel C has 50 teeth and meshes with wheels D and E, each of which has 60 teeth.

Find the speed and direction of E when A rotates at 200 rev/min if: (a) D is fixed, and (b) D rotates at 100 rev/min in the same direction as A.

In both cases, find the ratio of the torques transmitted by the shafts of the wheels A and E, the friction being neglected. (U. Lond.)

(*Answer:* (a) 800 rev/min, opposite to direction of A;
(b) 300 rev/min, opposite to direction of A; 4; 4)

20. Figure 12.25 shows an epicyclic gear in which the internally toothed annulus A is fixed. Wheels B and G are free to rotate on the arm J. Wheels F and H are keyed to shafts D and E respectively. The numbers of teeth in the gear wheels are as follows: A, 120; C, 60; H, 50; B, 30; F, 50; G, 30.

If D is driven at 80 rev/min and C is driven at 50 rev/min in the same direction, find the speed and direction of rotation of the arm J and also of the shaft E.
(U. Lond.) (*Answer:* $+50/3$ rev/min; $-140/3$ rev/min)

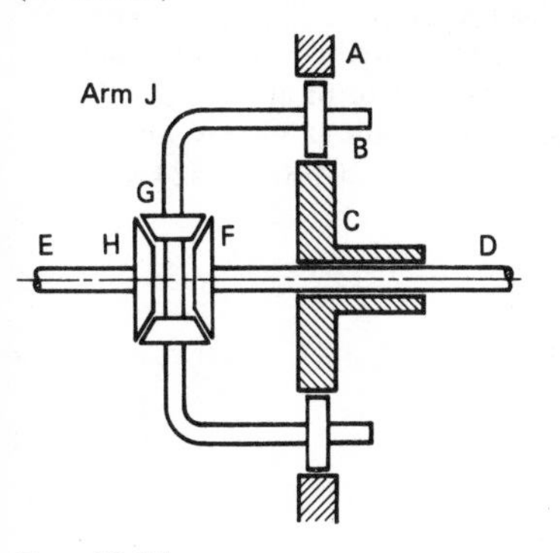

Fig. 12.25

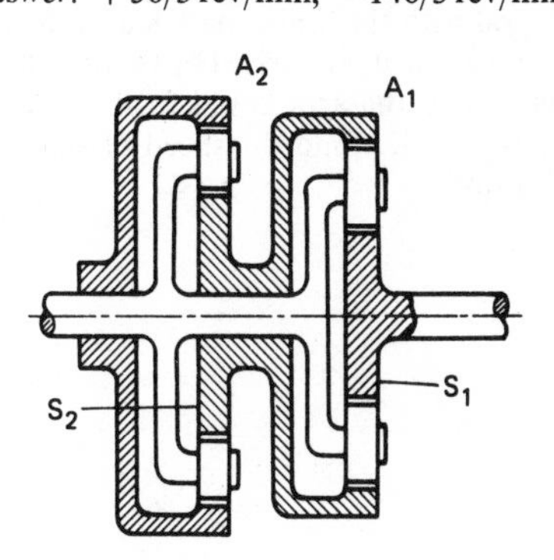

Fig. 12.26

21. In the epicyclic gear shown in Fig. 12.26, the annulus A_1 and the sun wheel S_2 are compound, the sun wheel S_1 rotates with the input shaft, and both planet cages rotate with the output shaft. The numbers of teeth on the wheels are: S_1, 24; A_1, 72; S_2, 40; A_2, 80.

(a) Find the gear ratio of the system when the annulus A_1 is held fixed.
(b) Find the gear ratio of the system when the annulus A_2 is held fixed.
(c) With the arrangement (a) the torque required to hold the gear box, which contained the epicyclic gear, fixed was found to be 347·5 N m when the output power was 45 kW at 900 rev/min. Find the efficiency of power transmission. (U. Lond.)

(*Answer:* 1/4; $-1/5$; 91·8 per cent)

22. Figure 12.27 shows an epicyclic gear box, the input shaft of which drives the sun wheel S_1 which has 31 teeth. The planets P_1 have 18 teeth and are carried on a rotating cage which is fixed to the output shaft. The annulus A_1 is compound with the sun S_2 and the planets P_2 which also have 18 teeth are carried on a cage also fixed to the output shaft. The annulus A_2 has 59 teeth.

Calculate the velocity ratio of the gear when the annulus A_2 is held stationary and the torque required to hold A_2 when the gear is transmitting 12 kW at 600 rev/min with negligible friction losses. (U. Lond.) (*Answer:* $-0{\cdot}22$; 1·057 kN m)

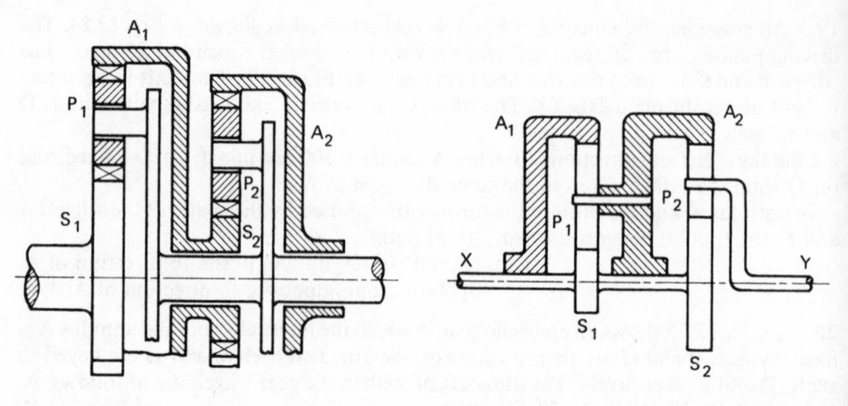

FIG. 12.27

FIG. 12.28

23. A compound epicyclic gear is shown in Fig. 12.28. Sun wheels S_1 and S_2 are integral with the input shaft X and the annular wheel A_1 is fixed. The planet wheel P_1 rotates about a pin carried by the annulus A_2 and the planet wheel P_2 rotates about a pin attached to the arm keyed to the shaft Y. The numbers of teeth are: S_1, 30; P_1, 20; P_2, 18; S_2, 32. Find the speed of shaft Y when shaft X rotates at 100 rev/min. (U. Lond.)

(*Answer:* 524 rev/min)

13
Introduction to automatic control

13.1 Introduction

Engineering control systems are used for the control of temperature, flow rates, liquid levels, chemical composition, speed of prime movers, position of ships and aircraft, radar guidance, machine tool operations, etc. The advantages over human control are speed of response, consistency, accuracy, predictability, maintenance of adequate control continuously over long periods in most environments and the ability to control large amounts of energy from low power sources.

Control system elements include various temperature and pressure measuring devices, springs, levers, gears, valves, gyroscopes, compressors, accumulators, bellows, motors, resistors, relays, transistors, etc. Transducers, which convert such quantities as pressure, temperature and acceleration into electrical signals are commonly used and the microprocessor, based on the 'chip', is now involved in sophisticated control of medical equipment, engine ignition systems and domestic appliances such as the microwave oven.

13.2 Open- and closed-loop systems

If the performance requirements of a system are preset into a controller, the machine or process will carry out the required function, regardless of the resulting output. Thus a dish-washing machine will continue with the cycle of operations for the preset conditions, irrespective of the state of the contents. Such systems are called *open-loop* or open-path, since no information is fed back to the controller to vary the output. The output of an open-loop system will vary with the load upon it and the appropriate setting of the controls is at the judgement of the operator.

When the difference between the desired and actual performance is continuously monitored and the difference between these values is used to correct the output, the system has *feedback* and is called *closed-loop**. Such systems may be *discontinuous* or *continuous*. An example of a discontinuous system is a thermostat controlling the temperature of a hot-water cylinder – this is merely an 'on–off' control which maintains the temperature

* An open-loop system becomes closed-loop if the operator manually adjusts the controls during the cycle.

between two limits, depending on the sensitivity of the thermostat. Other examples are to be found in car engine coolant systems, refrigerators and central heating systems.

Continuous control applications are water cisterns with ball-valve control, engine governors and automatic pilots for ships and aircraft, where the correcting action depends on the deviation at any instant. If excessive correction is applied, the system may 'hunt' or oscillate about the desired position and damping may be required. A system which will not settle in the required position is *unstable* and the conditions for stability form an important part in the design of a closed-loop control system, which is beyond the scope of this book.

13.3 System representation

A system consists of a number of *elements* and the relation of these elements is shown schematically by a *block diagram*, in which each component is represented by a box. Thus the essential features of a car drive consisting of a fuel system, engine, gear box and differential would be represented by a block diagram as shown in Fig. 13.1.

Since the control of the fuel supply is determined by the driver, the system is closed-loop 'with operator' but if the car is fitted with a *cruise control* (a device to maintain a steady, desired speed), the fuel control is determined by the speed

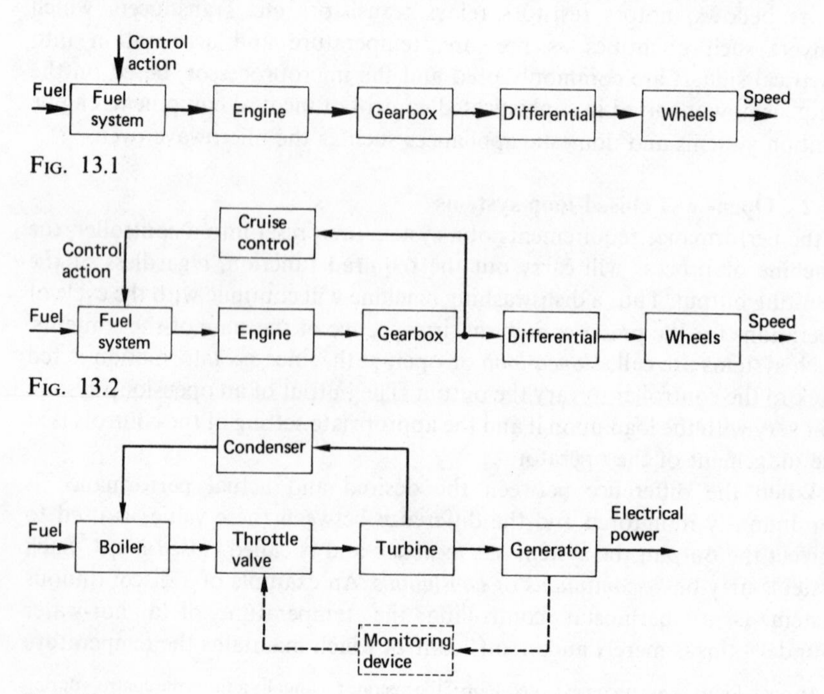

FIG. 13.1

FIG. 13.2

FIG. 13.3

of the car and the system becomes automatic closed-loop. The modification of the block diagram is shown in Fig. 13.2.

A further example of a block diagram is the steam power plant shown in Fig. 13.3. The system is open-loop but would become closed-loop if the throttle valve is actuated by a sensor from the load, as shown dotted.

13.4 Essentials of a control system

The *input* to a control system is the desired or reference value for some variable and is denoted by θ_i. The *output* or controlled variable is the actual response of the system and is denoted by θ_o. The difference between θ_i and θ_o is termed the *error* or *deviation* and is denoted by θ_e. The function of a control system is to maintain the value of θ_o as nearly as possible to θ_i and to correct any error as rapidly as practical.

The essentials of such a system are (a) a *controller* which operates on the energy input to the plant to adjust the output, (b) a *monitoring device* to measure the output and feed this information back to (c) a *comparing element*, which detects the error between the desired and actual output. The error signal from the comparing element is fed into the controller to initiate the control action.

The arrangement of the system is shown in Fig. 13.4.

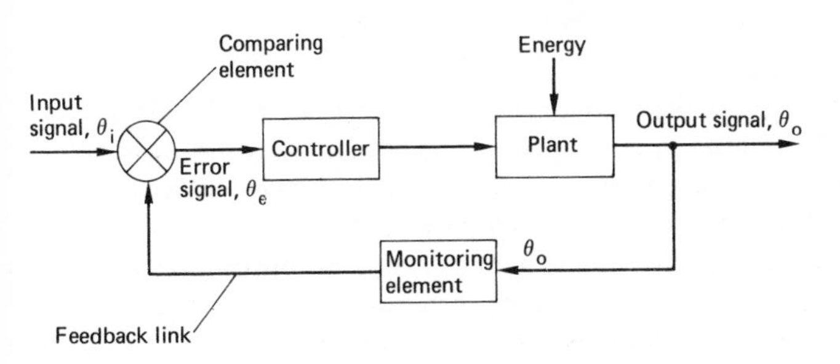

FIG. 13.4

If the control action requires power amplification, such as for the movement of a ship's rudder, this is provided by a *servomechanism* or servo-motor. Such mechanisms require energy from an external source and are said to have *power-gain*. The monitoring element in the feedback path may also include an amplifier.

The error signal supplied to the controller represents the difference between the desired and actual outputs and the system has negative feedback; this is a necessary (but not sufficient) requirement for a stable system.

In control work, the terms 'input' and 'output' have wide meanings and may represent variables such as movements, voltages, pressures and vibration frequencies. Furthermore, inputs and outputs need not be similar quantities.

13.5 Classification of control systems

In addition to open-loop and closed-loop types, systems may be further classified as servomechanisms, regulators and process control systems. The design and speed of response of the various types differ greatly but the general principles are similar.

Servomechanisms, or kinetic control systems, are usually used for mechanical applications such as to control direction, speed or acceleration in automatic guidance systems or gun position systems, and for machine tool operations.

Regulators are used to maintain constantly controlled values of such quantities as speed, voltage and temperature.

Process controllers maintain flow rates of liquids, liquid levels, chemical composition, etc. Some process control systems are basically regulators and may include servomechanisms.

13.6. The transfer operator

The ratio of the output of an individual element or complete system to the input is called the *transfer operator* or *transfer function**. This may depend on values and rates of change of variables so that the transfer operator is usually a function of time,

i.e. $$\text{transfer operator} = \theta_o/\theta_i = f(\mathrm{D})$$

where D represents $\mathrm{d}/\mathrm{d}t$.

The transfer operator depends on the characteristics of the element or system and is a differential equation defining its behaviour, being independent of the form of the input. The ratio θ_o/θ_i is not usually dimensionless; the units will depend on those of θ_i and θ_o, which may be different.

When the transfer operator for a system has been established, the response to different input signals such as sudden, progressive or harmonic changes may be investigated. This response is determined by the solution of the differential equation, which consists of two parts: (a) the complementary function, which represents the transient or natural response, approaching zero as time increases, and (b) the particular integral, which represents the steady-state response, persisting after the transient response has died away.

13.7 Types of controller response

When a system has 'on–off' response, the controller moves immediately from fully-on to fully-off or vice versa but if the control is continuous, the controller produces a response dependent upon the error. In some systems, this response may leave the output with a steady lag or the response may not be sufficiently rapid. Thus, in addition to providing a correction proportional to the error, a controller may also respond to the rate of change of error in order to anticipate the change in output. Such action is called *derivative response*. A further

* This term is usually used in conjunction with Laplace transforms.

refinement is *integral response*, where the correction also depends on the time for which the error existed and integral action is used to improve steady-state response.

In general, derivative action is used to improve transient response. These actions are reflected in the transfer operator for the controller.

13.8 Block diagram algebra

The block diagram for a system shows the interconnection of the various elements, with the values of the individual transfer operators shown in the respective boxes. The overall transfer operator for the system is then obtained from the combination of those of its elements. In this work, the transfer operators for elements in forward paths are denoted by G and those in feedback paths are denoted by H.

Elements in series or cascade, Fig. 13.5 The output of each element is enhanced by its transfer operator and so the overall transfer operator is the product of the individual values,

i.e. $$\frac{\theta_o}{\theta_i} = G_1 \times G_2 \times G_3$$

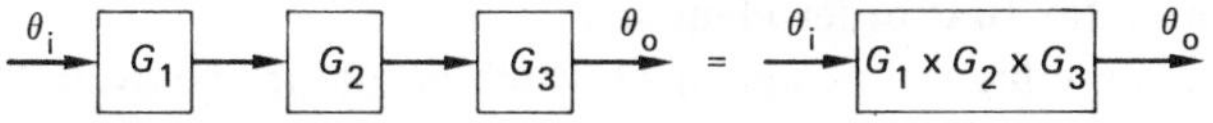

FIG. 13.5

Elements in parallel, Fig. 13.6 The three elements are each fed by the same input θ_i and the output is then the sum of the outputs from the separate elements,

i.e. $$\frac{\theta_o}{\theta_i} = G_1 + G_2 + G_3$$

The symbol ⊗ indicates a summing point, together with the sign of the signals entering it.

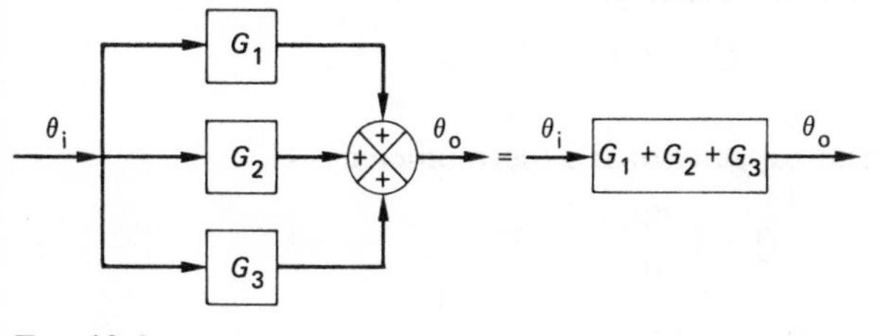

FIG. 13.6

Unity feedback system, Fig 13.7 The signal θ_o is fed-back to the input side of the system and only the difference $\theta_i - \theta_o$, denoted by θ_e, is enhanced by the element.

Thus $$\theta_o = G\theta_e = G(\theta_i - \theta_o)$$

from which $$\frac{\theta_o}{\theta_i} = \frac{G}{1+G}$$

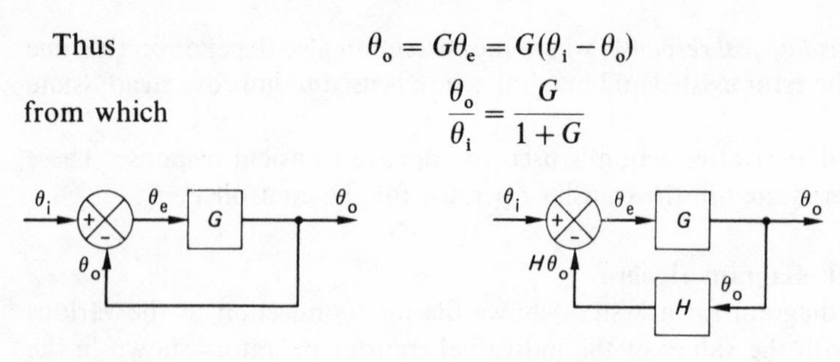

FIG. 13.7

FIG. 13.8

Feedback loop with element, Fig. 13.8 The signal θ_o in the feedback loop is modified by the element to give a signal $H\theta_o$ at the summing point. The error signal θ_e fed into the forward element is then $\theta_i - H\theta_o$.

Thus $$\theta_o = G\theta_e = G(\theta_i - H\theta_o)$$

from which $$\frac{\theta_o}{\theta_i} = \frac{G}{1+GH}$$

The analysis of more complex arrangements is illustrated in Examples 13.1 and 13.2.

13.9 Transfer operator for first order elements

The *order* of a system is the highest power of D which occurs in the overall transfer operator. The scope of this chapter is limited to first order linear systems, where displacements, forces, flow rates, etc, are directly proportional to the stimulus in the steady state.

The following examples show the determination of the transfer operator for simple elements.

(a) *Thermometer*

The rate of heat flow to a thermometer is proportional to the difference in temperature between the thermometer and its surroundings. Thus, if the heat flow rate is Q, Fig. 13.9, and the temperatures of the thermometer and surroundings are θ_o and θ_i respectively,

$$Q = \alpha(\theta_i - \theta_o)$$

where α is a constant.

Also, if the specific heat capacity of the thermometer is C,

$$Q = C\frac{d\theta_o}{dt} = CD\theta_o$$

$$CD\theta_o = \alpha(\theta_i - \theta_o)$$

from which $$\frac{\theta_o}{\theta_i} = \frac{\alpha}{CD+\alpha} = \frac{1}{1+CD/\alpha} = \frac{1}{1+\tau D}$$

where $\tau = C/\alpha$.

τ has the units of time and is called the *time constant* of the element. The significance of the time constant of an element or system is discussed in Section 13.10.

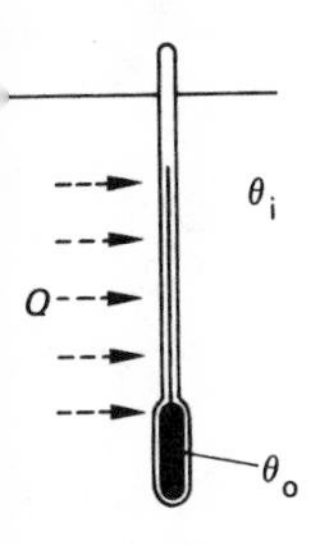

FIG. 13.9

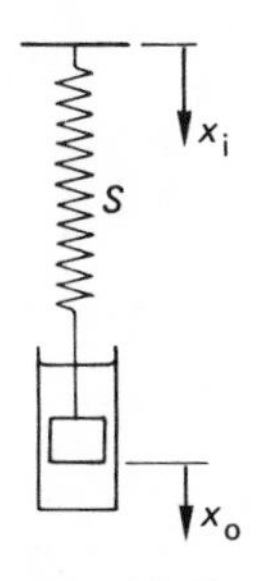

FIG. 13.10

(b) *Spring and dashpot*

Figure 13.10 shows a spring of stiffness S connected to a dashpot plunger of negligible mass. The dashpot is filled with a viscous liquid, so that the resistance to motion may be assumed to be proportional to the speed.

If the upper end of the spring is given a displacement x_i and the resulting displacement of the dashpot plunger is x_o, then

$$\text{compressive force in spring} = S(x_i - x_o)$$

If the resistance of the plunger is c per unit velocity,

$$\text{resisting force on plunger} = c\frac{dx_o}{dt} = cDx_o$$

$$\therefore \; cDx_o = S(x_i - x_o)$$

from which

$$\frac{x_o}{x_i} = \frac{S}{cD+S} = \frac{1}{1+cD/S} = \frac{1}{1+\tau D}$$

where $\tau = c/S$ and is again the time constant of the element.

(c) *Water tank*

Figure 13.11 shows a tank with inflow Q_i and outflow Q_o; the depth of water, h, in the tank is to be maintained approximately constant.

If the cross-sectional area of the tank is A, then

$$Q_i - Q_o = A\frac{dh}{dt} \tag{1}$$

The flow of a liquid through pipes, orifices, etc, is proportional to the square root of the head causing flow,

i.e.

$$Q_o = ch^{1/2}$$

where c is a constant.

$$\therefore \; \frac{dQ_o}{dt} = c \times \tfrac{1}{2}h^{-1/2}\frac{dh}{dt}$$

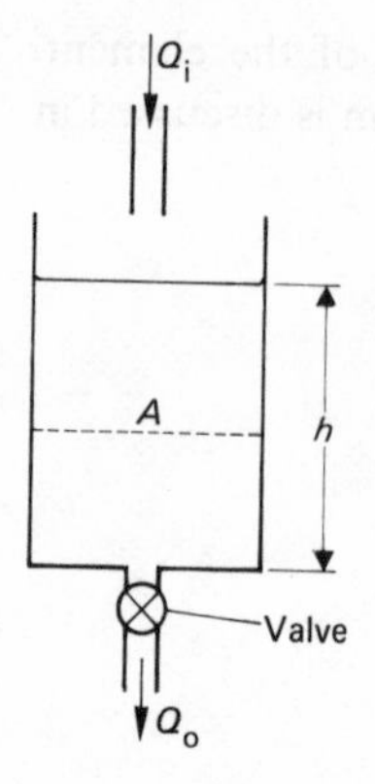

FIG. 13.11

or
$$\frac{dh}{dt} = \frac{2\sqrt{h}}{c}\frac{dQ_o}{dt} = \frac{2\sqrt{h}}{c}DQ_o$$

If h is approximately constant, $2\sqrt{h}/c$ may be replaced by another constant R.

Then
$$\frac{dh}{dt} = RDQ_o$$

Hence
$$Q_i - Q_o = ARDQ_o \qquad \text{from equation (1)}$$

or
$$\frac{Q_o}{Q_i} = \frac{1}{1+ARD} = \frac{1}{1+\tau D}$$

where the time constant $\tau = AR$.

In these examples, inputs and outputs have been similar quantities and have been applied directly to the element, leading in each case to the transfer operator $1/(1+\tau D)$. In general, however, the transfer operator for a first order linear system will be of the form $k/(1+\tau D)$, where k is a constant.

If the input and output are similar quantities but are enhanced or attenuated by a linkage or other device, as in Examples 13.7 and 13.8, then k will be a simple numerical constant but if the input and output are dissimilar quantities, k will have units corresponding with those of the output/input ratio, as in Example 13.5.

13.10 Response of first order elements to various inputs

The variation of an input to a system may be due to some external or internal disturbance, such as a change in steam demand or a voltage drop, or it may be due to a deliberate or pre-determined action. Thus the sequence of events may be of a random nature or it may be governed by a mathematical relationship. Of the latter type, the most common inputs for analysis are of step, ramp or harmonic form and the response to each type of input of first order elements with transfer operators of the form $1/(1+\tau D)$ will now be determined. This involves the formation of a differential equation, the solution of which gives the output θ_o in terms of t, τ and the parameters of the input function.

(a) Step input A step input is a sudden displacement which is maintained at its initial value, i.e. the input control is suddenly moved to a new position. If the value of this displacement is a, Fig. 13.12, then $\theta_i = a$ and

$$\theta_o = \frac{a}{1+\tau D}$$

The solution is $$\theta_o = Ae^{-t/\tau} + a$$

where A is a constant.

Since $\theta_o = 0$ when $t = 0$, $$A = -a$$

$$\theta_o = a(1 - e^{-t/\tau})$$

This is an exponential function which approaches a as t approaches infinity. Theoretically the output never reaches the input value but $\theta_o \simeq 0{\cdot}99\,\theta_i$ when $t = 5\tau$.

The slope of the graph, $d\theta_o/dt = ae^{-t/\tau}/\tau$ and hence the gradient at the origin is a/τ. Thus the tangent to the curve reaches the value a at time τ and so τ represents the time at which the output θ_o would equal the input θ_i if the initial rate of response were maintained.

At time τ, $$\theta_o = \theta_i(1 - e^{-1}) = 0{\cdot}632\,\theta_i$$

The time constant is a measure of the speed of response of the system and may be regarded as the time taken for the output to reach 63·2 per cent of the step change.

Applications of this type are a thermometer plunged into hot water and the charging of a condenser from a d.c. supply.

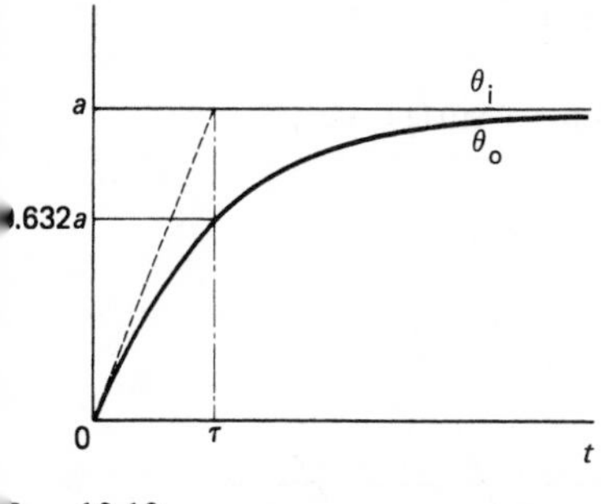

FIG. 13.12

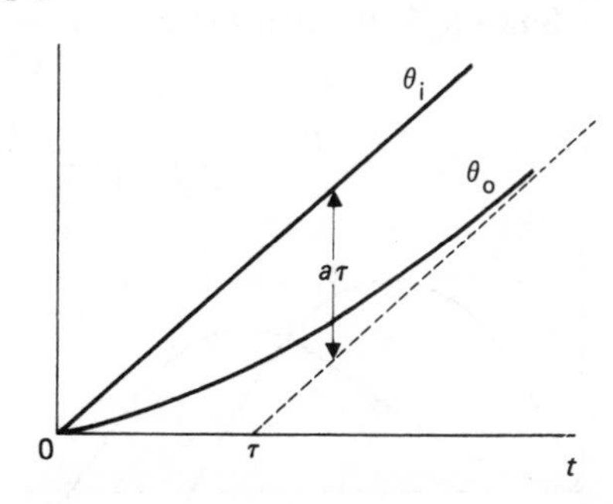

FIG. 13.13

(b) Ramp input A ramp input is a displacement which increases linearly with time. If this displacement is given by $\theta_i = at$, Fig. 13.13, then

$$\theta_o = \frac{at}{1+\tau D}$$

The solution is $$\theta_o = Ae^{t/\tau} + a(t-\tau)$$

where A is a constant.

Since $\theta_o = 0$ when $t = 0$, $A = a\tau$

Thus $$\theta_o = a\{t - \tau(1 - e^{t/\tau})\}$$

As t approaches infinity, θ_o approaches $a(t-\tau)$. The response is therefore asymptotic to a line representing a steady-state lag $a\tau$, shown dotted in Fig. 13.13.

Applications of this type are the temperature of a component in a furnace which is being heated at a constant rate and the displacement of a body being pulled through a viscous medium by an elastic member, the other end of which is moving at constant velocity.

Notes

(1) If the value of a is 1 in the foregoing cases, the input is referred to as a *unit step input* or *unit ramp input*.

(2) Since a step input is the differential of a ramp input ($\mathrm{d}[at]/\mathrm{d}t = a$), the response for the step input may be obtained by differentiating that for a ramp input.

(c) Sinusoidal input This is a harmonic displacement of the form $\theta_i = a \sin pt$, Fig. 13.14. Thus

$$\theta_o = \frac{a \sin pt}{1 + \tau D}$$

The solution is $$\theta_o = Ae^{-t/\tau} + \frac{a}{\sqrt{(1 + p^2\tau^2)}} \sin(pt - \alpha)$$

where A is a constant and $\alpha = \tan^{-1} p\tau$.

Since $\theta_o = 0$ when $t = 0$, $A = \dfrac{a}{\sqrt{(1 + p^2\tau^2)}} \sin \alpha$

Thus $$\theta_o = \frac{a}{\sqrt{(1 + p^2\tau^2)}} \{e^{-t/\tau} \sin \alpha + \sin(pt - \alpha)\}$$

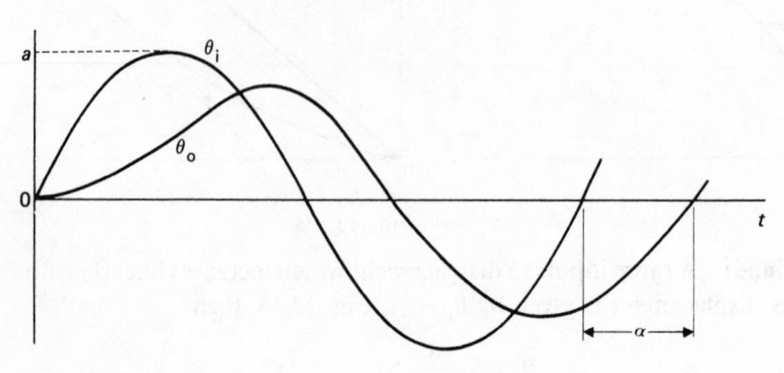

FIG. 13.14

The form of this response is shown in Fig. 13.14; θ_o has a smaller amplitude than θ_i and lags behind θ_i. In the steady-state condition, i.e. when $t = \infty$,

$$\theta_o = \frac{a}{\sqrt{(1 + p^2\tau^2)}} \sin(pt - \alpha)$$

The amplitude of θ_o is therefore $a/\sqrt{(1 + p^2\tau^2)}$ and the angle of lag, α, is $\tan^{-1} p\tau$.

Applications of this type are a capacitor in series with a resistor connected to an a.c. supply and a dashpot of negligible mass operating in a viscous medium at one end of an elastic member while the other end is given a sinusoidal displacement.

An approximate example is the depth of water in a tidal estuary.

The above cases have been applied to the transfer operator $1/(1 + \tau D)$ but in the general case of the transfer operator $k/(1 + \tau D)$, the output θ_o will be k times that for the unit transfer operator.

13.11 Practical control elements

Two simple servomechanisms in common use are based on the hydraulic spool valve and the pneumatic flapper valve, which are briefly described below.

Spool valve Figure 13.15 shows the arrangement of a hydraulic relay, which gives a large force application for a small input force.

The spool valve, shown in mid-position, is controlled by a lever which is pivoted to the valve and ram rods. The input signal is the movement of the lever and the output signal is the movement of the ram. If the input movement is to the right, the valve is immediately moved to the right since the ram does not move. Oil under pressure then passes to the right side of the ram, causing it to move to the left and oil from the left of the ram is displaced back to the sump. When the ram begins to move, the resulting movement of the valve cuts off the oil supply and the ram then stabilizes.

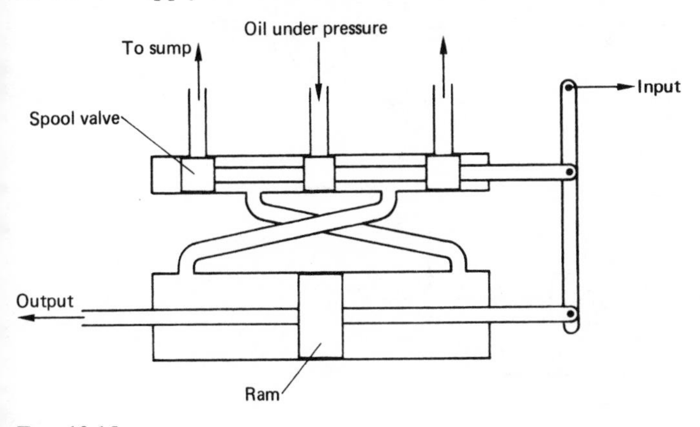

Fig. 13.15

Flapper valve In the simple flapper valve shown in Fig. 13.16, the output pressure θ_o depends on the movement of the arm θ_i, controlling the opening of the exhaust nozzle. A feedback action is provided by the bellows and the valve gives approximately linear response for small movements of the arm.

Provided that certain assumptions are made, both the spool valve and the flapper valve may be regarded as first order linear elements, having transfer operators of the form $k/(1+\tau D)$. (See Examples 13.7 and 13.8.)

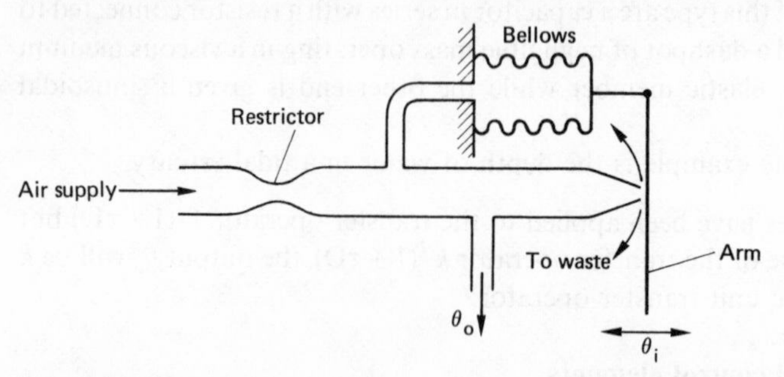

FIG. 13.16

Examples

Example 13.1 Find the transfer operator for the system shown in Fig. 13.17.

Solution

$$\text{Output from element } G_1 = G_1\theta_i$$

and
$$\text{output from element } G_4 = G_4\theta_i$$

Elements G_2 and G_3 may be combined to give a transfer operator G_2G_3.

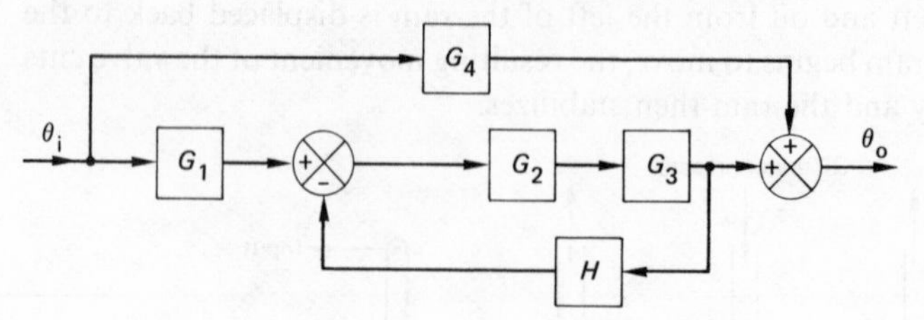

FIG. 13.17

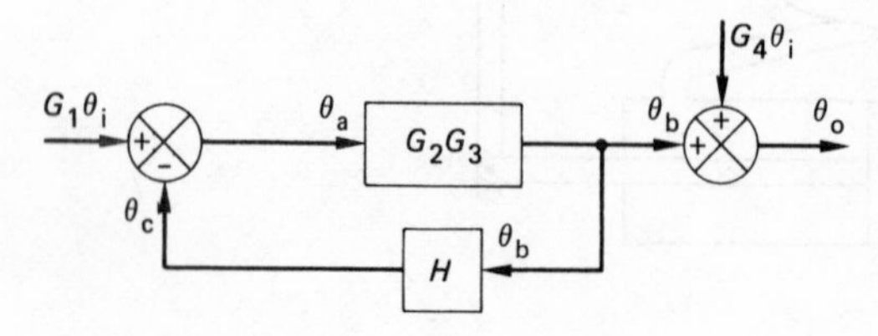

FIG. 13.18

The system may then be reduced to that shown in Fig. 13.18.

$$\begin{aligned}\theta_b &= G_2G_3\theta_a \\ &= G_2G_3(G_1\theta_i - \theta_c) \\ &= G_2G_3(G_1\theta_i - H\theta_b)\end{aligned}$$

from which
$$\theta_b = \frac{G_1G_2G_3\theta_i}{1+G_2G_3H}$$

Then
$$\begin{aligned}\theta_o &= \theta_b + G_4\theta_i \\ &= \frac{G_1G_2G_3\theta_i}{1+G_2G_3H} + G_4\theta_i\end{aligned}$$

from which
$$\frac{\theta_o}{\theta_i} = \frac{G_1G_2G_3 + G_4 + G_2G_3G_4H}{1+G_2G_3H}$$

Example 13.2 Figure 13.19 shows a system with inputs θ_i' and θ_i''. Obtain an expression for the output of the system, θ_o.

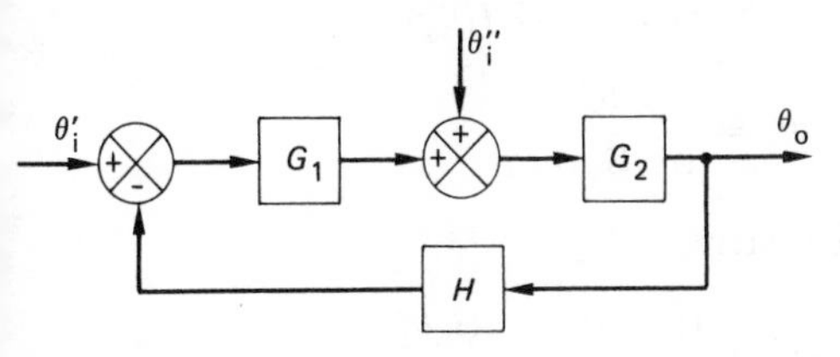

FIG. 13.19

Solution

The output θ_o is the sum of those outputs due to θ_i' and θ_i'', by the principle of superposition.

(a) Let θ_i'' be zero and let θ_o' be the output due to θ_i' alone. The system is then simplified to that shown in Fig. 13.20.

The elements G_1 and G_2 in series may be replaced by a single element of transfer operator G_1G_2, which reduces the system further to that shown in Fig. 13.21.

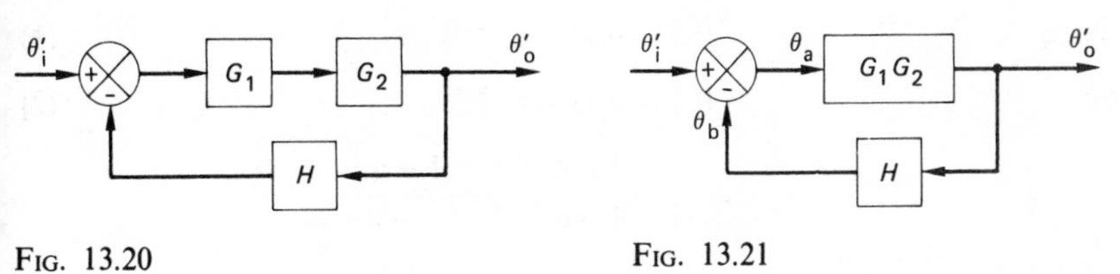

FIG. 13.20

FIG. 13.21

Then
$$\begin{aligned}\theta_o' = G_1G_2\theta_a &= G_1G_2(\theta_i' - \theta_b) \\ &= G_1G_2(\theta_i' - H\theta_o')\end{aligned}$$

from which $$\theta_o' = \frac{G_1 G_2 \theta_i'}{1+H}$$

(b) Let θ_i' be zero and let θ_o'' be the output due to θ_i'' alone. The system is then simplified to that shown in Fig. 13.22.

The elements H and G_1 in series may be replaced by a single element of transfer operator $G_1 H$, which reduces the system further to that shown in Fig. 13.23.

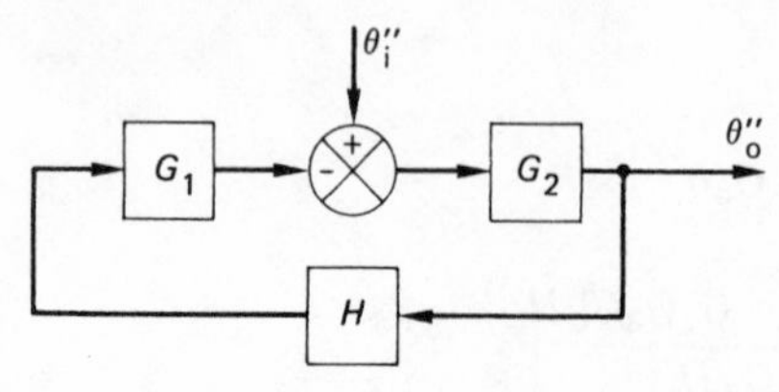

FIG. 13.22

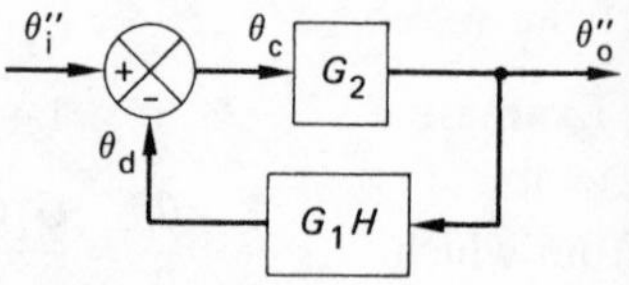

FIG. 13.23

Then $$\theta_o'' = G_2 \theta_c = G_2(\theta_i'' - \theta_d)$$
$$= G_2(\theta_i'' - G_1 H \theta_o'')$$

from which $$\theta_o'' = \frac{G_2 \theta_i''}{1 + G_1 H}$$

Hence the total output of the system,

$$\theta_o = \theta_o' + \theta_o'' = \frac{G_1 G_2 \theta_i'}{1+H} + \frac{G_2 \theta_i''}{1+G_1 H}$$

Example 13.3 In a servosystem controlling the angular position of a rotor, a step input results in an output of 0·5 rad after 4 s and 0·7 rad after a further 4 s. Find the time constant for the system and the total time to achieve an output of 0·8 rad.

Solution

For a step input, $\theta_o = \theta_i(1 - e^{-t/\tau})$

When $t = 4$ s, $\theta_o = 0{\cdot}5$ rad, and when $t = 8$ s, $\theta_o = 0{\cdot}7$ rad.

Hence $$0{\cdot}5 = \theta_i(1 - e^{-4/\tau}) \qquad (1)$$

and $$0{\cdot}7 = \theta_i(1 - e^{-8/\tau}) \qquad (2)$$

$$\therefore \quad \frac{7}{5} = \frac{1 - e^{-8/\tau}}{1 - e^{-4/\tau}} = \frac{1 - n^2}{1 - n}$$

where $n = e^{-4/\tau}$.

This reduces to $$n^2 - 1{\cdot}4n + 0{\cdot}4 = 0$$

$$\therefore \quad n = 1 \quad \text{or} \quad 0{\cdot}4$$

The solution $n = 1$ corresponds with $\tau = \infty$.
The solution $n = 0{\cdot}4$ corresponds with $e^{-4/\tau} = 0{\cdot}4$,

from which $$\underline{\tau = 4{\cdot}37\,\text{s}}$$

From equation (1),

$$0{\cdot}5 = \theta_i(1 - e^{-4/4{\cdot}37})$$

$$\therefore \quad \theta_i = 0{\cdot}833\,\text{rad}$$

When $\theta_o = 0{\cdot}8$ rad, $$0{\cdot}8 = 0{\cdot}833\,(1 - e^{-t/4{\cdot}37})$$

from which $$\underline{t = 14{\cdot}1\,\text{s}}$$

Example 13.4 A mercury-in-glass thermometer, indicating an ambient temperature of 20°C, is plunged into a liquid which is at a constant temperature. The reading is observed to vary in the following manner:

Time (s)	0	5	10	15	20	25	30
Thermometer reading (°C)	20	60	77	84	87	89	90

Estimate the time constant for the thermometer.

The same instrument is later used to take measurements in a liquid whose temperature is known to vary sinusoidally. When the period of the fluctuations is 25 s, the recorded temperature is observed to vary between the limits of 37·6 °C and 45·4 °C. Determine the actual maximum and minimum temperatures of the liquid. (U. Lond.)

Solution
From the graph, Fig. 13.24, the maximum temperature achieved is approximately 90 °C. Hence

$$\theta_i = 90° - 20° = 70\,°\text{C}$$

This is a step input and hence, from Section 13.10 (a),

$$\theta_o = \theta_i(1 - e^{-t/\tau})$$

When $t = \tau$, $$\theta_o = 0{\cdot}632\theta_i = 0{\cdot}632 \times 70 = 44{\cdot}2°\text{C}$$

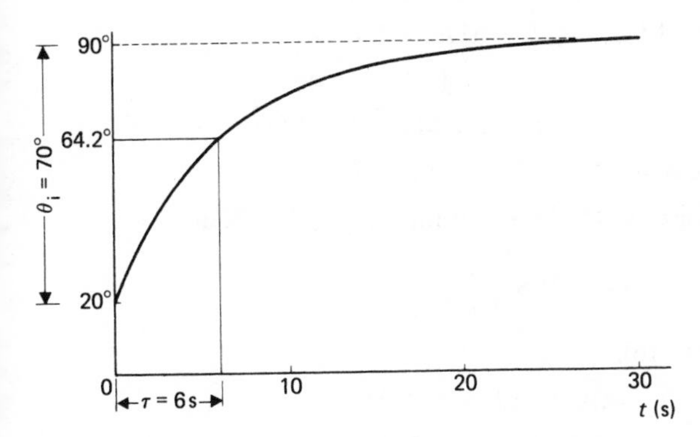

FIG. 13.24

By setting off a temperature of 44·2 °C above the datum temperature of 20 °C, it is found that the corresponding time is 6 s,

i.e. time constant, $\tau = \underline{6\,\text{s}}$

From Section 13.10 (c),

$$\text{steady-state amplitude of } \theta_o = \frac{a}{\sqrt{(1+p^2\tau^2)}}$$

$$\text{Amplitude} = \frac{45{\cdot}4 - 37{\cdot}6}{2} = 3{\cdot}9\,°\text{C}$$

$$p = 2\pi/25 = 0{\cdot}2513\,\text{rad/s}$$

$$\tau = 6\,\text{s}$$

$$\therefore\quad 3{\cdot}9 = \frac{a}{\sqrt{[1+(0{\cdot}2513\times 6)^2]}}$$

from which $a = 7{\cdot}06\,°\text{C}$

$$\text{Mean temperature} = \frac{37{\cdot}6 + 45{\cdot}4}{2} = 41{\cdot}5\,°\text{C}$$

$$\therefore\ \text{maximum temperature} = 41{\cdot}5 + 7{\cdot}06 = \underline{48{\cdot}56\,°\text{C}}$$

and $\text{minimum temperature} = 41{\cdot}5 - 7{\cdot}06 = \underline{34{\cdot}44\,°\text{C}}$

Example 13.5 A shaft carries a rotor of moment of inertia 5 kg m^2 and a dashpot which provides a torsional resistance of 2 N m per rad/s. If a steadily increasing torque, given by $T = 1{\cdot}2t$ N m is applied to the rotor, what speed will be reached after 3 s?

Solution

Referring to Fig. 13.25,

$$\text{accelerating torque} = I\frac{d\omega}{dt} = 5\text{D}\omega\ \text{N m}$$

and $\text{damping torque} = c\omega = 2\omega\ \text{N m}$

$$\therefore\ \text{applied torque, } T = 5\text{D}\omega + 2\omega\ \text{N m}$$

from which $\dfrac{\omega}{T} = \dfrac{1}{2+5\text{D}} = \dfrac{0{\cdot}5}{1+2{\cdot}5\text{D}}$

Thus time constant, $\tau = 2{\cdot}5$ s and $k = 0{\cdot}5$

When the rotor is subjected to the ramp input, $T = 1{\cdot}2t$ N m

$$\omega = 1{\cdot}2t \times \frac{0{\cdot}5}{1+2{\cdot}5\text{D}}$$

Hence, from Section 13.10 (b),

$$\omega = ak\{t - \tau(1 - e^{-t/\tau})\}$$
$$= 1{\cdot}2 \times 0{\cdot}5\{t - 2{\cdot}5(1 - e^{-t/2{\cdot}5})\}$$

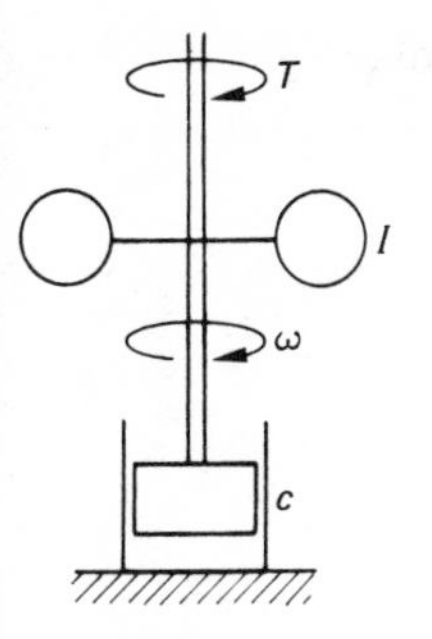

FIG. 13.25

Therefore, when $t = 3$ s,

$$\omega = 0{\cdot}6\{3 - 2{\cdot}5(1 - e^{-3/2{\cdot}5})\} = \underline{0{\cdot}752 \text{ rad/s}}$$

Example 13.6 Figure 13.26 shows a tank with inflow Q_i and outflow Q_o. The level of water is controlled by a valve, the setting of which is governed by a float with a screw adjustment. The input flow is proportional to the movement of the float and the output may be taken as proportional to the head in the tank if the variation in the head is small.

Obtain the relation between the actual head and the desired head when the screw adjustment is altered.

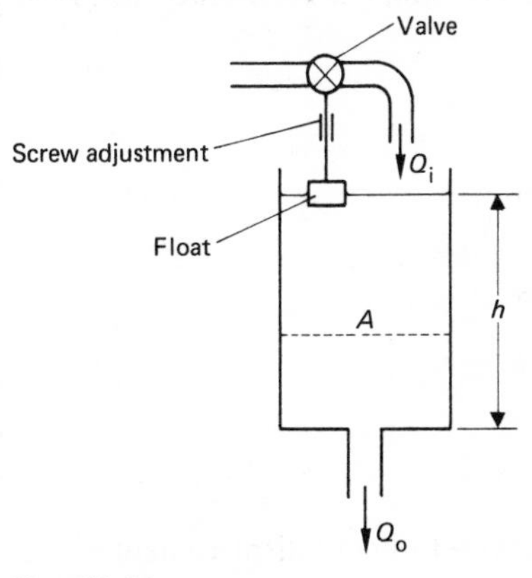

FIG. 13.26

Solution

Let the actual head be h_o and the desired head be h_i. Then

$$\text{change of head} = h_i - h_o$$

Since the inflow is proportional to the float movement,

$$Q_i = c_1(h_i - h_o)$$

where c_1 is a constant.

Also the outflow is proportional to the head in the tank,

i.e.
$$Q_o = c_2 h_o$$

where c_2 is another constant.

From continuity of flow,
$$Q_i - Q_o = A\frac{dh_o}{dt} = ADh_o$$

where A is the cross-sectional area of the tank.

Hence
$$c_1(h_i - h_o) - c_2 h_o = ADh_o$$

from which
$$\frac{h_o}{h_i} = \frac{c_1}{c_1 + c_2 + AD} = \frac{1}{1 + \frac{c_2}{c_1} + \frac{A}{c_1}D}$$

Example 13.7 Figure 13.27 shows a hydraulic ram controlled by a spool valve. When the valve is in its mid-position, it blocks off flow to and from both ends of the cylinder. The cross-sectional area of the ram is 0·003 m² and when the valve is moved from its mid-position, the rate of flow of oil into the cylinder is 0·01 m³/s per metre of valve measurement.
(a) Show that the transfer operator is of the form $k/(1 + \tau D)$, stating the assumptions made, and find the values of k and τ.
(b) If the end of the lever is suddenly moved 20 mm from the mid-position, determine the limiting displacement of the ram and the time taken to move through 85 per cent of this displacement.

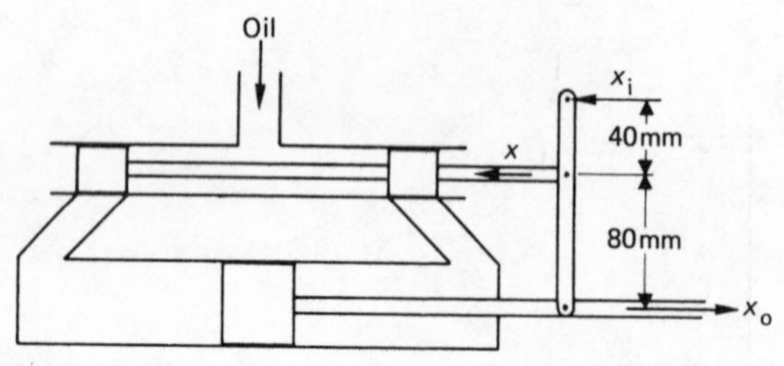

Fig. 13.27

Solution

(a) Let Q be the rate of flow through the ports and x be the displacement of the valve.

Then
$$Q = 0{\cdot}01\,x\ \text{m}^3/\text{s}$$

But
$$x = \frac{80}{40 + 80} \times x_i - \frac{40}{40 + 80} \times x_o$$
$$= \tfrac{1}{3}(2x_i - x_o)$$

$$\therefore\quad Q = \frac{0{\cdot}01}{3}(2x_i - x_o)\ \text{m}^3/\text{s}$$

Also $$Q = 0{\cdot}003\frac{dx_o}{dt} = 0{\cdot}003Dx_o$$

$$\therefore \quad \frac{0{\cdot}01}{3}(2x_i - x_o) = 0{\cdot}003Dx_o$$

from which $$\frac{x_o}{x_i} = \frac{2}{1+0{\cdot}9D}$$

This is of the form $k/(1+\tau D)$ where

$$\underline{k = 2} \quad \text{and} \quad \text{time constant, } \underline{\tau = 0{\cdot}9\,\text{s}}$$

The assumptions are that the inertia of the moving parts are negligible and that leakage and compressibility of the oil are ignored.

(b) When $x_i = 20\,\text{mm}$,

$$x_o = 20 \times \frac{2}{1+0{\cdot}9D}$$

Hence, from Section 13.10 (a),

$$x_o = 20 \times 2(1 - e^{-t/0{\cdot}9})$$

Thus the limiting displacement is $\underline{40\,\text{mm}}$ when $t = \infty$.

When $x_o = 0{\cdot}85 \times 40$ mm,

$$40(1 - e^{-t/0{\cdot}9}) = 0{\cdot}85 \times 40$$

from which $$\underline{t = 1{\cdot}71\,\text{s}}$$

Example 13.8 Figure 13.28 shows a pneumatic servomechanism in which the flow of air into the cylinder is controlled by a flapper valve operated by a lever. The movement of the valve, y, is half the movement of the link, x, and the area of piston is $1600\,\text{mm}^2$. The flow of air into the cylinder is given by $Q = 0{\cdot}01y\,\text{m}^3/\text{s}$, where y is in metres.

(a) Obtain the transfer operator for the mechanism and determine the time constant, neglecting the area of the piston rod.

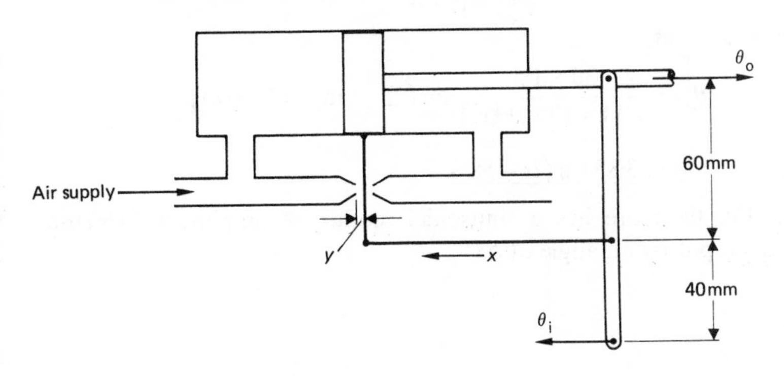

FIG. 13.28

(b) If a sinusoidal input given by $\theta_i = 30 \sin 2t$ mm is applied to the lever, obtain an expression for the steady-state motion of the piston.

Solution

(a) From the geometry of the linkage,

$$x = \frac{60}{100} \times \theta_i - \frac{40}{100} \times \theta_o = \frac{3\theta_i - 2\theta_o}{5} \text{ m}$$

$$\therefore\ y = \frac{x}{2} = \frac{3\theta_i - 2\theta_o}{10} \text{ m}$$

$$\therefore\ Q = 0{\cdot}01\left(\frac{3\theta_i - 2\theta_o}{10}\right) \text{m}^3/\text{s}$$

Also $$Q = 1600 \times 10^{-6}\frac{d\theta_o}{dt} = 0{\cdot}0016 D\theta_o$$

$$\therefore\ 0{\cdot}01\left(\frac{3\theta_i - 2\theta_o}{10}\right) = 0{\cdot}0016 D\theta_o$$

or $$3\theta_i - 2\theta_o = 1{\cdot}6 D\theta_o$$

from which $$\frac{\theta_o}{\theta_i} = \frac{3}{2 + 1{\cdot}6D} = \frac{1{\cdot}5}{1 + 0{\cdot}8D}$$

Hence time constant, $\tau = \underline{0{\cdot}8 \text{ s}}$

(b) If $\theta_i = 30 \sin pt$, then

$$\theta_o = 30 \sin pt \times \frac{1{\cdot}5}{1 + 0{\cdot}8D}$$

The transfer operator is of the form $k/(1 + \tau D)$ where $k = 1{\cdot}5$ and hence, from Section 13.10 (c),

$$\theta_o = \frac{ak}{\sqrt{(1 + p^2\tau^2)}} \sin(pt - \alpha)$$

where $\alpha = \tan^{-1} p\tau$,

i.e. $$\theta_o = \frac{30 \times 1{\cdot}5}{\sqrt{[1 + (2 \times 0{\cdot}8)^2]}} \sin[2t - \tan^{-1}(2 \times 0{\cdot}8)]$$

$$= \underline{23{\cdot}85 \sin(2t - 58°)}$$

The piston therefore has a sinusoidal motion of amplitude 23·85 mm, lagging the input by an angle of 58°.

Problems

1. Obtain the transfer operator for the system shown in Fig. 13.29.

(*Answer:* $G_1G_2/(1+G_1H)$)

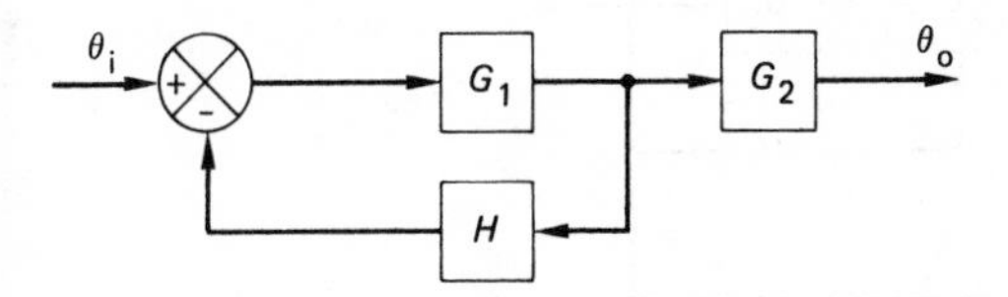

FIG. 13.29

2. The system shown in Fig. 13.30 (a) has been rearranged by moving the summing point behind the element G, Fig. 13.30 (b). Show that the transfer operator in each case is given by $G/(1+GH)$.

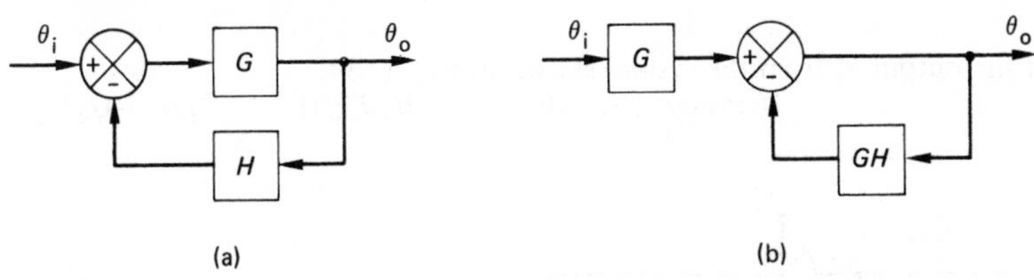

FIG. 13.30

3. Obtain the transfer operator for the system shown in Fig. 13.31.

(*Answer:* $G_1G_2/(1+G_1G_2+G_2H)$)

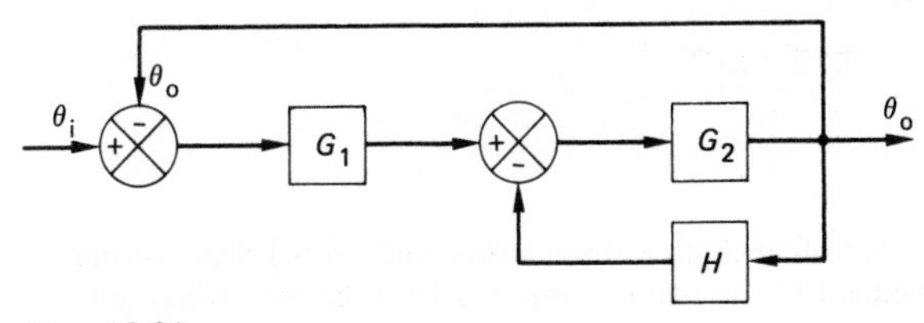

FIG. 13.31

4. Find, using the principle of superposition, the output θ_o for the system shown in Fig. 13.32, which is subjected to two inputs, θ_i' and θ_i''.

(*Answer:* $(G_1G_3\theta_i'+G_2G_3\theta_i'')/(1+G_3H)$)

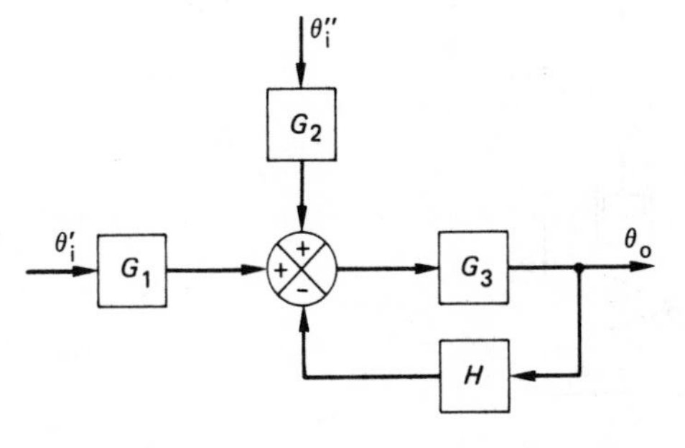

FIG. 13.32

5. Obtain the transfer operator for the system shown in Fig. 13.33.
(*Answer:* $G_1G_2/(1+G_1G_2H_2+G_1G_2H_3+G_2H_1)$)

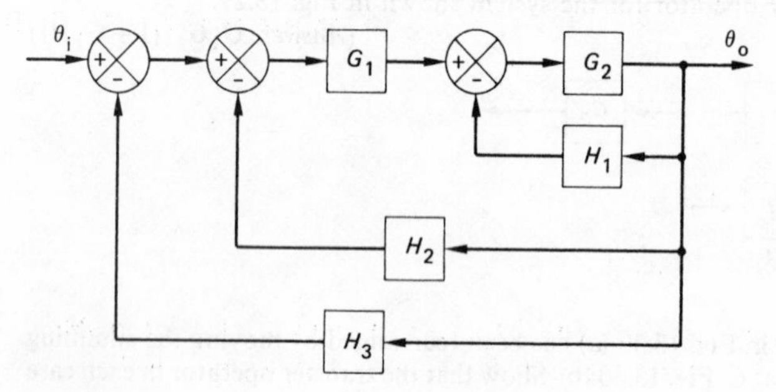

FIG. 13.33

6. Find the output θ_o from the system shown in Fig. 13.34.
(*Answer:* $(G_1G_2\theta_i' - G_1G_2H_1\theta_i'')/(1+G_1G_2H_1+G_2H_2)$)

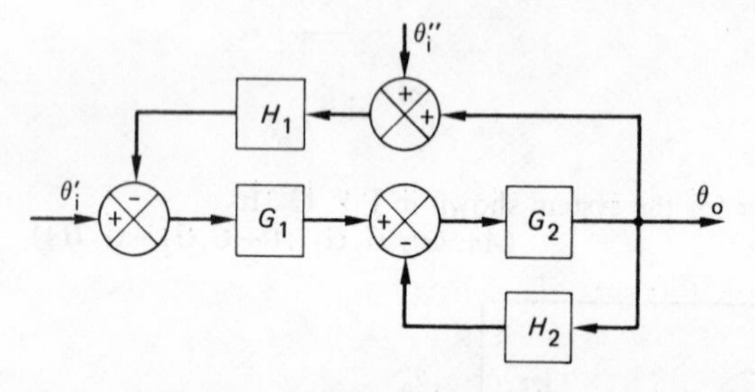

FIG. 13.34

7. Figure 13.35 shows a plate of negligible mass supported on a helical spring of stiffness 200 N/m. The movement of the plate is opposed by a dashpot which gives a resistance of 50 N/(m/s).
(a) Obtain the transfer operator and time constant for the system.
(b) Determine the amplitude of the motion of the plate when an axial periodic force $F = 20\sin 8\pi t$ is applied to it. (*Answer:* 0·005/(1 + 0·25D); 0·25 s; 15·7 mm)

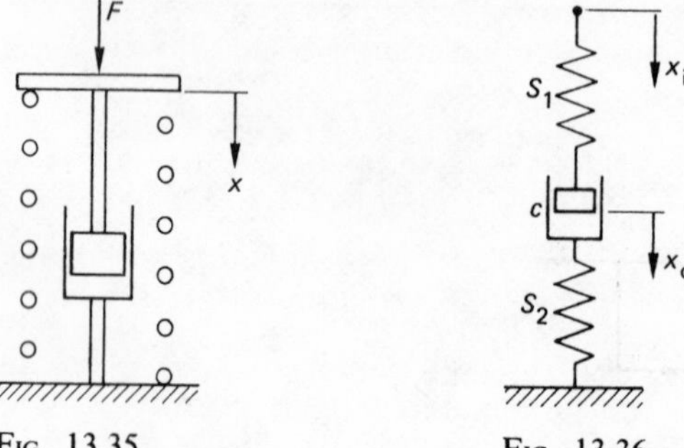

FIG. 13.35 FIG. 13.36

8. Obtain the transfer operator for the system shown in Fig. 13.36.

(*Answer:* $S_1(cD+S_2)/[cD(S_1+S_2)+S_1S_2]$)

9. The angular position of a radar aerial is controlled by a servomechanism. The control wheel is given a sudden angular displacement which causes the aerial to turn through 5° after 3 s and 10° after 7·3 s.

Determine: (a) the time constant for the system, (b) the step input rotation, and (c) the time taken for the aerial to rotate through 12°. (*Answer:* 10 s; 19·3°; 9·73 s)

10. The system shown in Fig.13.37 consists of a light rod, hinged at A, and a dashpot, of viscous resistance 40 N/(m/s), is connected to the other end, B. A spring of stiffness 1 kN/m supports the rod at C.

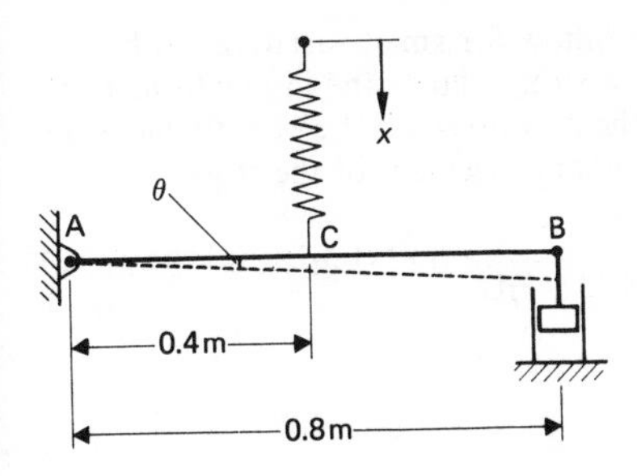

FIG. 13.37

(a) A vertical movement x of the rod results in an angular displacement θ of the rod. Find the ratio θ/x for small displacements.
(b) If the upper end of the spring is given a vertical sinusoidal motion of amplitude 10 mm at a frequency of 3 Hz, determine the steady-state amplitude of the dashpot plunger. (*Answer:* $2{\cdot}5/(1+0{\cdot}16D)$; 6·3 mm)

11. A thermometer with a time constant of 6 s is taken at a room temperature of 15 °C and plunged into a liquid at 60 °C, What would the thermometer read after 5 s?

(*Answer:* 40·5 °C)

12. A thermometer at room temperature, 15 °C, is subjected to a sudden rise in temperature; it shows a reading of 55°C after 3·5 s and after a further 3·5 s, the reading is 75 °C. Find the time constant for the thermometer and the applied rise in temperature. (*Answer:* 5·04 s; 90·9 °C)

13. A gas thermometer has a thermal conductivity of 0·02 W/°C and a thermal capacity of 0·1 J/°C.
(a) Determine the time constant for the thermometer.
(b) The thermometer is at a temperature of 20 °C. Calculate the reading 10 s after being plunged into a liquid at 80 °C.
(c) The thermometer is at a temperature of 20 °C. Calculate the reading after 10 s when the surrounding liquid is heated at the rate of 2 °C per second from an initial temperature of 20 °C. (*Answer:* 5 s; 71·9 °C; 31·35 °C)

14. The heat flow into a boiler is $1{\cdot}5(\theta_i-\theta_o)$ kW, where θ_i is the controller setting and θ_o is the temperature of the boiler. The thermal capacity of the boiler is 150 kJ/°C.
(a) Obtain the transfer operator and time constant for the boiler.
(b) If the boiler is operating steadily at 100 °C and the controller is suddenly reset to 200 °C, how long will it take for the boiler temperature to reach 150 °C?

(*Answer:* $1/(1+100D)$; 100 s; 69·3 s)

15. Referring to the spool valve in Example 13.7, the input x_i is moved steadily from its equilibrium position at 5 mm/s.

Find: (a) the movement of the ram 4 s after the action commences, and (b) the time taken for the ram to move 20 mm. (*Answer:* 31·1 mm; 2·863 s)

16. Referring to the spool valve in Example 13.7, the input x_i is given a sinusoidal motion of amplitude 20 mm and frequency 1 Hz.

Find the amplitude of motion of the ram and the angle of lag.

(*Answer:* 6·97 mm; 80°)

17. A small lake, of surface area 10^4 m², is fed by a stream and the outflow is measured by a weir, the flow rate being given by $Q = 5h^{3/2}$ m³/s, where h is the head of water over the weir in metres.

(a) Obtain a relation between the outflow and inflow for small variations in h.

(b) When conditions are steady and h is 0·1 m, a storm causes the inflow to increase progressively by 10^{-4} m³/s every second. Find the flow rate over the weir: (i) under the steady conditions and (ii) 20 minutes after the commencement of the storm.

$$\left(\textit{Answer:}\ \frac{Q_o}{Q_i} = \frac{1}{1 + (4000/3\sqrt{h})D};\ 0{\cdot}1581\ \text{m}^3/\text{s};\ 0{\cdot}1737\ \text{m}^3/\text{s} \right)$$

Appendix

Solution of differential equations
The differential equations which occur in this book may be generalized by

$$\frac{d^n x}{dt^n} + k^n x = f(t)$$

where k is a constant.
This may be written

$$(D^n + k^n)x = f(t)$$

where D represents d/dt.
The *order* of the equation is represented by the value of n.
This equation may be expanded to

$$(D^n + k^n)x = 0 + f(t)$$

and the solution consists of two parts:

(a) the value of x which satisfies the equation $(D^n + k^n)x = 0$, this being the *complementary function*;
(b) the value of x which satisfies the equation $(D^n + k^n)x = f(t)$, this being the *particular integral*.

The complete solution is then the sum of these two parts.

First order differential equations The forms to be considered are

$$(D + k)x = a$$
$$(D + k)x = at$$

and

$$(D + k)x = a \sin pt$$

where a is a constant.

Complementary function

$$(D + k)x = 0$$

The solution is of the form

$$x = Ae^{mt}*$$

where A and m are constants.

* This is the only type of function which can be differentiated any number of times without change of form.

Substituting in the equation $Dx + kx = 0$,

$$mAe^{mt} + kAe^{mt} = 0$$

$$\therefore\ m = -k$$

$$\therefore\ x = Ae^{-kt}$$

Particular integrals

For the form $$(D+k)x = a$$

$$x = \frac{a}{D+k} = \frac{a}{k\left(1+\dfrac{D}{k}\right)}$$

$$= \left(1+\frac{D}{k}\right)^{-1}\frac{a}{k}$$

$$= \left(1-\frac{D}{k}+\frac{D^2}{k^2}-\cdots\right)\frac{a}{k}$$

$$= \frac{a}{k} \text{ since all the other terms are zero}$$

For the form $$(D+k)x = at$$

$$x = \frac{at}{D+k} = \frac{at}{k\left(1+\dfrac{D}{k}\right)}$$

$$= \left(1+\frac{D}{k}\right)^{-1}\frac{at}{k}$$

$$= \left(1-\frac{D}{k}+\frac{D^2}{k^2}-\cdots\right)\frac{at}{k}$$

$$= \frac{at}{k}-\frac{a}{k^2} \text{ since all the other terms are zero}$$

$$= \frac{a}{k}\left(t-\frac{1}{k}\right)$$

For the form $$(D+k)x = a\sin pt$$

$$x = \frac{a\sin pt}{D+k} = \frac{(D-k)a\sin pt}{D^2-k^2}$$

Since $D^2 \sin pt = -p^2 \sin pt$, D^2 may be replaced by $-p^2$, and

$$(D-k)\sin pt = p\cos pt - k\sin pt = -(k\sin pt - p\cos pt)$$

$$= -\sqrt{(p^2+k^2)}\sin(pt-\alpha)$$

where $\tan\alpha = p/k$.

from which
$$x = \frac{-a\sqrt{(p^2+k^2)}\sin(pt-\alpha)}{-p^2-k^2}$$
$$= \frac{a}{\sqrt{(p^2+k^2)}}\sin(pt-\alpha)$$

Second order differential equations The forms to be considered are
$$(\mathrm{D}^2+k^2)x = a$$
and
$$(\mathrm{D}^2-k^2)x = a$$

Complementary functions
The solution is of the form
$$x = C\mathrm{e}^{mt}$$
where C and m are constants.

For the form $(\mathrm{D}^2+k^2)x = 0$
substituting for x gives
$$m^2C\,\mathrm{e}^{mt}+k^2C\,\mathrm{e}^{mt} = 0$$
$$\therefore\ m = \pm \mathrm{j}k$$
$$\therefore\ x = C_1\mathrm{e}^{\mathrm{j}kt}+C_2\mathrm{e}^{-\mathrm{j}kt}$$
$$= C_1(\cos kt+\mathrm{j}\sin kt)+C_2(\cos kt-\mathrm{j}\sin kt)$$
$$= A\cos kt+B\sin kt$$
where $A = C_1+C_2$ and $B = \mathrm{j}(C_1-C_2)$.

For the form $(\mathrm{D}^2-k^2)x = a$
substituting for x gives
$$m^2C\,\mathrm{e}^{mt}-k^2C\,\mathrm{e}^{mt} = 0$$
$$\therefore\ m = \pm k$$
$$\therefore\ x = C_1\mathrm{e}^{kt}+C_2\mathrm{e}^{-kt}$$
$$= C_1(\cosh kt+\sinh kt)+C_2(\cosh kt-\sinh kt)$$
$$= A\cosh kt+B\sinh kt$$
where $A = C_1+C_2$ and $B = C_1-C_2$.

Particular integrals
For the form $(\mathrm{D}^2+k^2)x = a$
$$x = \frac{a}{\mathrm{D}^2+k^2} = \frac{a}{k^2\left(1+\dfrac{\mathrm{D}^2}{k^2}\right)}$$
$$= \left(1+\frac{\mathrm{D}^2}{k^2}\right)^{-1}\frac{a}{k^2}$$

$$= \left(1 - \frac{D^2}{k^2} + \frac{D^4}{k^4} - \cdots\right)\frac{a}{k^2}$$

$$= \frac{a}{k^2} \quad \text{since all the other terms are zero}$$

For the form $\quad (D^2 - k^2)x = a$
a similar analysis gives

$$x = -\frac{a}{k^2}$$